Y0-BCK-888

Protein
Biosynthesis
and
Membrane
Biochemistry

There has been a tremendous growth of interest in protein synthesis in the last decade. This book develops the theme that a thorough understanding of the process of cellular protein biosynthesis will require the study of cell structure in general, and membrane structure in particular.

The author examines the concepts of protein biosynthesis in the light of knowledge at the time of their development and reconsidered in terms of newer information that caused their modification or rejection. The history of protein synthesis is reviewed and limitation of current knowledge of the subject is stressed. Emphasis is on the involvement of cell membranes in addition to the well-accepted ideas involving nucleic acids. Other fields of biochemistry covered in the book are: cell structure, membrane structure, oxidative phosphorylation, active transport, facilitated diffusion, and hormone action.

Protein Biosynthesis and Membrane Biochemistry can be used by graduate students in biochemistry and by professional biochemists.

Protein Biosynthesis
and Membrane Biochemistry

Protein Biosynthesis and Membrane Biochemistry

RICHARD W. HENDLER

National Heart Institute
National Institutes of Health
Bethesda, Maryland

John Wiley & Sons, Inc.
New York · London · Sydney · Toronto

Copyright © 1968 by John Wiley & Sons, Inc.

All rights reserved. No part of this book may
be reproduced by any means, nor transmitted,
nor translated into a machine language without
the written permission of the publisher.

Library of Congress Catalog Card Number: 68-30912
SBN 471 37050 9
Printed in the United States of America

To the Memory
of
Reba Gordon Hendler

PREFACE

Many books are now appearing on the subject of protein synthesis and I should like to state my aims and purposes for presenting this particular work. I think that it is clear to many observers in this field that strong trends have gathered many followers but very few dissenters. As a consequence most published works treat the subject in a somewhat similar manner insofar as they tend to emphasize the data that fit and less often to deal with certain aspects not readily treated by current concepts.

Our present knowledge of protein synthesis is extensive but incomplete. We still do not know, for example, how a cell turns out its wide spectrum of unique proteins and why cell destruction produces acellular fragments that function with low quantitative and qualitative efficiency. Many investigators in the field are aware of these problems, but most books continue to emphasize only the positive aspects. The major message to be conveyed by this book is that the missing parts of the knowledge necessary to a thorough understanding of the process of protein synthesis will be acquired in the study of cell structure in general and membrane structure in particular.

Students entering the field of protein synthesis today are confronted with a neatly worked out system of reactions that can explain the process in almost complete detail. This does not mean that all the questions have been answered, but the unanswered questions seem to concern smaller details. Indeed, this over-all view of protein synthesis has been well presented in two comprehensive reviews in recent years (Chapter 2, Refs. 1, 2). There is, however, a history of protein synthesis that seems to have been largely discarded, even though it does not extend back far in time. This is quite unfortunate because it is interesting to see how some of our fundamental concepts were derived and developed. What preliminary experiments laid the foundations that support the current structure of knowledge? How was the path followed from initial observations to our modern picture of protein synthesis?

When I first became interested in the problem of protein synthesis and started working actively in the field in 1949 as a graduate student under Dr. David Greenberg at California, the view of protein synthesis that I learned was also a modern view. In fact, as the field of protein synthesis was developing and changing, each new concept was initially part of a modern view.

Looking back over this time, which is rather brief in terms of years but which constitutes almost the entire history of the active study of the problems of protein biosynthesis, an interesting perspective becomes apparent. It is easy now to see how wrong our ideas were about certain things that looked so convincing at the time. It is also interesting to note how we worried so much about things that turned out to be really not so important and how much we neglected to see things that really mattered. Although the true historical perspective of our contemporary picture will be available to us only in the years to come, I hope that I can convey by analogy to previous examples the idea that the modern view of protein synthesis is really a pliable, changing picture which represents a phase in the historical development of the problem. In order to present the history as a series of momentarily contemporary concepts, I make frequent use of the present tense in describing older work. Now and then I take advantage of newly gained knowledge to comment on various developments in light of more up-to-date information.

A purely impersonal history would consist of masses of data and dates, but even the compilation of that kind of material causes the historian to exercise a degree of selectivity. Furthermore, history, pure and impersonal, can be dull and tedious to read. In this work I will describe the developments as they happened and as I saw them. I realize the danger that the autobiographical aspects of this device may tend to confer on me an exaggerated importance in connection with the material that I treat. This is not my intention and I hope not to convey that impression. Facts are stated as facts and my interpretations are clearly stated so that the reader will have no difficulty deciding who is responsible for the views expressed.

In the first three chapters I describe many of the problems involved in studying protein synthesis and how current views were obtained by experimentation and the interpretation of these experiments. These chapters also point up the deficiencies and gaps that seem to exist.

Chapter 4 provides a descriptive treatment of the development of current concepts of cell structure. Because this discipline has escaped much of the controversy and uncertainty characterizing other topics treated in this book, it will make lighter reading than those chapters that cover more complicated areas of biochemistry.

The subject of membrane structure is paramount in the book. At the moment of writing, however, this is a field marked by uncertainties and a deficiency of fundamental information. A clear, concise picture of biological membranes does not exist. In Chapter 5, therefore, I attempt to trace the development of recent and current concepts, and I try also to evaluate the existing divergent views of membrane structure.

In Chapters 6 and 7 I try to emphasize how membranes seem to be used by living cells to accomplish their most complicated functions. Although this is an open-ended and fast-growing area of biochemistry, I have chosen to demonstrate this complexity by evaluating the fields of oxidative phosphorylation, membrane transport, the function of certain organelles, and hormonal control of cell function. The purpose in discussing these topics in a book whose major emphasis is devoted to protein synthesis is twofold. First, I wish to illustrate how cells employ membranes in active phases of metabolism, some of which involve integration of many different enzymes and cofactors, and, second, to emphasize that the localization of various metabolic activities in a membrane matrix allows the possibility of a coordinated integration of function and control of different systems. In the discussion of hormone action it is shown that single hormone substances can regulate the production of energy by oxidative phosphorylation, the uptake of exogenous metabolites including amino acids, and the biosynthesis of polymers, including proteins.

Future developments in biochemistry, it seems to me, will center on explaining the role of cell membranes in accomplishing these activities which are basic characteristics of living cells.

Finally, in Chapter 8 I attempt to bring together those positive experimental results that show that cell membranes *per se* provide the answer to questions raised in the first three chapters about protein biosynthesis.

For graduate students and biochemists working in other areas of biology and biochemistry the book provides an extensive and critical introduction to an active and vital field of research. A broad view of each area covered in the book can be obtained by a general reading of pertinent sections in which detailed evaluation of individual data and experiments are either skipped or skimmed.

The intensive treatment of protein synthesis, however, is intended for investigators actually concentrating in the field. Many experiments are carefully dissected and data and conclusions are viewed from divergent points of view. Although this treatment may tend to over-emphasize my scepticism at the present "state of the art," it will tend to balance other books on this subject that seem to me at this time to be overly sanguine about what we really know about this important subject.

To a large extent the continuity and clarity of the final version of this manuscript is due to the close reading and valuable suggestions of several of my colleagues. I am indebted to them for their time and their help. Particular acknowledgment is made to Dr. Raymond Scharff, Dr. Edward L. Kuff, and Dr. Bernard T. Kaufman for thorough readings of the early manuscript and to Dr. Richard B. Roberts, Dr. Joseph E. Rall, Dr. Edward D. Korn, Dr. Christian B. Anfinsen, Dr. Joseph Fruton, Dr. Harriet Maling, and Dr. Bernard M. Babior for suggestions made in its later stages.

RICHARD W. HENDLER

Bethesda, Maryland
May 1968

CONTENTS

Protein Biosynthesis
and Membrane Biochemistry

Chapter 1

THE PROBLEM OF PROTEIN BIOSYNTHESIS

Every living thing starts out as a single cell, often of microscopic dimension, which is largely indistinguishable from other cells in appearance or in the general composition of its contents. In the case of multicellular organisms, this first cell divides and the daughter cells continue the division process. At various stages, as the clump of cells continues dividing, particular areas become recognizably different from the rest. Much later various parts differentiate into individual tissues. Not only does the process predictably culminate in the form of an elephant, tree, or a human, but the individual resembles its individual parents in many characteristics not shared by other members of the species. The rate of this process of cellular biosynthesis and the quantities of protein produced in the early stages of growth are quantitatively the greatest of any time throughout life.

Consider the relative growth in nine months from a single cell of insignificant weight to a newborn infant of seven to ten pounds. The continued rate of growth from birth to teenager is still impressive but does not call for any particular comment except for the fact that during late adolescence, this high rate of growth so dependably slows down. The total weight of the organism which has been rapidly increasing for so long reaches a plateau level and thereafter maintains this level closely. Biosynthesis does not stop at this point but continues at a rate entirely and exactly balanced by the rate of degradation. If at any time an injury is experienced by which tissue is destroyed, such as a cut or severe gash, something happens to the synthetic processes in the cells along the damaged area. Where previously synthesis proceeded only to the extent of degradation, now synthesis exceeds degradation until the destroyed tissue is replaced and the old balance is restored. Consider

1

also the development of a plant as the stem grows. Not only are particular leaf shapes reproduced one by one, but the exact characteristic location of leaves along the stem is reproduced and at some places, instead of leaves, flowers are formed.

All of the precise information and instructions for reading and carrying out the complicated synthetic operations and instructions for control of their relative rates are contained in the space encompassed by the starting cell. This information is written in the language of molecular configurations. All that we are is largely prescribed by what is written in this original cell. The disturbance of this perfect balance which, far from being constant, continually readjusts itself throughout life, can lead to cancer. This disease is believed to be the result of a defect in some aspect of cellular balance mechanisms.

We may be able to influence or to control these processes once we understand them. At the present state of our knowledge in this field such possibilities still seem remote. As an ultimate possibility in the light of new knowledge, such predictions may prove accurate.

Energy is required to link amino acids together in peptide chains. In fact, according to current concepts, the minimum amount of free energy necessary for a cell to make 100 g of protein containing 100 amino acids would be the same as that needed to bring 240 ml of water from 0° to the boiling point.[1] It is possible that the number may even be three times greater than this amount.[2]

A given cell simultaneously produces a wide spectrum of different proteins—some for export, as in the case of serum albumin and beta lipoprotein produced by the liver for the blood; some for cell structure in the replacement and maintenance of cytoplasmic and nuclear organelles; some for enzymatic purposes, as pepsin in the stomach or aldolase in the muscle; and some for hormones such as insulin in the pancreas or ACTH in the pituitary. Hundreds of different proteins for different functions may be produced simultaneously in one cell at any time. The cell must draw from the stock of individual amino acids, it must amass tremendous quantities of energy, it must simultaneously read the blueprints for all the various proteins, and it must honor messages to scale up or scale down or completely turn off or turn on the synthesis of different varieties of proteins. All of these factors must be brought to the synthesizing site in the right order, in the right quantities, and at the right time. At the same time that the protein synthetic machinery

[1] These calculations are based on a free energy of hydrolysis of ATP of 9000 cal at pH 7.0 and the relative concentrations of ATP, ADP, phosphate, and magnesium existing in the cell under usual *in vitro* conditions.
[2] That is, if the terminal configuration of sRNA is replaced (see page 66).

is in use, new machinery is synthesized and old machinery is torn down. The process is so well arranged spatially and so efficiently automated that mistakes seldom occur. Loftfield has shown that in the case of chicken ovalbumin the amino acid isoleucine is mistakenly substituted by the related amino acids valine or leucine only one time out of 3000 (1).

There are many specific questions about protein synthesis that have been posed and experimentally examined; for example, what are the chemical stages through which an amino acid must pass enroute to becoming a part of a particular finished protein? Between the time that the amino acid enters the synthesizing area and emerges as a finished product, does it enter into combination with other amino acids to form intermediary peptides? This question of peptide intermediates in protein synthesis was, until recently, the most important of research questions. The route that an amino acid follows to become part of a finished protein needs definition not only in chemical terms but also cytologically (i.e. the manner in which cell structure participates in the process). This thought will be foremost in the discussion of the problem that follows.

If the information to make all proteins that the adult of the species will ever need is contained inside of the starting first cell, in what manner is it stored? Matters related to this question have come to be known as the coding problem. This problem has long intrigued many scientists in and out of biological disciplines. It was first proposed in modern terms by George Gamow (2). Starting with the knowledge that the information for genetic continuity is stored in deoxyribonucleic acid (DNA), and assuming that the basic message serves to specify the linear sequence for the 20 different amino acids that make up proteins, we can state the problem with only one additional basic assumption, which in the light of recent experiments seems quite sound—that all of the information is stored in the linear sequence of the four nucleotide bases which comprise DNA.

The basic question, then, is how can a language with four letters store sufficient information to specify messages in a twenty-letter language? It is evident that the four-letter alphabet is not sufficient to relate to the 20-letter alphabet on a one-for-one basis. If two nucleotides were used to specify each of the amino acids, then 4×4 or 16 specifications would be possible. This number is also insufficient but it is clear that three nucleotide combinations would allow $4 \times 4 \times 4$, or 64 possibilities, which would be more than ample. Therefore the combination of three nucleotides has become known as a coding triplet or codon. This conclusion (as well as many others of equally fundamental importance) was reached and firmly believed long before the evidence was forthcoming to support it. It is gratifying that more recent evidence tends to establish this important piece of intuitive reasoning.

Problems of a higher order of complexity must also be included in the broad definition of protein synthesis. So far I have described the problem concerning determination of the linear sequence of amino acids in proteins (the problem of the primary structure). A given protein owes its uniqueness not only to its primary structure, but also to higher orders of structural integrity. A string of amino acids is biologically inert unless it is folded in a precise manner. If 200 amino acids were strung together and numbered sequentially from one end to the other, a given biological activity might result because the folding was such as to bring amino acids No. 3, 28 and 142 into proximity. This would be accomplished by other than peptide bonding; for example, two cysteines in different positions in the chain could be linked by disulfide bonds. Other bonds of importance in forming secondary and tertiary structures are shown in Figure 1. When amino acids are strung together they frequently assume a coiled helical structure, due to restraints imposed by bonds in the peptide chain backbone and to interactions among amino acid residues. The formation of particular areas of coiling and the spatial characteristics of such coiled and uncoiled areas comprise aspects of the secondary structure of proteins. These coils of linked

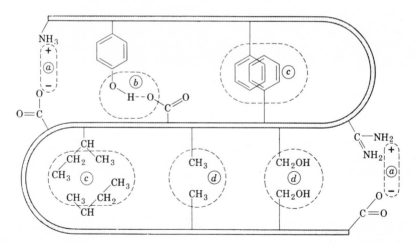

Figure 1 Some types of noncovalent bond that stabilize protein structure: (*a*) electrostatic interaction; (*b*) hydrogen bonding between tyrosine residues and carboxylate groups on side chains; (*c*) interaction of nonpolar side chains caused by the mutual repulsion of solvent; (*d*) van der Waals interactions. From Christian B. Anfinsen, *The Molecular Basis of Evolution,* Wiley, New York, 1963.

amino acids are further folded in a highly precise and unpredictably tortuous manner to produce pretzel-like structures. Figure 2 shows these structures (tertiary structures) for (a) myoglobin and (b) hemoglobin. Quaternary structures are built in a precise manner in which individual properly folded protein subunits are brought together to produce a more complicated structure.

An interesting extension of these problems considers the kind of information that must be stored to make cell structures such as membranes, nuclei, and mitochondria. Not only do the various cell membranes have over-all differences of lipid and protein composition, but within a single type of membrane such as the endoplasmic reticulum (Chapters 4 and 5) there may be specifically differentiated areas that would comprise ribosomal attachment sites or areas for phospholipid biosynthesis or electron transport. The structure of a mitochondrion is in itself amazingly complex with a fairly smooth outer membrane and an inner membrane specifically involuted to form cristae (Chapter 4). The inner membrane is periodically studded with knoblike projections. Interacting and related enzyme systems are specifically built into the structure. Would all of this specificity automatically follow from a code that specified only the linear sequence of amino acids? It seems hard to believe, but this is a problem we must deal with in stages. We are still in the first stage. The evidence that the linear sequence of bases in a nucleic acid specifies the linear order of amino acids in a polypeptide is convincing. The linear order of amino acids in a polypeptide may by itself specify higher orders of structure. These problems are further considered in later chapters.

It is not an oversimplification to say that the study of protein synthesis is the study of how a cell produces complete and biologically functional molecules. Many research efforts are centered on one particular aspect which is characteristic of protein synthesis but need not be entirely representative of the whole process: the incorporation of a radioactive amino acid into undefined precipitable proteinaceous material. Thus, a problem arises when attempts to study protein synthesis involve systems less organized than the intact cell.

A little understood but recognizable consequence of the destruction of cell structure is that the uniqueness and completeness of protein synthetic processes are generally lost. Superimposed on this fact is the observation that usually more than 98% of the cell's quantitative ability for protein synthesis is also lost, and furthermore, that new and artifactual mechanisms of incorporation arise. Some prime examples of this disturbing unnatural synthetic ability can be seen in early reports from the literature (see pages 22–25). These examples are not necessarily typical and the acellular approach to increase our understanding of the mechanisms of protein synthesis is essential. Through this approach alone, however, and especially when undefined products are being studied, a true solution to the problem cannot be obtained. All acellular findings must be restudied and reestablished in the context of the organized cell which is turning out a spectrum of unique proteins with amazing efficiency and few mistakes.

There is an important question of research philosophy here. The early triumphs of biochemistry were obtained with clear juices pressed from broken cells. The glycolytic pathways were defined in such systems. When cells are broken, various of their minute subparticles can be sus-

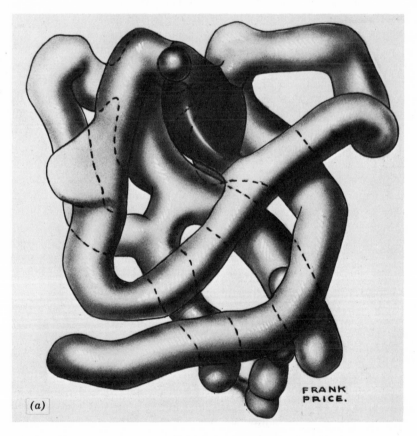

(a)

FRANK
PRICE.

Figure 2 Three dimensional models for myoglobin and hemoglobin molecules based on X-ray crystallographic data. (a) Myoglobin. The dark disc at the top represents the porphyrin prosthetic group. Dr. J. C. Kendrew of Cambridge University kindly furnished this photograph. (b) Hemoglobin. Dr. M. F. Perutz of Cambridge University kindly furnished this photograph.

Figure 2 (*continued*)

pended as desired. Problems of inhomogeneities which are always present in intact cell systems are minimized or eliminated in acellular systems. Furthermore, large molecules such as ribonuclease can be introduced. Highly electrically charged molecules such as ATP and polynucleotides can be added. In intact cell systems approaches involving the use of such additions are often denied because of the selective permeability barrier (cell membrane) which surrounds the cell. There is a comforting feeling associated with using an acellular, almost completely defined system that the investigator can manipulate as he chooses. The value of acellular systems are obvious and need not be extolled for anyone working in this field. What may need explanation and justification is the importance of intact cell studies for protein synthesis. Both approaches are vital to our understanding, but the final proving ground for any information obtained must be with a system actively and efficiently engaged in the synthesis of authentic proteins—the intact cell. The past successes with acellular systems do not justify the feeling that similar successes will be achieved in the problem of protein synthesis. The earlier research achievements concerned fairly isolated single-step metabolic conversions. The problem of protein synthesis as defined in this section, and which includes the intricate interplay of many series of complex reaction systems, is probably the most involved activity of the cell. In fact, it is believed to be the basis for cellular individuality, differentiation and propagation. The intact cell approach is admittedly limited in its application. A radioactive amino acid or nucleotide is introduced into a system where all of the required pools and structural integrity are intact. By stopping the incubation at particular times, the radioactive material can be isolated in different chemical combinations and located in particular cell organelles. By sampling at rapid time intervals we can trace the rate of passage of these radioactive substances through stages suspected of being intermediates for the formation of protein.

The acellular and cellular approaches should be used to complement and corroborate each other; for example, through studies involving acellular systems, amino acids were found to become associated with molecules of soluble ribonucleic acid. The natural interpretation of the role of such a compound would be in the intermediate stages of protein synthesis. Later work showed this to be so. The real question is whether this intermediate is on the predominant path for the intact cell. Once this chemical entity was discovered in the acellular system, it was important to move to an intact cell system to establish that the behavior of such an intermediate during the natural formation of proteins was consistent with its suspected role. In the particular example, a reading of the published papers shows that the relevance to normal protein synthesis was accepted before adequate testing in intact cells. This subject is covered again in detail in Chapter 3.

The total reliance on cell-free systems, and studies on isolated aspects of protein synthesis in systems producing nondefined atypical proteinaceous endproducts is not unlike an approach illustrated in an anecdote told by Dr. D. E. Green at a meeting that I attended several years ago: a fine Swiss watch was laid on a table and demolished by a single well-placed blow from a heavy hammer. The various pieces, as many as could be recovered, were swept up and placed in a bag. This bag of parts was taken to a small group of highly skilled and ingenious mechanics who, incidentally, had never built a watch before.

After many days of concentrated effort, thought, and experimentation, they produced from some of the parts a meshed interlocking set of gears, springs, and hammers. The gears, although they meshed with others in the reconstructed fragment, could with equal facility be arranged in totally different assemblies. The analogy is admittedly imperfect. Although the heavy hammer is a good comparison with the normal method of breaking cells, the cell itself is infinitely more intricate than the fine Swiss watch. Although more pieces of the cell are likely to be destroyed and altered, the biochemists are every bit as ingenious as the mechanics in the story. With their limited information, to be obtained from limited systems, complex assemblies can be constructed.[3] But which, if any, will keep time? That is the important question.

REFERENCES

1. R. B. Loftfield, *Biochem. J.,* **89,** 82 (1963).
2. G. Gamow, *Nature,* **173,** 318 (1954).
3. E. Baldwin, *Dynamic Aspects of Biochemistry,* 2nd ed., Cambridge University Press, London, 1952, p. 199.

[3] I have since learned that Dr. Green's story, which apparently I have freely paraphrased, concerned a baby Austin, the parts of which could be assembled by an ingenious mechanic into a perambulator, a pressure pump, or a hairdryer of sorts (3).

Chapter 2

THE MODERN PICTURE— CONTEMPORARY VIEWS 1947 TO 1955[1]

THE MODERN PICTURE—1947

The field of protein synthesis has recently been reviewed in considerable depth by J. H. Northrop (3). As Northrop points out, evidence is rapidly accumulating to show that all enzymes and at least some viruses are proteins. It is possible to consider all of the suggested mechanisms for protein synthesis as falling within one of the following categories:

1. Purely catalytic synthesis—no energy added.
   a. Equilibrium point shifted in favor of synthesis by change of concentration or other condition (plastein formation).
   b. Removal of protein as soon as it is formed. Insoluble proteins— plastein (?); synthesis of anilides (Bergman and Fruton, 1941). Protein surface films (Langmuir and Schaefer, 1938).
2. Energy added to shift equilibrium in favor of protein.
   a. Coupled reaction (cf. Borsook and Dubnoff, 1940, hippuric acid).
   b. Synthesis from building stones other than amino acids.

Examples of each of the preceding cases are briefly discussed.

CASE 1a. Since enzymes are catalysts and can catalyze a reaction from either direction, proteolytic enzymes are indeed capable of catalyzing the synthesis of protein. Consider a protein that releases 100 amino acids by hydrolysis.

[1] As explained in the preface, the present tense is used in the historical passages to convey the feeling and views existing in the field at the time. Indicated references are left for historical effect. Precise references can be obtained from original articles.

At equilibrium, we have an equilibrium constant, $K = $ [aa]^{100}/[protein]. By simply increasing the concentration of amino acid tenfold, the equilibrium concentration of unhydrolyzed protein would be increased 10^{100} times.

It has been known for quite some time that when pepsin is added to a concentrated solution of the products of pepsin digestion, a precipitate (plastein) is formed. This problem was carefully investigated some 17 years ago (1930) by Wastneys and Borsook. More recently, Haddock and Thomas (1942) have shown that plastein is also formed by the action of trypsin or papain on a solution of insulin which has been previously digested by pepsin. The real question is what relation these products have to authentic proteins. From a few reports it seems that the molecular weight of the product does not exceed 1000. A crucial test of the hypothesis has been the attempt to resynthesize the enzyme trypsin or pepsin from the hydrolysis products. Such experiments have produced products with no enzymatic activity (Northrop, 1947). It is possible, however, that conditions for authentic resynthesis were not right.

CASE 1b. The synthetic reaction can be driven by removing the product. A prime example of this possibility is illustrated by the formation of insoluble anilides, a reaction discovered by Bergmann and Fruton (1944). Furthermore, Langmuir, Gorter, and Schaefer, in independent investigations (1937–1939), showed that protein becomes insoluble at air-water interfaces. Such a process operating at some cell surface, if the insoluble protein could be removed, would ensure its continuous synthesis. If the protein could serve as a nucleus for orienting the "crystallization" of amino acids from the cell solution, then both the energy and specificity problem could be solved. This process, as Northrop has pointed out, deserves the most careful consideration.

CASE 2a. There are many recent examples of energy requiring reactions which will not proceed alone, but will occur if another reaction which liberates energy takes place simultaneously (Meyerhof, 1944, and Kalckar, 1944). Bergmann and Fruton (1944) have proposed that phosphoric acid esters may take part in the synthesis of proteins. Schoenheimer's remarkable experiments (1942) have shown that proteins are continuously being synthesized and hydrolyzed so that it is quite possible that part of the energy for protein synthesis is obtained from the hydrolysis of other proteins or peptides.

CASE 2b. An attractive hypothesis stems from the work of Alcock (1936) who states that plants can form proteins from sources other than amino acids. Since it is unlikely that animals would have an entirely different mechanism for protein synthesis, Alcock has suggested that animals may first change the amino acids into the hypothetical building stones common to plants and animals.

Northrop has pointed out that the reaction: protein + water ⇌ amino acids may favor hydrolysis for denatured proteins only. It is possible that the reaction will proceed spontaneously to the synthesis of native proteins without the addition of energy. The problem of specificity of synthesis is still perplexing, but one reasonable solution would be if all cells contained as a "genonema" at least one molecule of every protein needed. Then, by an autocatalytic crystallization-type process, specificity could be achieved.

Northrop has suggested a workable mechanism for protein synthesis which appears to be supported by many different lines of investigation too numerous to be described here. For additional detail the reader should refer to Northrop's article. In essence, it is proposed that each species or organ has a unique type protein that autocatalytically directs its own specific synthesis. This specific parent protein or proteinogen may contain nucleic acid. Spiegelman and Kamen (1946) have suggested that the energy for protein synthesis may be supplied by the phosphoric acid groups of nucleic acid. This proteinogen, which could be synthesized by a single organ like the liver, would circulate and be transformed to specific proteins in a manner similar to the conversion pepsinogen → pepsin, an enzyme being formed only in response to the presence of its substrate or related molecule. Since all proteins are in equilibrium with the proteinogen, they are in equilibrium with one another.

Viruses

Viruses could be proteins formed from virus proteinogens existing in the cell and uniquely folded in the presence of introduced virus. This process requires no new source of energy other than that used for normal protein synthesis, whereas building a virus *de novo* from building blocks (possibly amino acids) would require a new energy supply. There are a few recent papers which do not readily fit this picture of virus formation unless a few more assumptions are made. Cohen (1947) has reported that phosphorus and nitrogen for phage synthesis in *E. coli* come entirely from the medium, not from the host cell. Price (1947) has shown that Staphylococcus phage production requires some substances present in yeast extract, whereas the host cell does not require this substance. Therefore the phage proteinogen, instead of being present before infection, would have to be formed afterward.

These same concepts seem to cover antibody and adaptive enzyme formation. Burnet et al. have proposed a mechanism whereby normal proteinases in a cell convert the proteinogen to normal globulins and other proteins. When antigen (or inducer) is introduced, the proteinase is modified to produce antibody or adaptive enzyme. As the antigen or inducer is used up, the specific proteinase is reduced in quantity and the unique synthesis decreases. Sevag assumes that the antigen itself is the specific proteinase and not just the coenzyme which modifies the proteinase.

Notes from 1967 on the first amino acid-incorporation studies (1947–1949)

It was the work of Schoenheimer and his colleagues which first anticipated the experimental assault that began some five years after the appearance of his classic monograph on the dynamic state of body constituents. By use of stable isotopes it was conclusively demonstrated that, even in the adult animal, proteins, instead of representing a static pool of fixed amino acids, were in a continuous flux of breakdown and resynthesis.

Starting in 1947, the problem of protein biosynthesis entered an age of sophisticated experimentation and hypothesizing which has continued to grow ever since. This development grew out of the wartime Manhattan Project, in the form of the atomic pile at Oak Ridge, Tennessee. Although C^{14} was discovered at the turn of the decade in experiments with the Berkeley cyclotron (4), it was not available for general biological experimentation until the atomic pile was fully active; for example, to produce a millicurie of C^{14} by normal cyclotron methods, at least five efficient cyclotrons running continuously for a year (2000 total days of bombardment) at a final cost of around a million dollars would have been required. In 1947 the United States Atomic Energy Commission at Oak Ridge offered $BaC^{14}O_3$ for sale at \$50.00/mc at a specific activity of 3 mc per milliatom. $H_2S^{35}O_4$ was listed at \$2.40/mc. With the availability of these isotopes, and new ingenious modifications in organic synthetic procedures, to push up the yield of product based on the isotopic starting material, several laboratories produced and tested the first radioactive amino acids.

Several papers appeared in 1947 which described the uptake of radioactive amino acids by various tissue systems. Melchior and Tarver reported the uptake of cystine (S^{35}) by slices and homogenates of rat liver (5). Most of their incorporation, however, was due to disulfide bonding, and an important note of caution for the interpretation of *in vitro* uptake experiments was sounded by these authors. In an accompanying paper the same authors reported a small uptake of S^{35}-methionine by slices and virtually no uptake by a homogenate (6). Part of the label introduced as methionine was recovered in the form of cysteine. In the same year Winnick, Friedberg, and Greenberg reported the incorporation of C^{14} glycine into the proteins of a crude tissue mince of rat intestines (7). In these studies it was found that uptake was prevented by boiling the tissue, by homogenization and by 2×10^{-3} M sodium azide.

The year 1947 also marked the beginning of a long and fruitful series of pioneering investigations from the laboratories of the group at Harvard in association with Dr. Paul C. Zamecnik. Frantz, Loftfield, and Miller described the incorporation of C^{14}-alanine into proteins of rat liver slices (8). The important observation, that uptake occurred under an oxygen atmosphere but not under nitrogen, along with the earlier cited effects of sodium azide, signaled the coupling of protein synthetic reactions to those of oxidative energy production. Another of the early observations of *in vitro* amino acid incorporation was that of Anfinsen and coworkers, who discovered that $C^{14}O_2$ incubated with rat liver slices found its way into glutamic and aspartic acids subsequently released from tissue proteins by complete acid hydrolysis (9).

A very important first step was taken by Friedberg, Winnick, and Greenberg who showed that, contrary to the experience obtained initially, homogenates (broken cell preparations) of the rat spleen and liver, when properly prepared, could incorporate C^{14}-glycine (10). Boiling the tissue abolished incorporation, and homogenates of the intestine were inactive. The following year Zamecnik, Frantz, Loftfield, and Stephenson studied the question of whether malignant cells would show an enhanced rate of amino acid incorporation (11). Hepatomas were induced in rats by oral administration of p-dimethylaminoazobenzene. Normal liver tissue from the same and also from untreated rats was used for comparison. The uptake of C^{14}-alanine and glycine into tissue slices was studied. In this careful investigation a correction technique for self-absorption was thoroughly described. It was found that incorporation was significantly greater in the malignant tissues. Aware of the difficulties of interpretation involved if some of the radioactivity introduced as one amino acid were converted to and incorporated in the form of other amino acids, the authors isolated various amino acids by the tedious procedure of adding carrier and recrystallizing to constant specific activity. An attempt at quantitative estimation of the extent of incorporation indicated that 0.2% of the protein-bound alanine came from free alanine in the incubation medium. It was also found that fetal liver preparations and regenerating liver gave enhanced rates of uptake. A search was made for stimulating factors from tissue extracts of the various active liver preparations; the results were negative. The effects of the thickness of the slice, and the addition of a structural analogue (benzoylalanine), were studied and found to be minimal. Incorporation was also studied in relation to the rate of oxygen uptake, and it appeared that hepatoma made more protein per microliter of oxygen consumed than normal liver. Finally, in a perfusion experiment with an intact animal it was found that the hepatoma incorporated 50% more alanine than the surrounding normal tissue.

Friedberg, Schulman, and Greenberg reported in 1947 that homogenates of fetal rat liver were much more active for C^{14}-glycine incorporation than were homogenates from older animals (12). In a short note of the same year, Frantz, Zamecnik, Reese, and Stephenson showed that in oxygen, 2,4-dinitrophenol inhibited alanine incorporation into proteins in both normal and malignant liver slices at concentrations (up to 10^{-4} M) where oxygen uptake was not affected (13). Since at these concentrations it had been shown (14) that phosphorylation was uncoupled from oxidation, a link between amino acid uptake and the use of high energy resulting from substrate oxidation was strongly implied.

The first experiment in molecular biology (i.e., the use of *E. coli* to study protein synthesis) was reported in 1948 by Melchior, Mellody, and Klotz (15), who studied the uptake of S^{35}-methionine by nonproliferating cells. Azide, cyanide, and fluoride exerted pronounced inhibitory effects, whereas penicillin G had a minimal influence. Sulfanilamide was inhibitory and the effect was not reversed by p-amino benzoic acid, contrary to the effects of these agents on growth in a nutrient medium.

THE DEVELOPMENT OF ACELLULAR AMINO ACID-INCORPORATING SYSTEMS

A more detailed study of the conditions which influence homogenate incorporation systems was published by Winnick, Friedberg, and Greenberg (16). Unfortunately, glycine was the amino acid tested and later studies revealed that under the conditions used an appreciable amount of such incorporation involved nonpeptide bonds. Nonetheless, the influence of various factors and conditions on homogenate-incorporation was studied and this work constituted the basis for the future investigations with acellular systems, from which almost our entire concept of protein synthesis has been developed. The uptake of glycine was concomitant with oxygen consumption and both processes were drastically reduced by boiling the homogenate. The uptake was increased by increasing the concentration of glycine in the solution, although the efficiency of uptake was greatest at the lowest concentrations. Of particular interest was the finding that an appreciable portion of the incorporation was associated with a sedimentable fraction of "insoluble cell particles." These particles showed a broad optimal pH range for incorporation of around 7.5. The importance of some kind of structural integrity was indicated by the loss of activity upon homogenization in distilled water, lyophilization, and repeated freezing and thawing. The washed particles were used to test the importance of various additions on the incorporation process. Thus Ca^{2+}, glucose, and citrate were beneficial, whereas cytochrome c, ATP, phosphoglyceric acid, and pyridoxal were without significant effect, and a mixture of amino acids was slightly inhibitory.

Many points of difference were emphasized between this incorporation process and p-amino hippuric acid synthesis. At that time, and for many years, thereafter, it was believed that a similarity in mechanism existed for protein synthesis and for the formation of other peptide-like compounds. In an accompanying paper Winnick, Moring-Claesson, and Greenberg reported isolating various amino acids from a hydrolysate of the homogenate-labeled protein obtained after incubation with glycine (17). They found that only 12% of the radioactivity was in the form of glycine. Approximately 60% was in the form of serine, and small amounts were present as glutamic and aspartic acids and arginine. In another study Borsook and collaborators introduced a concept which was to become a major preoccupation with investigators in this field for the next 10 to 15 years—that of peptide intermediates in protein synthesis (18).

If single amino acids are destined for inclusion in molecules in which several hundred are linked together, would it not be most reasonable for this process to proceed in a stepwise fashion involving intermediate peptides? Incorporation of C^{14}-lysine in guinea pig liver homogenates was studied. A peptide was isolated from the protein-free solution that exhibited a rate of radioactive lysine turnover much greater than that which occurred in the protein fraction. If a peptide intermediate existed between free lysine and protein-bound lysine, then the relative turnover rates observed would be consistent with the suspected role of the isolated peptide described in this work. A similar finding with C^{14}-leucine, discovered in another suspected peptide intermediate, was also reported in 1948 (19), along with the discovery of the existence of other peptides in these homogenates.

An important added refinement, that of observing the incorporation of radioactive amino acid into a pure isolated protein, was introduced by Zamecnik, Loftfield, Stephenson, and Williams, who demonstrated the incorporation of radioactive glycine and alanine into silk protein produced by silkworms (20). It was also found that a crude silk fraction produced in vitro by isolated silk glands became labeled upon incubation with radioactive amino acids.

Publications following soon after these first demonstrations extended the observation of in vitro amino acid incorporation to other tissues and other radioactive amino acids. Thus Borsook et al. showed the incorporation of radioactive glycine, leucine, and lysine into the proteins of bone marrow cells (21) and of rat diaphragm (22). Shemin, London, and Rittenberg showed the uptake of N^{15}-histidine into crude hemoglobin of duck red blood cells (23). Peters and Anfinsen reported the labeling of pure serum albumin with radioactive glutamic and aspartic acids by incubating chicken liver slices with radioactive CO_2 (24). Shortly thereafter Peters and Anfinsen found a net formation of serum albumin in this system (25).

*Contemporary Views on the Mechanism of
Peptide Bond Formation (1949–1950)*

Fritz Lipmann has recently analyzed the current status of the problem of peptide biosynthesis (26). He points out that the mechanism of protein synthesis will most likely be similar to that of the synthesis of "peptidic"[2] bonds in simpler molecules. The animal body has fortunately supplied a number of model reactions for the study of the synthesis of peptidic links in two component systems. Lipmann's studies on the acetylation or aromatic amines and the participation of acetyl phosphate in these reactions leads to the generalization that carboxyl activation by phosphorylation may represent a major phase in biosynthesis.

Lipmann reviewed the recent studies of Speck (27) and Elliot (28) on the biosynthesis of glutamine and suggested that, although an acyl phosphate has not been identified, its presence is indicated in the formation of glutamhydroxamic acid by the system. A third model system, that of p-amino hippuric acid formation (29), by its enhancement due to added ATP also points to a thoroughly analogous situation; the energy-rich phosphate of ATP acts as an immediate energy source attaching itself primarily to the carboxyl component.

The formation of protein is certainly a much more complex problem, but here too the dependence of amino acid incorporation on reactions of oxidative phosphorylation is indicated by the recent studies of Greenberg and coworkers and Zamecnik and coworkers. The parallel inhibitory effect of dinitrophenol on phosphorylation and protein synthesis suggests that phosphate bond transfer is operative in the process of amino acid-incorporation into protein. In fact, it appears that a single energy rich phosphate bond is used to effect a single peptide link.

The Importance of Microsomes as the Major
Site of Protein Synthesis Is Established

Using the recently developed techniques of Hogeboom, Schneider, and Palade for tissue fractionation (30–32), Borsook *et al.* separated adult guinea pig liver into nuclei, mitochondria, microsomes and supernatant fractions (33a). Incubation of these fractions with C^{14}-labeled glycine, L-leucine and L-lysine was carried out. No consistent pattern of incorporation resulted and the microsome fraction was particularly inert. In light of more recent knowledge (see pages 22–25) it seems that these early studies were plagued by *in vitro* incorporation artifacts. Shortly thereafter, the same authors studied the distribution of radioactivity administered in the form of the same radioactive amino acids plus C^{14}-L-histidine after they were injected into the tail vein of intact guinea pigs (33b). The animals were killed 30 min after injection and the radioactive livers were homogenized and fractionated. Here the picture was clear-cut. The greatest incorporation for every amino acid

---

[2] Lipmann introduced the term "peptidic link" as a generic name for a —CO—NH- link between any amino and carboxyl group. The term "peptide link" is then reserved for the 1-carboxyl, 2-amino link between two alpha amino acids as it occurs in proteins.

was into the microsome fraction. In the same year Hultin (34) used tissue fractionation techniques of Claude (35) and Hogeboom, Schneider, and Palade (30–32) to obtain nuclei, mitochondria, microsomes, and supernatant fractions from chick livers. The chicks were first injected in the wing vein with N^{15}-glycine, and sacrificed at intervals from 10 min to 5 hr later, whereupon their livers were removed and fractionated. Hultin found that *in vivo*, N^{15} from glycine was incorporated into the microsome proteins earlier and at a more rapid rate than into the proteins of any other cell fraction isolated by differential centrifugation. The following year, Keller showed that with C^{14}-leucine injected into rats, the subsequent fractionation of the liver resulted in the microsome fraction being the most highly labeled (36). Similar observations were later reported by Allfrey *et al.* with mouse pancreas (37). Siekevitz and Zamecnik, in a preliminary note, reported that after the incorporation of alanine in a whole homogenate and subsequent fractionation of the tissue, the microsomes represented the most radioactive fraction (38). Lee, Anderson, Miller, and Williams observed that after an intraperitoneal injection of S^{35}-DL cysteine, an analysis of the subcellular components isolated from rat liver also showed the microsomes to be the most active fraction (39).

A later rather detailed study employing C^{14}-leucine and C^{14}-valine both *in vivo* and *in vitro*, using rats as the experimental animal, again firmly implicated microsomes as the major cytological fragments involved in protein synthesis (40). At about the time the first liver fractionations were being attempted along the lines of recognized cell organelles, as discussed, Peterson, Winnick, and Greenberg used an empirical fractionation method to obtain a particulate fraction from rat liver homogenates which was active in the uptake of C^{14}-labeled serine, phenylalanine, leucine, and glycine (41). The system, which contained mainly mitochondria, was rather thoroughly studied and will be referred to again later in this chapter in relation to the establishment of an ATP requirement and the demonstration of an *in vitro* incorporation artifact.

The definitive paper which established the basis for future acellular studies and clearly revealed the microsomal incorporating ability, its dependence on energy and soluble factors, and the precipitability at pH 5 of synergistic factors, was the study of Siekevitz (42). The incorporation of C^{14}-alanine into various combinations of liver-homogenate fractions was studied. Incubation of the homogenate with radioactive alanine and subsequent fractionation clearly showed the microsomal fraction to be the most active from the first point taken at 5 min and also later in the incubation. Although each fraction by itself was relatively inactive, mitochondria plus microsomes were 40% as active as the whole homogenate, and the addition of supernatant fractions brought this value to 90%. Mitochondria plus a pH 5 precipitate of the mitochondrial supernatant (which contained the microsomes) gave 100% of the activity of the whole homogenate. Boiling this pH 5 sediment destroyed its activity. It was further observed that a precipitate formed upon the addition of magnesium chloride to the mitochondrial supernatant was able to replace the pH 5 sediment. Since magnesium chloride is known to bring down nucleoprotein, Siekevitz in 1952 cautiously observed, "It is possible, as Caspersson and Brachet have suggested, that the ribonucleic acids are somehow concerned in protein synthesis." This paper, which also helped to define the energy requirement for protein synthesis, is referred to again in the discussion which follows.

Relationship between Amino Acid
Incorporation and Metabolic Energy

Concurrent with the establishment of the roles of the various cellular subfractions was the elucidation that the energy of oxidative metabolism was used for protein synthesis. Writing in 1948, Greenberg *et al.* stated that there was still no definite proof that phosphate bond energy was required for protein synthesis (43). Additions of ATP were not able to stimulate amino acid incorporation in liver homogenates. However, the authors felt that the question was open since oxygen seemed to be required for incorporation, and such poisons of oxidation and phosphorylation as 2,4-dinitrophenol, cyanide, and azide were strong inhibitors of amino acid uptake into proteins. The eventual experimental demonstration of the participation of ATP in amino acid-incorporation processes was anticipated by the results of Frantz *et al.* who showed that anaerobiosis and dinitrophenol inhibited the uptake of alanine by rat liver slices (8, 13); by Winnick *et al.* who found that oxygen consumption was correlated to glycine uptake in a rat liver homogenate and that the process was inhibited by anaerobiosis (17); and by Melchior *et al.* who found that the uptake of methionine by *E. coli* was inhibited by azide, fluoride, and cyanide (15). Similar findings were reported by Borsook *et ali.* (44). The first observation that ATP addition caused a stimulation of amino acid uptake was that of Peterson, Greenberg, and Winnick (45, 41). These investigators found that a particulate fraction isolated from a rat liver homogenate, after it was drained of supernatant fluid, was very low in incorporating ability with respect to C^{14}-glycine, phenylalanine, serine and methionine. The restoration of incorporating ability depended upon the presence of ATP in the suspending medium. Winnick promptly extended these findings to C^{14}-glycine and alanine in homogenates of fetal rat liver that had been depleted of small molecules by dialysis (46). These homogenates, which after dialysis lost most of their incorporating ability, were substantially reactivated by the addition of ATP with the maximum effect being attained at 1.5×10^{-3} M. A minimal activation by a mixture of amino acids was also noted. Siekevitz and Zamecnik used an acellular rat liver preparation of mitochondria plus microsomes to show a striking dependence of alanine incorporation on conditions which affected oxidative phosphorylation (38).

In a more complete study Siekevitz showed the quantitative relation between alanine incorporation and oxidation-dependent phosphorylation in the same system (42). Oxidation reactions were left intact but ATP formation was prevented by the addition of 2,4-dinitrophenol. Alanine incorporation was eliminated, showing that incorporation was linked to the utilization of energy released by oxidation and not to oxidation *per se*. Various techniques which interfered with oxidative phosphorylation had a correlative inhibitory effect on alanine incorporation. A thoroughly fascinating observation was that a soluble factor was produced by incubation of the mitochondria with an oxidizable substrate which, when added to the microsome preparation, stimulated the incorporation of alanine even under anaerobic conditions. The factor was stable to boiling for 7 min in 1 N hydrochloric acid and thus could not be ATP or any similar pyrophosphate. This finding has unfortunately never been subsequently pursued or explained and remains a mystery to this day. Peterson and Greenberg further described their acellular particulate preparation derived from rat liver and showed the dependence of incorporation of several amino acids on the presence of an oxidative phosphorylating system (47). A stimulation by added amino acids was also shown, although full stimulation was provided by a mixture of only six amino acids which would suggest more the function of supplying acceptable oxidative substrates rather than a complete mixture of amino acids to satisfy the needs of protein synthesis. This is especially true, for later work with acellular rat liver systems failed to show a dependence on added L amino acids. Inhibitors of respiration and phosphorylation as well as the presence of certain metal ions inhibited incorporation. Particularly potent inhibitors were toluene or n-octanol at a concentration of three parts per hundred. Similar findings were published by Kit and Greenberg with the same system but employing other amino acids (48).

The further evolution of the microsomal incorporating system and clarification of the ATP requirement was achieved in the excellent study by Zamecnik and Keller with fractions derived from rat liver homogenates (49). Until this study, incorporation seemed to require an aerobic environment. In a system consisting of mitochondria and microsomes it was shown that hexose diphosphate plus nicotinamide under anaerobic conditions caused a tenfold stimulation of C^{14}-leucine incorporation. It was learned that mitochondria could be eliminated from the system if conditions for glycolysis existed in the supernatant fluid and hexose diphosphate and nicotinamide were added. It was also observed that gentler homogenization resulted in more active preparations. This was one of the earliest indications of the importance of cellular structural integrity for the process of amino acid incorporation, although as will be discussed, subsequent experimentation moved in the direction of further structural fragmentation rather than preservation. Addition of DPNase[1] destroyed incorporating ability. The glycolytic system could be replaced by the endogenous high energy compounds, phosphocreatine, phosphorylenol pyruvate, or 3-phosphoglycerate. The participation of adenine nucleotides in the process when one of the aforementioned compounds was used was demonstrable only after the passage of the supernatant through Norite which lowered the endogenous level. It was again confirmed that microsomes were the most active fraction with respect to amino acid incorporation. The microsomes by themselves, however, were inactive, and in order to obtain activity it was necessary to add supernatant fraction. Both fractions were found to be heat labile, but the supernatant fraction was relatively stable and could be stored (frozen) with retention of activity for months. The microsome fraction, on the other hand, lost activity on standing even at $0°$ and also upon quick freezing. After dialysis the supernatant fraction required the addition of ATP and phosphocreatine for full activity. The interesting observation was made that ribonuclease was a potent inhibitor of incorporation. Deoxycholate was also found to be a potent inhibitor even when energy was supplied by anaerobic glycolysis. In experiments with radioactive leucine, alanine, and glycine in the presence or absence of other C^{12}-amino acids it was found that each amino acid was incorporated without benefit from the addition of other amino acids and in some instances a competition between combinations of C^{14}-amino acids seemed evident. The unnatural amino acids α-amino butyric acid or D-leucine were not actively incorporated. Protein that had become labeled with C^{14}-lysine, when subsequently subjected to a slow acid hydrolysis procedure, released radioactivity about as fast as it released amino acids in general, which provided good evidence that the incorporated leucine was held by bonds as strong as those holding the rest of the molecule together. Preliminary experiments were described which indicated that small peptides released by this partial hydrolysis procedure were being

[1] An enzyme which destroys diphosphopyridine nucleotide.

recovered and analyzed to provide further proof of incorporation of amino acid into the matrix of protein chains.

Some Dangers of Acellular Systems Are Discovered

In the manner just traced many pioneers charted the route for the experimental inquiry into the complex processes involved in amino acid-incorporation and provided the foundation for future developments which are outlined in the pages that follow. The studies of Zamecnik and Keller already discussed, together with that of Siekevitz, serve as a particularly solid support for later work. Nonetheless, the experimental route laid out in the painstaking experiments just described was dangerous, fraught with deceptive turns, and should be explored only by those who are aware of its limitations and pitfalls. Because much of our current knowledge has been obtained by studies with these systems and by many investigators who take the reliability of such systems for granted, it would be useful to reconsider some of the dangers which have been recognized in the past.

It was 1949 when I joined Dr. Greenberg's department as a graduate student at the University of California at Berkeley. The first experimental phase in the study of the problem of protein synthesis—that of examining the incorporation of radioactive amino acids into proteins of various intact and acellular systems, was drawing to a close. Winnick, Friedberg, Schulman, Melchior, and Siekevitz were gone from the campus. Peterson and Kit were finishing their thesis work with acellular liver incorporating systems. Tarver and Greenberg were continuing the pursuit of the problem with their newer graduate students.

Although it was not long ago as time goes, many changes in procedure have since come about. At that time after completing an experiment it was necessary to sit in front of a counter and manually assay every sample. Some of the counters had a compartment which would pregas the next planchet to be counted and thus save a considerable amount of time. Nonetheless, many hours were spent in the counting room and this gave time for reading and frequent discussions. One subject of major concern was the question of how related were the phenomena of amino acid-incorporation in acellular systems and actual protein synthesis. Melchior and Tarver had shown that sulphur-containing amino acids were "incorporated" into proteins in such systems by disulphide rather than peptide bonds (5). The early workers in this field rapidly became aware of these dangers by uncovering several instances of artifactual incorporations. It was disturbing that a severe and unyielding critic was on the scene at the time—Dr. Albert Keston. Keston was often mentioned by the graduate students and the comments that Keston made were not easily dismissed, even though that is what we tried to do. Although Keston was not himself working in the field, and there are no published accounts of his observations in this connection, he had a marked influence on developments at the time. Winnick, in one of his early papers, tells of a personal conversation with Keston during which Keston told him that, after incorporating C^{14}-glycine into the proteins, of a rat liver homogenate, Keston could release most of the radioactivity by dialyzing the protein against ammonium hydroxide (50). This unpublished observation of Keston was also mentioned by Peterson and Greenberg. Greenberg and associates had confirmed this finding. Winnick showed that by dissolving the trichloroacetic acid (TCA) precipitate in dilute sodium hydroxide and reprecipitating with TCA the same result could be accomplished. This occurred only for proteins labeled in a homogenate, *not* for liver proteins labeled by injecting the rat with C^{14}-glycine. I heard that at one meeting where various incorporation studies were being discussed, Keston asked Greenberg whether the term "incorporation" could be used to properly describe the uptake of C^{14}-phenylalanine by activated charcoal.

Peterson and Greenberg described another artifact which they discovered in acellular incorporations which principally involved glycine, but which also occurred with other amino acids. It was possible to remove up to two thirds of the incorporated radioactivity by employing agents that broke disulfide bonds such as thioglycolic or performic acids. The explanation was that glycine was incorporated into glutathione which was then bound to proteins in the homogenate (but *not* in the intact cells) by disulfide bonds. Other amino acids, however, not present in glutathione, were subsequently found to be held by bonds susceptible to treatment with performic acid (48, 51).

As an example of the extent that Keston's views occupied the minds of some of the early workers in this field, the reader is referred to Cold Spring Harbor Symposia.[2] Zamecnik had just presented his paper entitled, "Peptide Bond Synthesis in Normal and Malignant Tissue." In the discussion period, Martin Schulman, recently of Greenberg's laboratory, volunteered a public answer to some of Keston's unpublished criticisms. It is interesting that Keston's criticisms were not at issue and therefore not mentioned in Zamecnik's paper. Schulman summarized the principal arguments supporting the concept that acellular incorporation represented true peptide bond synthesis and not adsorption or related nonenzymatic processes.

[2] Cold Spring Harbor Symposia, Vol. XIV, 1949, p. 208.

There were four points made:

1. The inhibition of incorporation by heating, lack of oxygen, or by azide or cyanide.

2. In experiments with carboxyl-labeled glycine, the labeled protein did not release $C^{14}O_2$ by treatment with ninhydrin but did after complete acid or alkaline hydrolysis.

3. Similarly, ninhydrin treatment released no $C^{14}O_2$ from partial enzymatic hydrolysates but did from enzymatic hydrolysis to the free amino stage.

4. The greater uptake ability of embryonic and regenerating liver was cited.

The controls mentioned as points (2) and (3) became standard tests to prove that the free amino acids were not simply adsorbed to protein, but that the carboxyl and/or amino group was involved in some type of chemical bond. Such tests were used by Peterson and Greenberg, and Zamecnik, as well as many others, even up to fairly recent times.

In 1952 a classic paper appeared on the subject of amino acid incorporation in an acellular liver system (52). Brunish and Luck fractionated a liver homogenate supernatant into crude albumin, globulin and ribonucleoprotein fractions. A deoxyribonucleohistone fraction was prepared from the residue. The homogenate either as a whole or in the form of fractions was incubated according to standard procedures with C^{14}-phenylalanine, glycine, alanine, and lysine, and the reaction was terminated with the addition of 10% TCA. The protein was isolated, washed, and assayed for radioactivity. The most active "incorporation" occurred with the deoxynucleohistone preparation. A zero time control had no radioactivity and incorporation was nearly linear for the 2 hr of incubation. A boiled control showed 50% activity. No incorporation occurred below 25°, but incorporation at 70° was more than double the incorporation at 37°. The pH range of incorporation was 5.8 to 9.0. It was found that histone itself, isolated from the deoxynucleohistone, was just as active and that the buffered medium could be replaced by distilled water with no loss of activity. The zero time control for this system showed no incorporation and incubation at 100° resulted in about seven times as much incorporation as at 37°. If the histone was heated 10 min at 100° before incubation at 100° for 7 hr, the incorporating ability was destroyed. The incorporation phenomenon showed a pH maximum at 8.8. As controls, one sample was mixed with an equal volume of 1 M nonradioactive amino acid and dialyzed 4 hr against a 1 M solution of nonradioactive amino acid. Another sample, after precipitation with TCA was dissolved in 0.5 N sodium hydroxide, allowed to stand 1 hr, reprecipitated with TCA and then washed in the usual way. These samples were compared with protein isolated and washed in the usual way. With three different radioactive amino acids, these drastic procedures removed little or no radioactivity. Finally, the most powerful control of all was used—for proteins labeled by incubation with carboxyl-labeled alanine or glycine the determination whether ninhydrin treatment at pH 2.7 would release $C^{14}O_2$ prior to hydrolysis. Such treatment released no $C^{14}O_2$ from the labeled proteins. These studies must surely justify cautious reservation for the interpretation of *in vitro* amino acid-incorporation into noncharacterized proteinacious products. Later experiments by Schweet showed that lysine could be bound through its epsilon amino group (53). A test developed in more recent years to establish that incorporation is not simply due to the addition of amino acids to terminal or epsilon amino groups within a peptide, is to show by use of dinitrofluorobenzene, followed by hydrolysis, that little of the amino acid is recovered in the form of the dinitrophenyl derivative. This reagent reacts only with amino acids at the end of a peptide. Even this test, however, has its limitations since a mechanism that adds amino acids to free N-terminal positions could cover up recently added amino acids by other amino acids and much of the activity would seem to be held by interior amino acids. In acellular systems, this tendency toward artifactual incorporation, combined with the great loss of incorporating ability (see page 297), and the general loss of ability to make characteristic proteins of the tissue, all combine to make such systems dangerous by themselves unless some correlation can be drawn with intact systems or with the formation of a normal protein.

*A Sophisticated Mechanism of Protein Synthesis is
Proposed and Supported by Experimental Data*

Transpeptidation. In 1950 a model for explaining protein synthesis was taking shape. The proposed model was supported by many experimental observations and was extremely attractive to workers in the field as evidenced by the many papers, reviews, and symposia held during the ensuing four or five years. The impetus behind this movement came principally from Cambridge and Yale Universities. The spokesman and leading figure at the time was Joseph Fruton at Yale. Additional support was supplied by Heinrich Waelsch at Columbia University.

Transpeptidation may be defined as the direct replacement of one amino acid (or peptide) involved in a peptide bond by another amino acid (or peptide) through an enzyme catalyzed reaction. Transamidation is the corresponding reaction in which either the original or final amide bond is not a true peptide bond. The compound possessing the original bond is the substrate or donor and the other participant is the replacing substance, or acceptor. Interest in reactions of this type was based upon the belief that they may be part of the integrated mechanism for the formation and modification of peptides involved in protein synthesis. The first reported transamidation reaction was the formation of hippurylanilide and ammonia from an incubation of hippuric acid amide and aniline in the presence of cysteine-activated papain (54). In this paper, Bergmann and Fraenkel-Conrat determined that the rate of formation of the anilide was greater than its formation by direct incubation of hippuric acid and aniline and therefore concluded that the reaction represented a transfer type, rather than hydrolysis followed by synthesis. The process of transpeptidation however, always occurs under conditions where the original bond is undergoing active hydrolysis and is in fact, believed to be catalyzed in an identical way by the same enzyme. Bergmann and Fraenkel-Conrat suggested that such transfer reactions may play a role in the biological synthesis of individual proteins. Transpeptidation reactions involving glutathione and other gamma glutamyl peptides have received so much attention and study that they are discussed separately.

Characteristics of the Reactions

Conditions. The most active laboratory in the study of transpeptidation and transamidation reactions has been that of J. S. Fruton and his collaborators at Yale University. In a series of papers from 1950 to 1953 he covers a substantial bulk of the evidence on which the possible significance of these reactions is based (55–61). The general observation was that in simple aqueous or alcoholic solutions consisting of the substrate, isotopic replacing agent, and activated enzyme it was possible to demonstrate the presence of isotope in a portion of the residual substrate after approximately 40% had been digested by the enzyme. The several papers extended this to many different substrates, replacing agents, and enzymes. The substrates were amide derivatives of amine substituted amino acids or peptides, the replacing substances were ammonia, hydroxylamine, amino acids, and peptides, and the enzymes used were cysteine and cyanide activated papain, ficin, chymotrypsin, and cathepsin C. The concentrations of reactants have usually been 0.05 M, or 0.025 M, but the effect was observed with hydroxylamine down to 0.01 M. Reactions have been as simple as replacing the NH_2 portion of an amide by $N^{15}H_2$ derived from $N^{15}H_3$, and as complicated as the successive addition of two different amino acid residues in a single incubation with two enzymes. Insoluble polypeptides containing four or five repeating dipeptide units have been formed when the same compound was able to serve as both a substrate and replacing agent. The demonstrations of suspected reaction products have been accomplished by chromatography on paper and on columns of Dowex 50. Although the extent of transamidation is usually quite small compared to the amount of substrate hydrolyzed, in certain complex systems which can lead to an insoluble polymer instances of greater transamidation than hydrolysis have been found. Increase of pH (within limits) was found to favor transamidation, whereas the extent of hydrolysis was decreased. The interpretation offered was that the active replacing agent was the uncharged base and an increase in pH increased its concentration. Analogous reactions have been observed in other laboratories, thus Frantz and Loftfield observed the replacement of terminal glycine in glycylglycine by isotopic glycine (62). Brenner *et al.* incubated esters of methionine or threonine, and after one or two days observed mixtures of polypeptides of the single amino acids in addition to the hydrolysis product (63). Waley and Watson observed the formation of new peptides during enzymatic hydrolysis of other peptides, amides, and esters (64).

Gamma Glutamyl Transpeptidation. In a stimulating paper, Hanes, Hird, and Isherwood formally stated the proposition of γ-glutamyl transpeptidation (65). Their speculations sought to solve two very important mysteries. First, they gave a good reason for the ubiquitous occurrence of glutathione, a unique peptide in that it utilizes the γ-carboxyl group of glutamic acid and also because it is the only peptide so abundantly distributed in nature. Second, they offered the first insight into the fine mechanism for amino acid incorporation into peptide sequences. In their paper the authors described incubations of glutathione with each of six amino acids in the presence of an enzyme system extracted from sheep kidney. In every case a new chromatographic spot appeared when these incubation mixtures were chromatographed on paper. These spots increased in intensity with longer incubation at the apparent expense of glutathione. In one case the spot was eluted and structural analysis indicated its identity with the suspected γ-glutamylphenylalanine. Attention was drawn to the reported stability of the γ-glutamyl bond in glutathione to peptidase action and the apparent stability conferred by this bond to the adjacent α-link of cysteinylglycine. These properties are favorable to the concept of such compounds as reservoirs of primary peptide bonds. It was postulated that glutamine might be a primary compound formed at the expense of ATP. By transfer of the γ-glutamyl moiety to cysteinglglycine, glutathione would be formed, which could interchange its γ-glutamyl radical with other amino acids or peptides. If the γ-glutamyl peptides could then form α-peptides which themselves could be rearranged by transpeptidation, this sequence of reactions would constitute a mechanism for the utilization of ATP to form a great variety of particular peptide sequences for protein synthesis.

A second paper by the same group further established and extended the evidence for the occurrence of γ-glutamyl transpeptidation (66). In this paper, however, it was shown that glutamine was inert as a donor of the γ-glutamyl radical to the other amino acids and that it could not be formed by transfer of the γ-glutamyl radical of a peptide to ammonia. The spectrum of donor compounds was enlarged to include five other γ-glutamyl dipeptides. Nine amino acids including glutamine and three γ-glutamyl dipeptides were shown to function as acceptors of the γ-glutamyl radical. With proline and arginine as acceptors, little reaction, if any, occurred. Positive identification was given for the γ-glutamyl peptides of phenylalanine and glycine. When γ-glutamylphenylalanine was added to a kidney enzyme extract, it was quite stable, whereas α-glutamylphenylalanine was rapidly hydrolyzed. This fact supported the postulation concerning the stability of these bonds to enzymatic hydrolysis.

In a later paper Hird and Springel provided quantitative data for the extent of reaction of glutathione with amino acids together with chromatographic evidence for the existence of the peptides indirectly measured by the liberation of cysteine (67). The effect of acceptor concentration was also studied. Of the amino acids tested, aspartic acid, isoleucine, and valine were exceptions in that their presence caused a decrease in liberation of cysteine from glutathione. These authors demonstrated reactivity for arginine which had been reported relatively inert by Hanes *et al.* It is important to note that Hanes *et al.* used γ-glutamylglutamic acid as the donor whereas Hird and Springel used glutathione.

Very strong support in favor of the possible significance of γ-glutamyl transpeptidation for protein biosynthesis can be found in the reviews and papers of Heinrich Waelsch and his coworkers at Columbia University (68–70). Binkley and Olson had observed earlier that the hydrolysis of glutathione as measured by the release of cysteine in a pig kidney system was stimulated by glutamine and to a much smaller extent by other amino acids (71). Binkley and Christenson later reinvestigated the activating effect of amino acids on this reaction and reported that it was significant (72). They interpreted the phenomenon in support of a γ-glutamyl transpeptidation whereby the γ-glutamyl radical was transferred to glutamine and to the other amino acids. Other systems which can transfer the γ-glutamyl radical from glutamine to $N^{15}H_3$ or NH_2OH have been studied. In these systems, however, amino acids did not accept the γ-glutamyl residue to form γ-glutamyl peptides. Many similar observations were reported following these studies, including the formation of γ-glutamylarginine in ox kidney extracts by Kinoshita and Ball (73), γ-glutamyl transpeptidase activity in *P. vulgaris* (74), and in *B. subtilis* (75), an organism which normally synthesizes a polypeptide of D and L-glutamic acid connected primarily by γ-linkages. This system was able to transfer the γ-glutamyl radical from ammonia in glutamine to glutamic acid, a most active transfer occurring between L-glutamine and D-glutamic acid. Williams and Thorne suggested that glutamyl radicals might be continually added to form a polymer similar to the natural polypeptide.

In 1950 I started the experimental work for my doctoral thesis. The very exciting paper of Hanes, Hird and Isherwood, demonstrating γ-glutamyl transpeptidation, had just appeared as had a stimulating discussion entitled, "The Role of Proteolytic Enzymes in the Biosynthesis of Peptide Bonds" by J. S. Fruton (55). For my doctoral thesis I had hoped to demonstrate the role of γ-glutamyl peptides in protein synthesis by a direct test of the possibility in systems synthesizing protein (or at least incorporating amino acids). It is interesting to note that the leading laboratories engaged in the consideration of this possibility confined their activities to enzymatic studies and did not directly examine their proposals in systems which were incorporating amino acids into proteins. My plan was to synthesize a model γ-glutamyl peptide containing a radioactive amino acid. In acellular and cellular systems incorporating this amino acid, if it were necessary for the amino acid to pass through a free γ-glutamyl peptide stage (these peptides were demonstrated in the protein free supernatant by Hanes et al.), the γ-glutamyl-bound amino acid would be incorporated into protein at least as readily as the free amino acid. Furthermore, in incubations with various radioactive amino acids in the presence of added γ-glutamyl compounds to serve as a trap for the suspected γ-glutamyl intermediates, I looked for the radioactive γ-glutamyl compounds. I also tested for the possible stimulation of incorporation of radioactive amino acids by supplying small quantities of γ-glutamyl donors. Further, I investigated the proposal that one function of the γ-glutamyl link was to protect a primary reservoir of peptide bonds from enzymatic hydrolysis.

Although I approached this study with much enthusiasm and anticipation, the picture which gradually unfolded was that no support whatever could be found for the suggested role of these compounds in amino acid incorporation processes. The radioactive peptide that I synthesized was γ-glutamylglycine, labeled in both carbons of the glycine. The incorporation of free glycine was compared with that of glycine introduced as γ-glutamylglycine in an acellular liver preparation, in intact spleen cell suspensions, spleen homogenate, intact cells of mouse Gardner Lymphosarcoma, and kidney homogenates. In every case free glycine was much more actively incorporated into the proteins than the peptide-bound glycine. Added γ-glutamyl compounds did not stimulate the incorporation of radioactive glycine, valine, or phenylalanine. Added unlabeled γ-glutamylglycine in the presence of free radioactive glycine undergoing incorporation into protein did not trap any radioactive γ-glutamylglycine. In fact, γ-glutamyl peptides corresponding to any of the tested radioactive amino acid could never be demonstrated. Although the γ-glutamyl bond was found to be stable against proteolytic hydrolysis in the spleen, it was hydrolyzed with ease in the liver and kidney preparations (76, 77).

CURRENT PICTURE—1952

The second International Congress of Biochemistry has just concluded in Paris. There was a special symposium on protein biosynthesis (78). Papers were presented by J. S. Fruton and H. Waelsch on the role of transpeptidation in protein synthesis, by H. Borsook on a general review of the problem of protein synthesis, by T. S. Work, P. N. Campbell, and B. A. Askonas on the existence of peptide intermediates in protein synthesis, by F. Haurowitz on some specific considerations regarding the synthesis of serum proteins and antibodies, by M. R. Pollock and J. Monod on induced enzyme formation, and by H. Chantrenne on the possible role of ribonucleic acids in protein biosynthesis.

In Fruton's paper, after he reviewed many instances of demonstrated transpeptidation reactions, he brought the situation up to date in his closing paragraphs, which are reprinted here:

"Although these findings indicate the ability of proteinases to catalyze the elongation of peptide chains in a specific manner under physiological conditions, it must be emphasized that, in the catalysis of replacement reactions, a proteinase acts at a preformed CO—NH bond. If a biological system is provided only with free amino acids from which it must make its proteins, some CO—NH bonds must be formed *de novo* in endergonic reactions coupled with energy-yielding processes. Studies on the metabolic oxidation of carbohydrates and fats have shown that much of the energy liberated in these processes is made available for the synthesis of the pyrophosphate bonds of adenosine triphosphate. Lipmann has made the important suggestion that the energy for the synthesis of the CO—NH bonds of natural peptides and of proteins is derived from exergonic transphosphorylation reactions involving adenosine triphosphate. Experimental evidence for this hypothesis has come from many laboratories; of especial importance have been the studies of Speck and of Elliott who showed that adenosine triphosphate is necessary for the formation (by pigeon liver homogenates) of L-glutamine from L-glutamic acid and ammonia, and of Johnston and Bloch on the biosynthesis of glutathione from the component amino acids (L-glutamic acid, L-cysteine, glycine) in the presence of suitable extracts of pigeon liver fortified with adenosine triphosphate. These results, together with studies in several laboratories, have led to the proposal of a working hypothesis which attempts to link the enzymatic catalysis of transamidation reactions to the role of adenosine triphosphate in the biosynthesis of glutamine and of glutathione. This hypothesis suggests that these ubiquitous amides (and possibly also asparagine) serve to link the specific formation of the peptide bonds of proteins to the exergonic reactions in the oxidation of carbohydrates and fats. Evidence for this speculation comes from experiments of Hanes, who showed that glutathione can participate in transpeptidation reactions catalyzed by enzymatic constituents of kidney extracts. Also, in 1949, Waelsch reported that extracts of *Proteus vulgaris* catalyze the formation of hydroxamic acids from glutamine, and a later publication from his laboratory showed this to be an enzyme-catalyzed transamidation reaction. An analogous replacement reaction was found by Stumpf with extracts of pumpkin seedlings. The results of Hanes, Waelsch, and Stumpf thus encourage the hope that some aspects of the hypothesis concerning the role of glutamine and of glutathione as funneling agents of energy for the elongation of peptide chains may be correct. It must be iterated, however, that this view is only a speculative one at present, and its ultimate validity will depend on the outcome of intensive experimental efforts now under way in several laboratories, including our own.

"On the basis of the available data, it would appear that when intact amino acids serve as the starting point for protein synthesis in biological systems, peptides are metabolic intermediates. The manner in which such peptides are formed, and the specific nature of the coupled reactions which lead to the utilization of the energy liberated in transphosphorylation reactions involving adenosinetriphosphate, are problems which still require solution. The subsequent metabolic fate of the peptides is also obscure. The possibility remains that, under suitable circumstances, some of these peptides may undergo condensation reactions with the removal of the synthetic product in an exergonic reaction. A more general metabolic pathway appears to be the participation of peptides in transpeptidation reactions leading to the elongation of peptide bonds. Suggestive evidence for the latter mechanism has come from microbiological studies by Gale and by my colleague, Dr. Sofia Simmonds, and it may be hoped that the continued examination of the metabolism of peptides in bacteria will provide more definite evidence on the occurrence of transpeptidation reactions *in vivo* and the role of proteolytic enzymes in the processes.

"In recent years, much attention has been devoted to the study of the incorporation of isotopic amino acids into the proteins of tissue slices, of tissue homogenates, or of microorganisms. The work of numerous investigators has shown that, under suitable circumstances, the incorporation of the isotope into the protein fraction of a tissue preparation or of bacteria is favored by the presence of adenosine triphosphate, or by the operation of oxidative processes which promote the synthesis of the pyrophosphate bonds of adenosine triphosphate. In some instances, substances such as dinitrophenol, which appear to inhibit oxidative phosphorylation, also prevent the incorporation of the isotope into the protein fraction. This has been taken by some workers to mean that a synthesis of peptide bonds was taking place with the *direct* participation of adenosine triphosphate. In view of the complexity of the enzymatic systems present in the biological material used for these studies, it would appear premature to draw these conclusions; for example, the intermediate occurrence of transpeptidation reactions has not been ruled out, and is a distinct possibility. It must be added, however, that the extension of this experimental approach may eventually lead to an analysis of the enzymatic mechanisms operative in the incorporation of amino acids into proteins, and thus provide important data on the problem of protein synthesis.

"In conclusion, the view may be offered that our knowledge is insufficient to permit us to cling to any one of the various current hypotheses concerning the mechanism of the biosynthesis of peptide bonds, to the exclusion of other plausible hypotheses. It may well be that there are, in fact, alternative pathways of peptide bond synthesis, and that different enzymatic mechanisms are available for the synthesis of small peptides such as glutathione, of larger peptides which serve as precursors of the completed protein molecule, and finally, of the native protein itself. At present, all we may justifiably ask of any hypothesis that bears on this baffling problem of the biosynthesis of proteins is that it provide a conceptual basis for further biochemical experimentation."

Fruton's paper was followed by a short appendix in which he reported that when carbobenzoxy glycinamide was the substrate and L-leucylglycine was the replacing agent in a papain catalyzed transamidation, 65% of the substrate which disappeared was converted to carbobenzoxy glycyl-L-leucylglycine, while only 35% underwent hydrolysis. Specificity towards selection of L-leucylglycine over D-leucylglycine in the reaction was shown. Other new chain elongation transpeptidation reactions were also described. Dr. Fruton closed his paper with the following remarks:

"These newer data focus attention on the probable importance, in protein synthesis, of differences in the rates of the component proteinase-catalyzed reactions. It may be suggested that the reproducibility of protein synthesis may be a consequence of the additive specificity of a number of enzymatic catalysts involved in the elongation of peptide chains. The coupling of specific steps in each of which preference is shown for a particular reaction may be expected to accentuate these preferences and, in a dynamic system, to favor the synthesis of a unique structure. If this speculation should find additional experimental support, it would provide a reasonable alternative to hypotheses involving a matrix or template in the biosynthesis of proteins."

Waelsch's paper summarized findings concerning the transfer of the γ-glutamyl residue and the possible relationship of these studies to protein synthesis. The closing paragraph to Waelsch's paper follows:

"The discovery of enzymes catalyzing the exchange of the substituents of the γ-carboxyl of glutamic acid in glutamine and glutathione and of the β-carboxyl in asparagine suggests a function of these compounds in the biosynthetic mechanisms of living matter. The present hypothesis which considers these substances as the funneling agents of biological energy into the synthesis of the peptide bond may be modified considerably by the experimental work of the future. However, already at the present stage, the evidence suggests the general importance of the principle strongly supported by observations made in the field of carbohydrate metabolism that the energy of the energy rich phosphate bond may be utilized for the formation of key intermediates which may enter further synthetic reactions by enzymatically catalyzed exchange of their components."

Borsook presented a clear discussion of some of the considerations involving the energetics of protein synthesis and a general discussion of the status of amino acid-incorporation reactions at the moment. Although transpeptidation reactions were covered, no position was taken by Borsook on the question of their possible involvement in protein synthesis.

The paper of Work, Campbell, and Askonas is pertinent to the question of the involvement of peptide intermediates in protein synthesis and will be discussed when this topic is developed further.

Haurowitz discussed the possibility that an expanded peptide film may act as a specific adsorbing surface, each of its amino acid residues adsorbing preferentially an identical amino acid. By the action of a nonspecific enzyme, peptide bond formation between the adsorbed amino acids might occur so that a peptide layer would be formed which was identical with the expanded template layer. Specificity would then be introduced by a second phase involving the specific folding of this peptide chain. Haurowitz thought that this process must take place in the mitochondria.

The papers of Pollock and Monod concerned a general consideration of enzyme induction and did not deal with specific mechanisms for amino acid incorporation. Chantrenne reviewed some of the correlations in the location of RNA and active protein synthetic capabilities. These ideas are discussed more fully later on.

In 1952 a symposium on phosphorous metabolism was held in Baltimore, sponsored by the McCollum-Pratt Institute of Johns Hopkins University. Because it was generally recognized that phosphate bond energy must in some way be involved in protein synthesis, a session on amino acid and protein metabolism was held. In the session on protein synthesis two papers were given; one by Hanes, Connell, and Dixon and one by H. Waelsch. In their presentation Hanes *et al.* compiled a table of 43 published cases of the formation of new peptides by transpeptidation reactions involving alpha amino acyl groups. They commented on the recent work of Fruton and collaborators in which, in a reaction catalyzed by the presence of two proteolytic enzymes, papain and cathepsin C, a new N-substituted tripeptide was synthesized, namely, carbobenzoxy, 1-alpha-glutamyl-L-phenylalanyl-L-arginyl-amide. An up-to-date review of γ-glutamyl transfer reactions was also presented, as well as a discussion of the conditions which seem to influence the reaction. Dr. Waelsch discussed the biological significance of the gamma glutamyl radical. He covered many instances of γ-glutamyl transfer reactions and suggested that when the γ-glutamyl residue is attached to the α-amino group of an amino acid, the α-carboxyl group of the amino acid becomes more reactive for further synthesis with the participation of high energy phosphate. Once a critical chain length has been formed, a true transpeptidation reaction can lead to the formation of higher amino acid polymers.

Notes from 1967

It would not be entirely accurate to leave the impression that in the early 1950's all voices were raised in accord with the significance of γ-glutamyl transpeptidation and transpeptidation in general, as an important part of the protein synthetic machinery. Henry Borsook, who wrote many reviews on the problems of protein synthesis, pointed out why it would not seem likely to have a major role played by such reactions from the standpoint of energies and logic. As regards γ-glutamyl transpeptidation in particular, Borsook discussed the work that Greenberg and I published, and he concluded that it seemed most unlikely that γ-glutamyl transpeptidation could be playing the role it was assigned (79).

CURRENT PICTURE—1954

The proceedings have just been published from the gluthathione symposium held in November 1953 at Ridgefield, Connecticut (80). The current status of the problem concerning the possible involvement of γ-glutamyl transpeptidation in protein biosynthesis was discussed in papers by Hanes, Dixon, and Connell, and H. Waelsch. Hanes *et al.* discussed several possibilites by which glutathione or other γ-glutamyl peptides might serve in biosynthetic reactions but was cautious in his summation to point out:

"It will be clear from what has been said that our knowledge of transpeptidation reactions, and of the participation of glutathione in such reactions is not sufficient at present to justify the conclusion that, by such participation, glutathione acts as an intermediary in the synthesis of proteins. This possibility is not excluded and various interesting avenues await exploration."

Waelsch has once more listed many instances of demonstrated and implied formations of γ-glutamyl peptides and discussed the possibility of the γ-glutamyl residue serving as an analogue to a carbobenzoxy group facilitating the involvement of the carboxyl group of a bonded amino acid in synthetic reactions. In a very brief communication Anfinsen told of experiments in progress with Martin Flavin to test the possibility that glutathione may transfer intact cysteinylglycine dipeptides to the proteins, ovalbumin, and ribonuclease. To date the results have not been in support of this possibility.

PEPTIDES AS INTERMEDIATES IN PROTEIN SYNTHESIS

At about the same time that ideas and experimentation were directed toward transpeptidation great interest was also building in the related question of peptide intermediates in protein biosynthesis. The question was posed as a choice between two alternatives: either amino acids were combined gradually through stages involving many free peptides to the degree of complexity of a complete protein molecule, or there had to be an "implosion" or near simultaneous joining of the individual amino acids into a complete protein. For some reason, the simple idea of adding amino acids one by one to a growing peptide chain was not discussed as a reasonable alternative.

As I was finishing my thesis work at Berkeley, California, I learned of the exciting experiments of Anfinsen and Steinberg on nonuniform labeling (81). Rather than work with nondescript trichloroacetic acid precipitates, these workers, after incubation of intact oviduct minces, isolated pure ovalbumin and studied fragments of the clean molecule released by a peptidase of limited specificity obtained from $B.$ $subtilis.$ In the first paper of a series of such studies Anfinsen and Steinberg reported that after incubating $C^{14}O_2$ with a mince of intact oviduct, crystalline ovalbumin could be isolated (81). This substrate was then enzymatically reacted with the $B.$ $subtilis$ enzyme. A hexapeptide was split away from the protein molecule leaving a protein that crystallized in the form of plates and which was known as plakalbumin. The hexapeptide and plakalbumin were separately hydrolyzed and aspartic acid was isolated as a barium salt and subsequently purified by chromatography on paper. The aspartic acid was decarboxylated by treatment with ninhydrin and the specific activities of the carboxyl groups were determined by assaying the $C^{14}O_2$ trapped as $BaC^{14}O_3$. The whole assay was rechecked and confirmed by isolating aspartic acid as the copper salt. The results showed that the aspartic acid released from the hexapeptide was 1.3 to 3.5 times more radioactive than the average aspartic acid released from the rest of the molecule. In the discussion the authors stated that these results were not compatible with a mechanism of synthesis involving the formation of ovalbumin molecules directly from free amino acids by means of a template process. They felt that protein synthesis by way of peptide intermediates differing in rates of replacement or pool size was compatible with these data, and pointed out the compatibility of these results with the data being accumulated on transpeptidation and the utilization of peptides for growth by bacteria.

I wrote to Dr. Anfinsen and asked if I might join his laboratory on a fellowship to participate in these studies, and arrangements were made to join the group in the summer of 1952. Steinberg and Anfinsen had just finished the next phase of their investigation (82). They confirmed the findings with aspartic acid, isolated after incubation with $C^{14}O_2$, and extended the results to glutamic acid which was also released by the enzymatic treatment. In experiments with labeled alanine, nonuniformity was also established between one of the three alanines contained in the hexapeptide and the average alanine contained in the plakalbumin. In another elegant approach to the problem the whole ovalbumin molecule, after labeling with alanine, was subjected to partial hydrolysis by pepsin and the resulting peptides fractionated on paper in butanol:acetic acid. Twenty-eight fractions were isolated and the alanine released by hydrolysis was assayed for radioactivity. Specific activities varied over a range from 170 cpm/μmole to 760 cpm/μmole with a mean value of 460 cpm/μmole.

This experiment showed that nonuniformity is not limited to any special small portion of the molecule, but appears to be quite general. It also was a superb demonstration of the incorporation of an added amino acid into the entire fabric of a *de novo* synthesized pure protein molecule. Finally, after injecting $NaHC^{14}O_3$ into a hen and isolating the ovalbumin, it was shown that nonuniformity resulted from *in vivo* as well as *in vitro* incubations. In the discussion a strong argument for interpreting these results in terms of the synthesis of ovalbumin by way of a series of peptide intermediates was presented. The possibility of a template mechanism[3] was considered to be most unlikely. Mention was made of recent experiments by Muir, Neuberger, and Perrone who reported studies on the formation of hemoglobin by the intact rat after injection of C^{14}-valine (83). The N-terminal valine was found to have the same specific activity as the average for the rest of the valines in the molecule. This finding in itself is not necessarily negative insofar

---

[3] That is, the amino acids would be selected from the pool and lined up on a surface (template) possessing a structural analogy to the protein being built.

as the valines lumped together in the average may individually have been nonuniform. A more important criticism of these experiments, however, was that the animals were sacrificed 24 hr after the injection of isotope. In that time, especially in an *in vivo* situation, an initial nonuniformity could have been "equilibrated."

When I arrived at Bethesda the problems posed were whether continued incubation would tend to reduce nonuniformity of labeling and also to determine the rate of change of the specific activities of the free amino acid pools of glutamic and aspartic acids during incubation with $C^{14}O_2$ and synthesis of ovalbumin. These problems, as well as some related problems concerning the nature of $C^{14}O_2$ fixation in general, and fixation into amino acids in particular, occupied my attention from 1952 to 1954. In the meantime, with others in the laboratory, Anfinsen extended the findings of nonuniformity to other proteins (insulin and ribonuclease), and to other amino acids (84, 85).

All during the 1950's, reports came from many laboratories showing either nonuniformity or uniformity (86–94). In a related but different approach T. S. Work and his collaborators studied the incorporation of labeled amino acids into the blood and milk proteins of lactating rabbits (78). Since the milk proteins were labeled faster than the blood proteins, they concluded that plasma protein was not a sole precursor to milk proteins. Similar results were obtained with a lactating goat. Prelabeled rabbit blood proteins reinjected into a lactating rabbit did not label the milk proteins. On hydrolyzing two milk proteins, however (casein and whey), each of four different amino acids had different specific activities in the two proteins. The authors concluded that this showed that both proteins could *not* have been made directly from one free amino acid pool, and they postulated the existence of peptide precursors, possibly coming from partial degradation of plasma proteins by transpeptidation reactions.

In an earlier paper Campbell and Work studied arterio-venous changes in amino acids and proteins of lactating goats and cows (95). It was decided that not enough amino acids were taken up to account for the synthesis of the milk proteins and it was suggested that a globulin fraction of the blood was utilized for milk protein synthesis without prior degradation to free amino acids. The authors concluded that milk proteins are synthesized primarily from plasma amino acids with a contribution of peptides from partial degradation of blood proteins. In a later paper this view was substantially modified (93). Radioactive casein and β-lactoglobulin isolated 3 and 4 hr after injection of radioactive amino acids were subjected to partial acid hydrolysis and isolation of peptides. It was found that radioactivity was uniformly distributed throughout the molecules. An interpretation relying on synthesis entirely from free amino acids as opposed to a contribution of preformed peptides from blood proteins was given.

There were other reports from time to time of suspected peptide intermediates found in tissues and of peptides more readily utilized for bacterial growth than free amino acids. Although during this time such reports sustained what seemed to be a most logical possibility that peptide intermediates function in protein biosynthesis subsequent work has given alternative explanations. For example, an amino acid contained in a peptide could be protected against destruction or could be taken into the cell more readily than the free amino acid.

Arguments against peptide intermediates were drawn from the work of Monod et al. (96), Halvorson and Spiegelman (97), Rotman and Spiegelman (98), Hogness et al. (99), Taliaferro and Talmage (100), and Geiger (101). These experiments have shown that in spite of the presence of prelabeled protein in a cell newly formed protein does not acquire radioactivity if the free amino acids are unlabeled and vice versa (i.e., no preformed peptides contributed to the newly synthesized proteins). In feeding experiments it was shown that all essential amino acids must be present for growth and that stored forms of the amino acid do not appear to be present. It was also shown that a single amino acid antagonist will interfere with protein synthesis, whereas if the needed amino acid could be stored as a peptide such inhibition would not be expected. Finally, it has been argued that despite intensive searches by many laboratories, the suspected peptide intermediates have never been found.

Today no one talks about peptide intermediates. This is not because any of the data cited above have shown that they do not exist. Actually, the presence of a quantitatively small but significant pool of peptides as obligatory intermediates in a rapid state of turnover between free amino acid and protein, is compatible with all of the experiments cited in the preceding discussion. If peptide intermediates existed these are precisely the characteristics they would be expected to have.[4] The idea of individual peptide intermediates simply wasted away all by itself. In the absence of positive evidence to support the possibility, and in the presence of a vast accumulation of facts interpreted in terms of a scheme not employing peptides, the idea, like the old soldier, did not die, but simply faded away.

From 1952 to 1954 I worked on the problem of nonuniform labeling

---

[4] An interesting exchange of views on this point can be found between R. B. Loftfield and myself in the book *Amino Acid Pools* (102).

in hen oviduct labeled by incubation with $C^{14}O_2$. I studied the change of nonuniformity with time, the change of the specific activities of the free amino acid pools with time, and the changes in nonuniformity ratios with induced changes in the specific activities of the free amino acids caused by flushing the medium with unlabeled $C^{12}O_2$ at some point during the incubation. Although initially I believed that nonuniformity did indicate individual peptide (or other) intermediates, the more I thought about the problem, the less convinced I became. My earlier experience in the field of γ-glutamyl transpeptidation involved a build-up phase of anticipation followed by a letdown and the birth of a critical view about theoretical interpretations. There were reports from Gale's laboratory which indicated that free amino acids might exchange with protein-bound amino acids. Subsequently it was learned that these data really did not demonstrate the phenomenon with proteins in general but were concerned with a more restrictive case involving a cell wall peptide.

But even in the absence of a clear-cut demonstration of this kind, other possibilities existed. If a protein were assembled on a template, then the newly formed protein in the vicinity of the template might be in an activated state in which the peptide bonds were labilized, making the amino acids available for exchange. These considerations were later developed in more detail by Borsook (103). Furthermore, as Dalgliesh explained, a series of growing peptide chains at different stages of completion on a template surface could result in a graded nonuniformity down the chain (104). This would be true even if the specific activity of the free amino acid pool were constant, although Dalgliesh and others introduced the idea that the specific activity of the free amino acid pool would be changing during synthesis.

There were three other developments of significance taking shape during this same period. One was a growing accumulation of observations pointing to a correlation between RNA and protein synthesis. A second development involved studies with induced enzyme-forming systems. It had been known for quite a long time that microorganisms could develop the ability to metabolize certain substrates if they were first exposed to the substrate for a certain amount of time. This period of adaptation was explained by the biosynthesis of an enzyme to utilize the particular substrate. When the substrate was removed, a decay in the ability to metabolize the substrate or, in other words, a gradual decrease in the concentration of the induced enzyme resulted. Such systems give the investigator a means of turning on and off the synthesis of particular proteins. The third development, and by far the most exciting at the close of the period under consideration, was the finding of Gale at Cambridge that fragments of a suspected RNA template could be added back to an acellular system to stimulate the incorporation of specific amino acids. These developments are briefly reviewed in the following discussion.

Observations Pointing to an Involvement of RNA in Protein Synthesis

The conclusion that RNA is somehow concerned with protein synthesis was reached independently by Caspersson (105) and Brachet (106) 27 years ago as a result of cytochemical observations. Caspersson used a technique involving ultraviolet microspectrophotometry, whereas Brachet worked with the basic dyes methyl green-pyronine in the form of Unna's stain and combined the use of these labels with specific digestion caused by RNAase.

Brachet listed an impressive collection of data showing that RNA is abundant in rapidly growing cells, whereas tissues having high physiological activity but not synthesizing large amounts of protein contain only small amounts (107). Microorganisms which multiply very rapidly are rich in RNA. Organs which synthesize large amounts of protein are rich in RNA. In embryonic development the synthesis of RNA precedes protein synthesis. The RNA content of bacteria appears to be correlated to its growth rate. Studies of Gale and Folkes showed that the presence of purines and pyrimidines in the medium enhanced protein synthesis in Staphylococci (108). Furthermore, as Brachet pointed out, interference with RNA synthesis by UV irradiation or RNAase treatment results in drastic interference with protein formation. The microsomes that are rich in RNA are active sites of protein synthesis.

Brachet discussed a theory of protein synthesis originated by Caspersson which gave the nucleus the central role in protein synthesis. Although this theory gave rise to much discussion for many years, it appears now to be incorrect in its details. At the same time, however, it was visionary in its appreciation of the problems of communication between the DNA of the nucleus and the protein-forming centers in the cytoplasm and in its postulation of a messenger to perform such a duty. In this theory cytoplasmic ribonucleoprotein read information carried from DNA in the form of histones to the perinuclear cytoplasm. The newly formed ribonucleoprotein then induced the synthesis of cytoplasmic proteins.

Brachet gave some support to Chantrenne's suggestion (109) that microsomes might represent early stages in the development of mitochondria, although he acknowledged that this possibility was not consistent with all the data.

Enzyme Induction

The phenomenon of induced or adaptive enzyme formation held great promise for providing a key to the elucidation of the basic mechanisms of protein synthesis. Work with these systems in the early 1950's, however, was concerned with understanding the particular phenomenon in itself and not much progress in increasing our understanding of basic mechanisms of protein biosynthesis resulted. Later developments in the study of such systems have served to shape our ideas on the regulation of protein synthesis. The status of the field as of 1955 was fairly well summarized by Jacques Monod (110) and Sol Spiegelman (111). Monod's summation of knowledge on induced enzyme synthesis consisted of four points.

1. Induced enzyme formation involves the complete *de novo* synthesis of the enzyme protein molecule from its elements or elementary building blocks (amino acids). For proof Monod cited experiments by Halvorson and Spiegelman with alpha glucosidase formation in yeast which showed that interference with the free amino acid pool inhibited enzyme formation (these experiments are discussed more fully later in this section), and also a similar dependence of induced formation of β-galactosidase in *E. coli* on the presence of all free amino acids (96). Also cited were elegant experiments by Hogness, Cohn, and Monod (112), and by Rotman and Spiegelman (113). Hogness et al. prelabeled amino acids and proteins of cells of *E. coli* with S^{35}, and Rotman and Spiegelman prelabeled cells of *E. coli* with C^{14}. The cells were then washed, suspended in an unlabeled medium, and induced to form β-galactosidase. The enzyme was isolated, purified, assayed for radioactivity, and found to be essentially unlabeled (less than 1%).

2. The induced synthetic process is virtually irreversible and the "finished" enzyme molecule is not renewed at a measurable rate (i.e., is not in a dynamic state) within the cells. The earlier cited observations that so little of the prelabeled cellular protein was reutilized for the induced formation of β-galactosidase were used to support this conclusion. In fact, Monod, with this information, questioned the whole concept of the dynamic state of mammalian proteins that had been established as a result of the investigations of Schoenheimer and his coworkers.

3. The process of enzyme induction is independent of the action of the enzyme induced (i.e., the functions of inducer and substrate are distinct, although inducers and substrates of a given enzyme always possess in common certain essential elements of molecular configuration).

This conclusion rests on the comparison of various compounds as substrates, specific binders, and specific inducers of β-galactosidase in *E. coli*. All of these properties are independent, and a given molecule may possess one or neither of the other two.

4. The inducer is not consumed in induction. It acts as a catalyst in the sense that one molecule of inducer may activate the synthesis of more than one enzyme molecule. For proof, the excellent experiments of M. Pollock with penicillinase induction in *B. cereus* were cited (114).

The other summary of the status of induced enzyme systems was presented by Spiegelman (111). His paper dealt with three questions: (a) the nature of the building blocks; (b) the nature of the enzyme-forming mechanism; and (c) the role of the inducer. In regard to the first question a whole series of intricate experiments were reported, all of which supported the interpretation that all 20 of the free amino acids had to be present to form an induced enzyme and that almost no material comes from stored complex precursors in the cell (115). The interpretation of these data was that peptide precursors were now eliminated from consideration. Actually, this conclusion is not supported by the data. A small supply of peptides as obligatory intermediates between free amino acids and proteins, if they are in a rapid state of equilibrium with the free amino acid pool, would be entirely compatible with these observations. Also, in a complex integrated series of reactions linking free amino acids to protein synthesis, a monkey wrench such as an amino acid analogue thrown into the mechanism could stop everything. The interpretation widely quoted, however, was that since an amino acid deficiency for a single amino acid inhibited the whole process of protein synthesis, this meant that a supply of the required amino acid in the form of a peptide precursor could not be present. The enzyme forming unit was deduced to consist of a complex between an RNA template, the inducer which activated the template and gave it specificity, and the enzyme(?) which caused protein formation.

Gale's Experiments Relating to the RNA Template[5]

Gale stated that the involvement of RNA in protein synthesis which had been suspected since the work of Brachet and Caspersson, was demonstrated directly for the first time with disrupted Staphylococci, where removal of nucleic acid resulted in cessation of protein synthesis and replacement of the extracted nucleic acid enabled the cell fragments to resume specific enzyme synthesis (116). He referred to the work of Spiegelman, Halvorson, Ben-Ishai

---

[5] A fuller discussion of templates in protein synthesis appears in pages 81–82. See also the footnote on page 36.

(117), and Creaser (118), who showed that adaptive enzyme formation in bacteria and yeast can be inhibited by the presence of purine and pyrimidine analogues, whereas the presence of specific combinations of nucleic acids and their precursors are necessary for the synthesis of particular enzymes in disrupted Staphylococci (119). When a single amino acid is incorporated into Staphylococci in the absence of other amino acids, no net synthesis of protein occurs and for reasons put forward elsewhere (120), the incorporation is believed to represent an exchange reaction between the added amino acid (or a metabolite thereof) and corresponding residues in the protein of the preparation. Incorporation of amino acids takes place also in sonically disrupted cells. Removal of nucleic acids from such a preparation results in loss of ability to incorporate certain amino acids. The ability is restored by adding back Staphylococcal RNA or DNA. The restoration phenomenon is species specific for the nucleic acid. Certain amino acids are more affected by slight removal of nucleic acids than others. It may be that each amino acid responds to the presence of a specific portion of the nucleic acid structure and that some portions are more readily damaged or replaced than others.

It was found that digestion of the RNA by RNAase does not abolish the reactivation phenomenon although the active material becomes dialyzable. Furthermore, RNA's from foreign sources which themselves are inactive, yield active fragments upon RNAase digestion. Such digests, when fractionated by paper chromatography and ionophoresis, yield fractions which could replace RNA for the incorporation of specific amino acids with activities at least 100 times greater than Staphylococcal RNA when compared on an optical density basis. In two cases the base composition of the active polynucleotide fractions have been tentatively determined. An RNAase digest of Staphylococcal RNA gave six bands of ultraviolet-absorbing material. It was found that band 5 was very effective in stimulating the incorporation of aspartic acid. Bands 4 and 6 had some activity. Each band contained a number of small polynucleotide fractions which could be subfractionated by ionophoresis. In such fashion band 5 was subfractionated into 4 or 5 components. The most active subfraction on an optical density basis was 100 to 120 times as effective as RNA and by hydrolysis yielded adenine and cytidylic acid in equimolar proportions. Its general properties agree with those given by Markham and Smith (121) for adenine-cytosine dinucleotides. Another active fragment present in smaller amounts gave the tentative analysis of ACC.[6] Free adenylic and free cytidylic acids give only partial activity. This aspartic acid incorporation factor is effective for aspartic acid incorporation only, although it has some activity (less than 20 to 25%) towards glutamic acid, glycine, or leucine. In a similar manner a subfraction was isolated which was effective for glutamic acid incorporation but not for aspartic acid. The factor corresponds to the trinucleotide GUU.[7] By these techniques separate factors which are effective in stimulating the incorporation

---

[6] Adenylcytidylcytidylic acid trinucleodide.
[7] Guanyluridyluridylic acid trinucleotide.

of each of seven amino acids have been isolated. Since active factors were isolated in each case from different fractions of RNA digest, it seemed highly probable that a different factor would be found for each amino acid studied. Native DNA was found to be more effective than native RNA in promoting amino acid-incorporation. The factor released from RNA by hydrolysis was also more effective than native RNA and this fact suggests that the activity of the factor is in some way restricted as long as it is part of the RNA complex. Proponents of the template theory of nucleic acid function suggest that specific groupings of nucleotides act as combining points for specific amino acids, so that the sequence of bases along the polynucleotide chain determines the sequence of amino acids in the protein which will be formed on that chain as template. Such a mechanism might underlie the experimental findings described here.

SUMMARY

This brief review has traced the development of our research and thinking on protein synthesis through the early days which showed that the problem could be taken into the laboratory and subjected to an experimental approach. Systems were defined, questions were posed, hypotheses were formed, and evidence was accumulated. The history has been dynamic if not actually turbulent. The center of attention was for a period located in the laboratories of Greenberg, Zamecnik, and Borsook, who with their many talented collaborators opened up the early paths. Attention was then shifted to the laboratories of Fruton, Hanes and Waelsch, who for many years influenced thinking in terms of transpeptidation and an involvement of proteolytic enzymes. The scene for a while then shifted to Bethesda where Anfinsen and collaborators stimulated much thought and experimentation in the direction of concepts involving peptide intermediates. Toward the end of this initial phase (up to 1955) attention was dramatically shifted both in geography and in approach. Although initially all work was with tissues of higher organisms, dramatic findings were now forthcoming from laboratories working with microorganisms as in the case of Spiegelman with yeast, Monod with *E. coli*, and Pollock with *B. cereus*. Then in the year 1955, the center of attention was intensely focused on the laboratory of E. Gale in Cambridge, England.

It was the purpose of this review of history not only to examine the strength of our early foundations for the continuing study of protein synthesis but also to show that each stage in the development was considered at the time to be essentially correct and to represent a modern view, and that in the perspective of time we can appreciate the transience of ideas which serve as important working hypotheses.

REFERENCES

1. J. D. Watson, *Science,* **140,** 17 (1963).
2. P. C. Zamecnik, *Biochem. J.* **85,** 257 (1962).
3. J. H., Northrop, in *Crystalline Enzymes,* 2nd ed., Columbia University Press, N.Y. 1948.
4. S. Ruben and M. D. Kamen, *Phys. Rev.,* **59,** 349 (1941).
5. J. B. Melchior and H. Tarver, *Arch Biochem.,* **12,** 301 (1947).
6. J. B. Melchior and H. Tarver, *Arch. Biochem.,* **12,** 309 (1947).
7. T. Winnick, F. Friedberg, and D. M. Greenberg, *Arch. Biochem.,* **15,** 160 (1947).
8. I. D., Frantz, R. B. Loftfield, and W. W. Miller, *Science,* **106,** 544 (1947).
9. C. B. Anfinsen, A. Beloff, A. B. Hastings, and A. K. Solomon, *J. Biol. Chem.,* **68,** 771 (1947).
10. F. Friedberg, T. Winnick, and D. M. Greenberg, *J. Biol. Chem.,* **171,** 441 (1947).
11. P. C. Zamecnik. I. D. Frantz, R. B. Loftfield, and M. L. Stephenson, *J. Biol. Chem.,* **175,** 299 (1948)
12. F. Friedberg, M. P. Schulman, and D. M. Greenberg, *J. Biol. Chem.,* **173,** 437 (1948).
13. I. D. Frantz, P. C. Zamecnik, J. W. Reese, and M. L. Stephenson, *J. Biol. Chem.* **174,** 773 (1948).
14. W. F. Loomis and F. Lipmann, *J. Biol. Chem.,* **173,** 807 (1948).
15. J. B. Melchior, M. Mellody, and I. M. Klotz, *J. Biol. Chem.,* **174,** 81 (1948).
16. T. Winnick, F. Friedberg, and D. M. Greenberg, *J. Biol. Chem.,* **175,** 117 (1948).
17. T. Winnick, I. Moring-Claesson, and D. M. Greenberg, *J. Biol. Chem.,* **175,** 127, (1948).
18. H. Borsook, C. L. Deasy, J. W., Dubnoff, C. T. O. Fong, W. D. Fraser, A. J. Haagen-Smit, G. Keighley, and P. H. Lowy, *Fed. Proc.,* **7,** 147 (1948).
19. H. Borsook, C. L. Deasy, A. J. Haagen-Smit, G. Keighley, and P. H. Lowy, *J. Biol. Chem.,* **174,** 1041 (1948).
20. P. C. Zamecnik, R. B. Loftfield, M. L. Stephenson, and C. M. Williams, *Science,* **109,** 624 (1949).
21. H. Borsook, C. L. Deasy, A. J. Haagen-Smith, G. Keighley, and P. H. Lowy, *J. Biol. Chem.,* **186,** 297 (1950).
22. H. Borsook, C. L. Deasy, A. J. Haagen-Smith, G. Keighley, and P. H. Lowy, *J. Biol. Chem.,* **186,** 309 (1950).
23. D. Shemin, I. M. London, and D. Rittenberg, *J. Biol. Chem.,* **183,** 757 (1950).
24. T. Peters and C. B. Anfinsen, *J. Biol. Chem.,* **182,** 171 (1950).
25. T. Peters and C. B. Anfinsen, *J. Biol. Chem.,* **186,** 805 (1950).
26. F. Lipmann, *Fed. Proc.,* **8,** 597 (1949).
27. J. F. Speck, *J. Biol. Chem.,* **168,** 403 (1947).
28. W. H. Elliot, *Nature,* **161,** 128 (1948).
29. P. P. Cohen and R. W. McGilvery, *J. Biol. Chem.,* **169,** 119 (1947).
30. W. C. Schneider, *J. Biol. Chem.,* **165,** 585 (1946).
31. G. H. Hogeboom, W. C. Schneider, and G. E. Palade, *Proc. Soc. Exptl. Biol. Med.,* **65,** 320 (1947).
32. G. H. Hogeboom, W. C. Schneider, and G. E. Palade, *J. Biol. Chem.,* **172,** 619 (1948).

33a. H. Borsook, C. L. Deasy, A. J. Haagen-Smit, G. Keighley, and P. H. Lowy, *J. Biol. Chem.*, **184**, 529 (1950).

33b. H. Borsook, C. L. Deasy, A. J. Haagen-Smit, G. Keighley, and P. H. Lowy, *J. Biol. Chem.*, **187**, 839 (1950).

34. T. Hultin, *Exptl. Cell. Res.*, **1**, 376 (1950).

35. A. J. Claude, *Exptl. Med.*, **84**, 51 (1946).

36. E. B. Keller, *Fed. Proc.*, **10**, 206 (1951).

37. V. Allfrey, M. M. Daly, and A. E. Mirsky, *J. Gen. Physiol.*, **37**, 157 (1953).

38. P. Siekevitz and P. C. Zamecnik, *Fed. Proc.*, **10**, 246 (1951).

39. N. D. Lee, J. T. Anderson, R. Miller, and R. H. Williams, *J. Biol. Chem.*, **192**, 733 (1951).

40. E. B. Keller, P. C. Zamecnik, and R. B. Loftfield, *J. Histochem. Cytochem.* **2**, 378 (1954).

41. E. A. Peterson, T. Winnick, and D. M. Greenberg, *J. Amer. Chem. Soc.*, **73**, 503 (1951).

42. P. Siekevitz, *J. Biol. Chem.*, **195**, 549 (1952).

43. D. M. Greenberg, F. Friedberg, M. P. Schulman, and T. Winnick, Cold Spring Harbor Symposium, *Quant. Biol.*, **13**, 113 (1948).

44. H. Borsook, C. L. Deasy, A. J. Haagen-Smit, G. Keighley, and P. H. Lowy, *Fed. Proc.*, **8**, 589 (1949).

45. E. A. Peterson, D. M. Greenberg, and T. Winnick, *Fed. Proc.*, **9**, 214 (1950).

46. T. Winnick, *Arch. Biochem. Biophys.*, **28**, 338 (1950).

47. E. A. Peterson and D. M. Greenberg, *J. Biol. Chem.*, **194**, 359 (1952).

48. S. Kit and D. M. Greenberg, *J. Biol. Chem.*, **194**, 377 (1951).

49. P. C. Zamecnik and E. Keller, *J. Biol. Chem.*, **209**, 337 (1954).

50. T. Winnick, *Arch. Biochem. Biophys*, **27**, 65 (1950).

51. M. L. Stephenson, K. V. Thimann, and P. C. Zamecnik, *Arch. Biochem. Biophys.*, **65**, 194 (1956).

52. R. Brunish and J. M. Luck, *J. Biol. Chem.*, **197**, 869 (1952).

53. R. Schweet, *Fed. Proc.*, **15**, 350 (1956).

54. M. Bergmann and H. Fraenkel-Conrat, *J. Biol. Chem.*, **119**, 707 (1937).

55. J. S. Fruton, *Yale J. of Biol. and Med.*, **22**, 263 (1950).

56. R. B. Johnston, M. J. Mycek, and J. S. Fruton, *J. Biol. Chem.*, **185**, 629 (1950).

57. R. B. Johnston, M. J. Mycek, and J. S. Fruton, *J. Biol. Chem.*, **187**, 205 (1950).

58. J. S. Fruton, R. B. Johnston, and M. Fried, *J. Biol. Chem.*, **190**, 39 (1951).

59. M. E. Jones, W. R. Hearn, M. Fried, and J. S. Fruton, *J. Biol. Chem.*, **195**, 645 (1952).

60. Y. P. Dowmont and J. S. Fruton, *J. Biol. Chem.*, **197**, 271 (1952).

61. J. S. Fruton, W. R. Hearn, V. M. Ingram, D. S. Wiggans, and M. Winitz, *J. Biol. Chem.*, **204**, 891 (1953).

62. I. D. Frantz, Jr., and R. B. Loftfield, *Fed. Proc.*, **9**, 172 (1950).

63. M. Brenner, E. Sailer, and K. Rufenacht, *Helv. Chim. Acta*, **34**, 2096 (1951).

64. S. G. Waley and J. Watson, *Biochem. J.*, **57**, 529 (1954).

65. C. S. Hanes, F. J. R. Hird, and F. A. Isherwood, *Nature*, **166**, 288 (1950).

66. C. S. Hanes, F. J. R. Hird, and F. A. Isherwood, *Biochem. J.*, **51**, 25 (1952).

67. F. J. R. Hird and P. H. Springel, *Biochem. J.*, **56**, 417 (1954).

68. H. Waelsch, *Adv. in Enzymol.*, **XIII**, 237 (1952).

69. P. J. Fodor, A. Miller, and H. Waelsch, *J. Biol. Chem.*, **202**, 551 (1952).

70. P. J. Fodor, A. Miller, A. Neidle, and H. Waelsch, *J. Biol. Chem.,* **203,** 991 (1953).
71. F. Binkley and C. K. Olson, *J. Biol. Chem.,* **188,** 451 (1951).
72. F. Binkley and G. M. Christenson, *Fed. Proc.,* **11,** 188 (1952).
73. J. H. Kinoshita and E. G. Ball, *J. Biol. Chem.,* **200,** 609 (1953).
74. P. S. Talalay, *Nature,* **174,** 516 (1954).
75. W. J. Williams and C. B. Thorne, *J. Biol. Chem.,* **210,** 203 (1954).
76. R. W. Hendler and D. M. Greenberg, *Nature,* **170,** 123 (1952).
77. R. W. Hendler and D. M. Greenberg, *Biochem. J.,* **57,** 641 (1954).
78. *Symp. sur la Biogénese des Protéines, IIe Congres des Biochemie,* Paris (1952).
79. H. Borsook, *Adv. Prot. Chem.,* **VIII,** 127 (1953).
80. *Glutathione Symposium,* Eds., S. Colowick, A. Lazarow, E. Racker, D. R. Schwarz, E. Stadtman, and H. Waelsch, Academic Press, New York 1954.
81. C. B. Anfinsen and D. Steinberg, *J. Biol. Chem.,* **189,** 739 (1951).
82. D. Steinberg and C. B. Anfinsen, *J. Biol. Chem.,* **199,** 25 (1952).
83. H. M. Muir, A. Neuberger, and J. C. Perrone, *Biochem. J.,* **49,** LV (1951).
84. M. Vaughan and C. B. Anfinsen, *J. Biol. Chem.,* **211,** 367 (1954).
85. M. Flavin and C. B. Anfinsen, *J. Biol. Chem.,* **211,** 375 (1954).
86. G. Gehrmann, K. Lauenstein, and K. I. Altman, *Arch. Biochem. Biophys.* **62,** 509 (1956).
87. V. Crekhovich, V. Shpikiter, O. Kasakova, and V. Mazourov, *Arch. Biochem. Biophys.,* **85,** 554 (1959).
88. K. Shimura, H. Fukai, I. Sato, and K. Saeki, *J. Biochem. Japan,* **43,** 101 (1956).
89. A. Yoshida and T. Tobita, *Biochim. Biophys. Acta.,* **37,** 513 (1960).
90. J. Kruh, J. C. Dreyfus, and C. Schapira, *J. Biol. Chem.,* **235,** 1075 (1960).
91. H. M. Muir, A. Neuberger, and J. C. Perrone, *Biochem. J.* **52,** 87 (1952).
92. M. Heimberg and S. F. Velick, *J. Biol. Chem.,* **208,** 725 (1954).
93. B. A. Askonas, P. N. Campbell, C. Godin, and T. S. Work, *Biochem. J.,* **61,** 105 (1955).
94. M. V. Simpson, *J. Biol. Chem.,* **216,** 179 (1955).
95. P. N. Campbell and T. S. Work, *Biochem. J.* **52,** 217 (1950).
96. J. Monod, A. M. Pappenheimer, and G. Cohen-Bazire, *Biochim. Biophys. Acta,* **9,** 648 (1952).
97. H. O. Halvorson and S. Spiegelman, *J. Bact.,* **64,** 207 (1952).
98. B. Rotman and S. Spiegelman, *J. Bact.,* **68,** 419 (1954).
99. D. S. Hogness, M. Cohn, and J. Monod, *Biochim. Biophys. Acta,* **16,** 99 (1955).
100. W. H. Taliaferro and D. W. Talmage, *J. Inf. Dis.,* **97,** 88 (1955).
101. E. Geiger, *Science,* **111,** 594 (1950).
102. R. W. Hendler and R. B. Loftfield, in *Amino Acid Pools,* Ed., J. T. Holden, Elsevier, Amsterdam, 1962, p. 738.
103. H. Borsook, *J. Cell. Comp. Physiol.* Suppl. 1, **47,** 35 (1956).
104. C. E. Dalgliesh, *Nature,* **171,** 1027 (1953).
105. T. Caspersson, *Naturwissenschaften,* **29,** 33 (1941).
106. J. Brachet, *Arch. Biol.* (Liege) **53,** 207 (1942).
107. J. Brachet, in *The Nucleic Acids,* Vol. II, Eds., E. Chargaff and J. N. Davidson, Academic Press, New York, 1955, p. 475.
108. E. F. Gale and J. P. Folkes, *Biochem. J.* **53,** 483 (1953).
109. H. Chantrenne, *Biochim. Biophys. Acta,* **1,** 437 (1947).

110. J. Monod, in *Enzymes—Units of Biological Structure and Function,* Ed., O. H. Gaebler, Academic Press, N.Y., 1956, p. 7.

111. S. Spiegelman, in *Proc. 3rd Int. Cong. Biochem.,* Ed., C. Liebecq, Academic Press, New York, 1955, p. 185.

112. D. J. Hogness, M. Cohn, and J. Monod, *Biochim. Biophys. Acta,* **16,** 99 (1955).

113. B. Rotman and Spiegelman, S., *J. Bact.,* **68,** 419 (1954).

114. M. R. Pollock, in *Adaptation in Microorganisms,* 3rd Symp. Soc. Gen. Microb., Cambridge University Press, London 1953, p. 150.

115. S. Spielgelman and H. Halvorson, in *Adaptation in Microorganisms* (3rd Symp. Soc. Gen. Microb.) Cambridge University Press, London, 1953, p. 98.

116. E. F. Gale, *Proc. 3rd Int. Cong. Biochem.* Ed. C. Liebecq, Asademic Press, New York, 1954, p. 345.

117. R. Ben-Ishai, in *Amino Acid Metabolism,* Eds. W. D. McElroy and H. B. Glass, Johns Hopkins University Press, Baltimore 1954, p. 124.

118. E. H. Creaser, *J. Gen. Microbiol.,* **12,** 288 (1955).

119. E. F. Gale and J. P. Folkes, *Biochem. J.,* **59,** 675 (1955).

120. E. F. Gale and J. P. Folkes, *Biochem. J.,* **59,** 661 (1955).

121. R. Markham and J. D. Smith, *Biochem. J.,* **52,** 538 (1952).

Chapter 3

THE MODERN PICTURE— CONTEMPORARY VIEWS 1955 TO 1967

The year 1955 marked a dramatic turning point in the history of the study of protein synthesis. Prior to this year, the majority of guesses, assumptions, interpretations, and hypotheses, had turned out to be incorrect. Starting in 1955, however, the most intricate hypothesis supported at the time a minimum of data, eventually turned out to acquire additional data to support it. Our present view of protein synthesis has been built steadily by placing one hypothesis upon another, with the acceptance of the scheme preceding the data which justified it. In my view the experience before 1955 was sufficiently full of unexpected turns and disappointments to throw immediate doubt on any new interpretation until the data to prove it were forthcoming. This natural tendency to skepticism was reinforced by studies with intact cells which gave only feeble support for certain parts of the generally accepted scheme.

Twice before, when the problem of protein synthesis was in the phases of study of gamma glutamyl transpeptidation and nonuniform labeling, I started with high hopes and, during the course of day-to-day contact with the problem, came to see that what seems true at one time can take on a different aspect much later. Therefore, as I trace each new development in the history from 1955 to 1967, I shall try to point out those parts of the scheme that appeared inadequate at the time. In most cases the proof that was missing at the time of acceptance of the idea was forthcoming at a later date. In some cases, however, weakness of proof is still an issue.

There is one other interesting aspect of the new study of protein synthesis. Many writers have already humorously referred to the current concept as the dogma built around the trinity of DNA, RNA, and protein. A prophet (or high priest) appears who says such mystical things as, "Let there be an adaptor," and there is an adaptor or, "Let there be a triplet codon," and there is a triplet codon. In a scientific discussion, the power of such words as "molecular biology," "tapes," "transcription," "cistron," and "information transfer," can hardly be doubted. With this brief introduction to the modern era in protein synthesis, let us now return to 1955.

AMINO ACID ACTIVATION

The beginnings of the new look in protein synthesis had their birth in the laboratory of M. B. Hoagland who was a member of the group working with P. C. Zamecnik in Boston. Hoagland observed that amino acids added to the supernatant fluid obtained from a rat liver homogenate would catalyze an interchange between P^{32}-labeled pyrophosphate and ATP (1). The exchange between inorganic phosphate and ATP, as well as the exchange between C^{14}-AMP and ATP, was unaffected. If the total amount of amino acid was held at 0.04 μmoles while the number of different amino acids making up this total was increased, the extent of PP^{32}-ATP exchange was significantly enhanced. This fact indicated that several different amino acid-recognizing sites were involved. Corresponding to studies of Maas and Novelli with pantoic acid activation (2), the following sequence was postulated:

(1) $Enz + ATP \rightarrow Enz - AMP - PP$

(2) $Enz - AMP - PP + aa \rightarrow Enz - AMP - aa + PP$

When a high concentration of hydroxylamine was present in an incubation involving leucine, a chromatographic spot identified as leucine-hydroxamic acid was found. It was suggested that an enzyme-bound AMP-amino acid was the activated form of the amino acid for protein synthesis and that hydroxylamine could serve as an acceptor in analogy to natural amino acids participating in peptide bond formation. A few months later P. Berg at Washington University in St. Louis published a note which solved another important mystery (3); that of the active form of acetate after reacting with ATP. He showed that in yeast acetate stimulated a PP^{32}-ATP exchange and then identified acetyl AMP as an intermediate in the formation of acetyl-CoA.[1] He also reported that

[1] The background of thinking and methodology concerned in the search for the active form of acetate which provided such tools as PP^{32}-ATP exchange and hydroxylamine as a trapping agent for activated carboxyl groups provided a basis for this and subsequent developments in the study of amino acid activation.

methionine catalyzed a similar PP^{32}-ATP exchange, and making reference to the just-published paper of Hoagland, discussed the possibility of having methionyl-AMP in the reaction mixture. At about the same time De Moss and Novelli, having learned of the work of Hoagland, reported a whole series of experiments with extracts from several microorganisms (4). By testing the individual amino acids they found that all of the PP^{32}-ATP exchange-catalyzing ability was associated with only eight amino acids: tryptophan, phenylalanine, histidine, methionine, tyrosine, valine, leucine, and isoleucine. The other amino acids were inert. They pointed out that such a situation would require clarification if the mechanism were a general means of energizing amino acid-incorporation into proteins. A more detailed paper by Hoagland, Keller and Zamecnik appeared early in 1956 (5). Since amino acids would not cause a net splitting of ATP unless a high concentration of hydroxylamine was present, it was proposed that amino acyl-AMP had a low rate of reactivity with water. No free amino acyl-AMP accumulated and it was therefore felt that this intermediate was tightly held by the enzyme. The extent of reaction of amino acids, as measured by the PP^{32}-ATP exchange method and by hydroxamic acid formation, differed and it was believed that this indicated a difference in the availability of the activation site to hydroxylamine. It was found that amino acyl hydroxamic acid formation was quite specific for ATP; the other nucleotides including GTP, were relatively inert. That different and specific activating enzymes were involved was indicated by a partial fractionation obtained by precipitation at different pH's and by preliminary studies on heat inactivation. It was also found that hydroxylamine inhibited amino acid-incorporation in an acellular system in a manner roughly parallel to its ability to form hydroxamic acid, and that hydroxylamine also inhibited PP^{32}-ATP exchange. De Moss, Genuth, and Novelli tested the reversibility of the proposed reaction sequence by synthesizing leucyl AMP and causing the formation of ATP upon incubation of this compound with pyrophosphate and the activating enzymes from *E. coli* (6, 7). Another extensive investigation by De Moss and Novelli was published in which they described their studies with microbial extracts (8). Their findings were essentially similar to those of Hoagland et al. (5), but they again found that only eight amino acids seemed to participate in the exchange reaction. Davie, Koningsberger, and Lipmann, who observed that tryptophan was the most active amino acid for the formation of a hydroxamate in a crude extract prepared from beef pancreas, proceeded to purify a tryptophan-specific activating enzyme (9). These authors pointed out that the participation of amino-acyl-AMP in protein synthesis would bear out an early proposition of

Hubert Chantrenne that a substituted acyl phosphate would make a better precursor for peptide formation than the acyl phosphate itself (10). Berg isolated a methionine-activating enzyme from yeast (11) and Schweet obtained a tyrosine-specific enzyme from pig pancreas (12). In a survey of this field in 1957 Novelli and De Moss again emphasized that in a variety of different microorganisms as well as rat liver, the same eight amino acids accounted for the full activity and that perhaps these were primarily activated and could activate the others by a transacylation reaction (13). They also pointed out that certain amino acids which did not stimulate PP^{32}-ATP exchange nonetheless formed hydroxamates and also that for amino acids participating in the exchange reaction there was no correspondence between the magnitude of the exchange and of hydroxamic acid formation.

At this point all the data strongly supported the following sequence of reactions:

(1)   Enz + ATP + R—CH—COOH $\rightleftharpoons$
　　　　　　　　　　　　　　　|
　　　　　　　　　　　　　　　NH_2

$$Enz\text{-}AMP\text{—}O\text{—}\overset{\overset{O}{\|}}{C}\text{—}CH\text{—}R + PP$$
　　　　　　　　　　　　　　　　　|
　　　　　　　　　　　　　　　　　NH_2

(2)   Enz-AMP—O—$\overset{\overset{O}{\|}}{C}$—CH—R + NH_2OH $\rightleftharpoons$

$$Enz + AMP + R\text{—}CH\text{—}\overset{\overset{O}{\|}}{C}\text{—}NHOH$$
　　　　　　　　　　　　|
　　　　　　　　　　　　NH_2

Additional strong support for this formulation came from an experiment reported by Hoagland, Zamecnik, Sharon, Lipmann, Stulberg and Boyer (14). They showed that O^{18} introduced in the carboxyl group of tryptophan, when subjected to the PP^{32}-ATP exchange reaction catalyzed by the purified tryptophan-activating enzyme, found its way into the oxygen of the phosphoric acid moiety contained in AMP isolated from the reaction mixture. The atom per cent excess found in AMP supported the interpretation that the direct condensation of tryptophan and AMP represented the major and probably sole reaction pathway for the amino acid in this reaction. Up to this point, although it was tempting to conclude that this activation of amino acids was involved in protein synthesis, there was little direct support for the conclusion. A more related observation was reported by Sharon and Lipmann (15). A series of six analogues of tryptophan was tested with a purified preparation of the pancreatic tryptophan activating enzyme. Four of the analogues were activated by the enzyme, as measured by hydroxamic acid formation and PP^{32}-ATP exchange. The other two, although they inhibited tryptophan activation, did not show any activation effects. These observations compared favorably to those of Pardee, Shore and Prestidge, who showed that in *E. coli* two of the four tested analogues that were activated by the pancreatic enzyme were also incorporated into protein (16). One of the two analogues that were not activated by the pancreatic enzyme was not incorporated into *E. coli* protein. Cole, Coote, and Work found amino acid hydroxamic acid formation catalyzed by extracts of different tissues from four different animals. They also achieved a partial purification for one fraction forming hydroxamates from serine and threonine and another which formed hydroxamate chiefly from tryptophan (17). Nisman, Bergmann, and Berg used a sedimentable preparation obtained from a lysate of *E. coli*-penicillin-spheroplasts and showed that 18 amino acids could catalyze the PP^{32}-ATP exchange reaction (18). There was a wide margin in abilities to catalyze the reaction which ranged from 2.3 μmoles of PP^{32} in ATP per mg protein in 15 min for valine, to 0.1 μmoles for glycine measured against a background (lacking amino acid) of 0.05 to 0.07 μmoles. By this time, many papers were appearing which showed that the addition of amino acids could lead to an enhancement of a PP^{32}-ATP exchange.

One reservation that I held at this time was that there could be many ways to stimulate a PP^{32}-ATP exchange. Berg very clearly showed the involvement of acetate in this reaction (3). How much of the observed stimulation could have been due to a further metabolism of the deaminated carbon skeleton of the amino acids? In cases in which hydroxamates were identified, clearly the activation of the amino acid-carboxyl group was involved. This reaction, however, occurred to a far lesser extent than PP^{32}-ATP exchange and was not directly quantitatively related to the exchange. In Federation meetings at this time, reports of the stimulation of PP^{32}-ATP exchange by added amino acid were accepted as synonymous with the first step in protein biosynthesis.

FURTHER CLARIFICATION OF THE NATURE OF MICROSOMES AND THEIR ACTIVITIES

During the two-year period (1955–1957) while the concept of amino acid activation was taking shape other significant developments were also happening. Before 1955 it was established that microsomes were the primary cytological unit concerned with protein synthesis; but what are microsomes? They were operationally defined as minute particles which sedimented in high centrifugal fields (up to 105,000 × gravity) and which contained the major fraction of cellular RNA and the amino acid-incorporating ability. In the hen oviduct, however, the major fraction of cellular RNA and amino acid-incorporating ability sedimented in 2 to 3 min at a few hundred times gravity (19). Similar atypical findings were reported for bacteria (20) and pea seedlings (21). Littlefield, Keller, Gros, and Zamecnik treated rat liver microsomes with the surface active agent sodium deoxycholate, and thereby released ribonucleoprotein granules which contained essentially all of the RNA and approximately one sixth of the original protein (22). These granules could be sedimented at 105,000 × gravity and were composed of approximately 50% RNA and 50% protein. They sedimented at 47s in the ultracentrifuge and in the electron microscope appeared as small particles of about 240 A in diamter. Whole microsomes contained similar particles in or on larger structural fragments. It was recognized that these particles corresponded to material previously described by Petermann et al. (23), and by Palade (24). It was found that, after intravenous injection of DL-C^{14}-leucine and fractionation of the rat liver microsomes into deoxycholate (DOC)-soluble lipoprotein and deoxycholate (DOC)-insoluble ribonucleoprotein granules, the ribonucleoprotein granules were initially labeled faster than the DOC-soluble fraction. After 10 to 15 min, however, the content of radioactivity in the DOC-soluble fraction surpassed that in the ribonucleoprotein (RNP) granule fraction. With a small dose of C^{14}-leucine, the RNP granule fraction showed a peak at 3 min and thereafter declined in radioactivity. After a large dose the RNP granule fraction reached a plateau at about 10 min and maintained this level. Radioactivity in the DOC-soluble portion continued to rise.

Palade, and Siekevitz were able to show that microsomes from liver (25, 26) and pancreas (27) were derived by a pinching-off process from the intracytoplasmic membrane system called ergastoplasm or rough surface endoplasmic reticulum (described more fully in Chapter 4). Thus microsomes isolated as separate particulate entities are really not present as such in the cell. In many types of intact cell there is a series of more or less continuous membranes composed of lipoprotein. Frequently, ribonucleoprotein granules (later called ribosomes) are found studded on vast areas of these membranes. These granules, or ribosomes, also occur unattached to membranes and those which are attached can be released by various treatments, the most extensively used being deaggregation of the supporting membrane with DOC.

Keller and Zamecnik found that amino acid-incorporation by the rat liver microsome system was dependent on guanosine di- or triphosphate in addition to ATP (28). This dependence was found by using a pH 5 precipitate from the soluble fraction, rather than the whole soluble fraction itself. Such precipitation freed the fraction of endogenous soluble nucleotides and under these conditions the requirement for the guanosine nucleotide was revealed. Although this requirement has been confirmed many times in succeeding years, the full explanation of the participation of these factors is still not available.[1]

By the year 1955 it was firmly established that microsomes represented the cells' major protein synthetic machinery and I was disturbed by the finding that in hen oviduct, the heavy, easily sedimentable fraction was the most active in amino acid-incorporation ability and that upon continued incubation in a medium of lowered specific radioactivity, radioactive proteins appeared to be transferred from this fraction to the soluble fraction. I thought that perhaps microsomes were derived from some larger cellular entity that fragmented during homogenization in other tissues (liver and pancreas) but not in oviduct, or that if microsomes existed as discrete entities they might occur adsorbed to larger structures in oviduct. At this time I became aware of the newer work of Palade (29) and Palade and Siekevitz (25–27). In the publication that described the high amino acid-incorporating ability of the easily sedimentable fraction of hen oviduct it was also reported that the major fraction of cellular RNA sedimented in this fraction (19). This difference of behavior in cell fractionation of different cell types was completely explained after a collaborative investigation involving biochemical, electron microscopic, and light microscopic techniques (30). We found that in this tissue the endoplasmic reticulum appeared as linings of sacs that were filled with amorphous material having the same resistance to the electron beam as the secretory protein found in the lumen. These sacs occurred in a variety of sizes from quite small to huge sacs near the apical end of the cell; they were filled to bursting with secretory protein. The pellet obtained from a homogenate by centrifugation at a few hundred times gravity clearly revealed the elements of the rough surfaced endoplasmic reticulum, and therefore these findings were in agreement with those in other tissues that the ribonucleo-protein studded-membrane system carried out the intense protein synthesizing function.

Rabinovitz and Olson showed after administering a pulse-dose of C^{14}-leucine in rabbit reticulocytes, that microsomes behaved kinetically with respect to C^{14}-content like a precursor to the soluble hemoglobulin (31).

[1] Most recent work seems to be resolving this mystery. See J. Lucas-Lenard and A. Haenni, *Proc. Nat. Acad. Sci.,* **59,** 554 (1968) and R. Ertel *et al., Proc. Nat. Acad. Sci.,* **59,** 861 (1968).

They became rapidly labeled and then decreased as the soluble protein continued to increase in radioactivity. Littlefield and Keller continued their studies with the dissection of the microsome and reduction of the amino acid-incorporating ability of the cell to its simplest elements (32). In an earlier paper concerning liver the RNP granules obtained with DOC were inactive in amino acid incorporation (22). This did not necessarily indicate a role for the membrane since DOC might be inhibitory to the incorporation system. In the later work with microsomes from mouse Ehrlich ascites cells, 0.5 M sodium chloride was used to liberate RNP particles from the membrane. Also, in this tissue, a greater proportion of the particles exist free (unattached to membranes). At this time the importance of controlling magnesium concentration for maintaining the aggregation of ribosomal subunits into 80s monosomes and polysomes was not appreciated (see page 77) and the particles present were 50 to 57s in the DOC preparations and 21s and 26s in the sodium chloride preparations. Using the whole cells and then separating the microsomes into DOC-insoluble RNP granules and DOC-soluble lipoprotein gave the same result as with liver, in that the DOC-insoluble granules showed the highest initial rate of incorporation. This system showed less dependence on added GTP than the liver microsome system. As was the case with the liver microsome system, the addition of a mixture of 17 unlabeled amino acids did not enhance incorporation. When the sodium chloride-insoluble particles were substituted for the whole microsomes, active incorporation into protein occurred and so it was concluded that the membranous component was not essential for incorporation.

It is important to state at this point that the membrane components removed by sodium chloride treatment were not essential for the extent and type of amino acid-incorporation process functioning in this preparation. Although the implication relates to the process of protein synthesis as defined in Chapter 1, it is important to bear in mind that the process actually observed in this and related systems is the incorporation of amino acids by peptide bonds into undefined positions in undefined products. The process may be entirely representative of that which goes on in organized systems, but unless actually shown to be such we should be aware that the relevance to protein synthesis is a point that must be experimentally demonstrated. In the case of incorporation of amino acids into the whole cell followed by fractionation into microsomal components it was shown that the initial rate of labeling of the RNP granules was sufficient to account for the flux of radioactivity passing into the total proteins of the cell. Therefore in the organized systems the RNP granules (attached or free) do appear to represent the focal point of the cell's protein synthetic machinery.

THE DISCOVERY OF sRNA

The year was 1957 and there was an obvious excitement beginning to build around the new discoveries linking protein synthesis to discrete RNP granules and more particularly around the findings concerning the reactions of amino acid activation involving ATP. Sometimes a great contribution to science occurs when the field is not ready to assimilate it, and years later the historians, in describing the discovery, say that it came before its time. This was definitely not the case for the discovery of sRNA (soluble ribonucleic acid—sometimes referred to as tRNA for transfer ribonucleic acid or acceptor ribonucleic acid). Once again attention was drawn to the laboratory of Hoagland, Zamecnik, and Stephenson (33). They showed that when C^{14}-leucine and ATP were incubated with a solution of activating-enzyme preparation from rat liver (prepared by precipitation of the supernatant fraction at pH 5) a low molecular weight species of RNA contained in this fraction (sRNA) became labeled. Addition of yeast RNA, microsomal RNA, and degraded microsomal RNA did not give additional labeling. The C^{14}-leucyl-sRNA was nondialyzable, charcoal and Dowex-1 adsorbable, acid stable, alkali-labile, nonexchangeable with C^{14}-leucine, and nonreactive toward ninhydrin. When incubated with hydroxylamine, the radioactivity was released and recovered as leucine hydroxamate. When incubated with microsomes and the complete system, radioactivity was transferred to protein, with GTP as a definite requirement, and the transfer was only slightly depressed in the presence of a great excess of C^{12}-leucine. Leucyl-sRNA prepared by the phenol method (34) transferred the amino acid to protein in the absence of added activating enzymes. This finding came just on the heels of a discovery by R. W. Holley that ribonuclease inhibited the alanine-catalyzed conversion of AMP to ATP in a pH 5 enzyme preparation from rat liver homogenate (35). In Holley's formulation the enzyme-bound AMP-amino acid passed its amino acid to an aceptor, X, and released the AMP. Therefore the conversion of AMP to ATP should depend on the integrity of X. Since RNAase inhibited this exchange, Holley proposed that an RNA would be the natural acceptor; but no amino acid other than alanine worked in this system. Nonetheless, the suggestion of an RNA as the natural acceptor of amino acid from amino acyl adenylate was strongly indicated and may have helped illuminate the way for Hoagland et al. The discovery that amino acid (alanine) appeared to become bound to a part of the RNA present in the pH 5 enzymes from rabbit liver was made independently by Ogata and Nohara whose report (36) appeared almost simultaneously with that of Hoagland et al. (33).

Unaware of these developments in the laboratories of Hoagland and Holley and Ogata, Crick, who was attending a symposium of the Biochemical Society in England, made some prophetic remarks in the discussion following a paper given by M. H. F. Wilkins on structural aspects of DNA in relation to protein synthesis (37). Crick stated that he could not see any aspect of the surface structure of DNA or RNA that might specifically adsorb amino acids. He proposed that the amino acid should combine with an "adaptor," which could then specifically form hydrogen bonds with a group of adjacent nucleotides. He also envisioned specific enzymes to attach the amino acids to their specific adaptors or coenzymes.

As we examine the evidence supporting the concept of the vital role of sRNA in protein synthesis, two distinct types of experiments should not be confused. In one case we shall be examining the involvement of sRNA in the process of amino acid-incorporation into undefined products as carried out in acellular systems. The quantitative extent of synthesis retained by the acellular systems compared to the intact cell is usually less than 1%. In the other case we shall examine the formation of characteristic cellular proteins by intact cells. The information we seek is an explanation of the latter process. The former process represents an experimental system employed to approach an understanding of the synthetic capabilities of the intact cell. Any finding obtained in the case of the acellular system must also be established for the intact cell system. With this in mind, we shall reexamine the data relating to the role of sRNA in protein synthesis as it appeared in the journal literature.

The initial paper of Hoagland et al. established that sRNA, contained in the pH 5 enzyme fraction of the supernate, became labeled with leucine and that this amino acid could be passed to hydroxylamine or microsomal protein in an acellular incubation (33). Weiss, Acs, and Lipmann reported that amino acid-incorporation into protein of pigeon pancreas microsomes was stimulated by a heat-stable nondialyzable factor (38). The active principle migrated with polynucleotide upon electrophoresis and showed a distribution of molecular weights between 20,000 and 40,000 by ultracentrifugation. In the absence of microsomes the incubation of C^{14}-leucine, supernate and concentrated factor resulted in radioactivity becoming associated with the isolated RNA. This radioactivity was lost by heating in TCA and therefore was primarily RNA-associated rather than peptide-bound. Direct transfer of radioactivity from the labeled factor to protein was not studied. Berg and Ofengand examined the same type of reaction in a sonic extract from $E.$ $coli$ (39). Incorporation of valine into a cold TCA-precipitable material was studied. Radioactivity could be removed by hot TCA, dilute alkali, hydroxylamine, or RNAase. The extent of incorporation was proportional to the amount of RNA added, and when leucine, valine, and methionine were tested together at saturation concentrations, the total amount of incorporation was the sum of the three amino acids tested individually. This pointed to separate sites for each amino acid. The reaction was specific for *E. coli* RNA. RNA's similarly extracted from yeast, *Azotobacter vinelandii*, turnip yellow mosaic virus (TYMV), tobacco mosaic virus (TMV), and rat liver, were inactive, as well as a portion of *E. coli* RNA precipitable with 1.5 M sodium chloride and also several synthetic polynucleotides. With the valine activating enzyme, of four amino acids tested, only valine was transferred to RNA. The same was true of a methionine activating enzyme. During a hundredfold purification of the valine activating enzyme, the ratio of PP^{32}-ATP exchange activity to that of RNA labeling remained constant. It was concluded therefore, that both activities were catalyzed by the one enzyme. Similar findings were simultaneously announced by Schweet, Bovard, Allen, and Glassman (40). These authors used standard enzyme purification techniques to achieve a separation of the RNA-labeling, PP^{32}-ATP exchange activities for tyrosine from that of leucine starting with a pH 5 preparation from guinea pig liver which was active for both amino acids. Activity was dependent on two factors, one heat labile (activating enzyme) and one heat stable. The heat-stable factor was RNAase-sensitive and could be replaced by an RNA preparation from the active fraction isolated by phenol extraction. Microsomal RNA was inert as well as RNA extracted from virus and yeast. RNA prepared from the pH 5 fraction from rat liver, however, was fully active. Furthermore, working at saturating concentrations for tyrosine and threonine, incorporation was strictly additive. The adddition of C^{12}-threonine did not diminish the labeling obtained with C^{14}-tyrosine nor did C^{12}-tyrosine affect the labeling with C^{14}-threonine. Clearly, specific sites on the RNA were indicated.

Zamecnik, Stephenson, and Hecht participated in a symposium on this subject which is reported in the *Proceedings of the National Academy of Sciences* (41). These authors, in my opinion, pointed out the single most important consideration governing the plausibility of the idea that amino acyl-RNA is a direct intermediate for protein synthesis. They wrote:

"If the leucine-labeled RNA is to be considered as a true intermediate in the over-all scheme rather than as a storage locus for activated amino acids, it should satisfy certain criteria of intermediates. It should, for example, be more rapidly labeled than are the peptide chains in the ribonucleoprotein particles. Such was found to be the case in short term experiments carried out on intact ascites tumor cells (42)."

The crucial question concerns whether this essential requirement has been fulfilled.

There are, to my knowledge, six published experimental evaluations of this vital facet of the problem: two by Zamecnik and coworkers, one by F. Gros, one by R. B. Roberts and coworkers, one by Hunter and Goodsall, and one by myself. Each of these experimental papers is considered in detail in the pages that follow. It will be apparent that the amino acyl-RNA has failed to meet this essential criterion in all of these cases and that this particular defect still stands as a major obstacle to accepting the idea that the *free* amino acyl-RNA represents *the sole or major path* for amino acids entering proteins *in intact living cells*.

One of the most important studies in support of the concept of the role of sRNA is in the paper entitled, "A Soluble Ribonucleic Acid Intermediate in Protein Synthesis" by Hoagland et al. (42). It was shown that the extent of labeling of sRNA was dependent on amino acid concentration, ATP concentration and the amount of added pH 5-RNA. Prelabeled pH 5 enzymes lost considerable radioactivity upon subsequent incubation in a medium lacking ATP. Pyrophosphate elevated and ATP reduced the extent of this loss. Thus it appears that the amino acyl RNA-forming reactions are reversible. The RNA-bound C^{14}-leucine did not react with ninhydrin but did with hydroxylamine to form leucine hydroxamic acid. The first test of the rate of labeling of sRNA in an intact cell compared to the rate of labeling of the primary protein incorporation site (the RNP granules released by sodium chloride treatment) is reported in this paper. The incubation was carried out at 25° in order to slow the reaction to more easily measurable rates. The results obtained are reproduced in Figure 1. Note the difference in scale of the ordinates for the two sets of curves. This experiment has often been thought to demonstrate the fact that sRNA in the intact cell is labeled faster than the protein. Figure 1, since it occupies such an important place in the development of the sRNA story, will be analyzed in detail.

1. The zero time point for the sRNA shows a value of 340 cpm/mg. As any worker in the field is aware, the isolation of a labeled product from the soluble fraction in the presence of radioactive amino acid entails the very real danger of adsorption of a minute quantity of highly radioactive C^{14}-amino acid. For this reason zero time values are always suspect and are usually subtracted as a background from the other values.

2. The initial rate of labeling of sRNA in this experiment depends

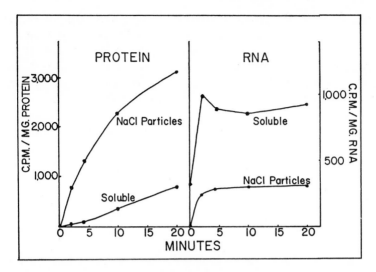

Figure 1   Time-curve of incorporation of L-leucine-C^{14} into the RNA and protein of the ribonucleoprotein particles and the soluble fraction in intact ascites cells. Ascites tumor cells (approximately 10 g. of packed cells) were incubated at 25° in 50 ml. of their own ascitic fluid fortified with glucose (0.04 M), Tris buffer, pH 7.6 (0.02 M), and containing 3 μmoles of L-leucine (3.5 × 10^6 cpm per μmole). Aliquots were taken at the time points shown; NaCl-insoluble and -soluble fractions were prepared from the 15,000 x g supernatant fraction. The specific activities of the RNA and protein of these fractions are shown. From M. B. Hoagland et al., "A Soluble Ribonucleic Acid Intermediate in Protein Synthesis," *J. Biol. Chem.*, **231**, 241 (1958).

on a single experimental point—that taken at 2 min. Accepting this point as valid by subtracting the zero time, the initial rate of labeling of the sRNA is 330 cpm/(min)(mg) RNA. (As will become clear in the following comments, this rate of labeling is inadequate for a major precursor role, even if we do not subtract the suspect zero time value.)

3. The initial rate of labeling of the sodium chloride-insoluble RNP granules was 385 cpm/(min)(mg) protein.

These data are both expressed per mg of protein or RNA. This is not the way to compare these rates. The question that is important is, "Can the sRNA in this cell handle the flux of radioactive amino acid that is observed to enter the protein?" From a previous paper of this laboratory (43) we know that the sodium chloride-insoluble particles contain essentially all the RNA of the microsomes and that this accounts for 70% of the RNA of the cell. From both papers we know that the sRNA accounts for 20% of the cell's RNA. This figure could actually be high since in the liver, much less of the cell's RNA is considered to carry out the functions of sRNA. Nonetheless, on a weight basis there is at least 3.5 times more RNA in the sodium chloride-insoluble particles than in the sRNA fraction. We also know from the previous paper that the weight of protein in the particles equals the weight of RNA and therefore that the cells contain at least 3.5 times as much RNP-protein as sRNA. Thus for the sRNA to accommodate the flux of radioactive amino acids that are being incorporated into proteins it must be labeled 3.5 times faster than the protein on a weight basis; that means at least 3.5×385 or 1350 cpm/(min)(mg) RNA. The observed value was 330 if the zero time correction was made, or 500 if no zero time correction was made. These data *do not* support the concept that in this intact cell all of the amino acid passes through the amino acyl-sRNA stage en route to the protein. It is possible, however, that a small fraction of the radioactive sRNA has the proper kinetics but that a large fraction that is relatively inert is obscuring the picture. This possibility, however, seems to be lessened by the other experiments reported in the paper. It was shown that when the C^{14}-leucine-sRNA fraction was incubated with microsomes 100% of the radioactivity leaving the sRNA entered the protein and the rate of loss of radioactivity from the sRNA mirrored the rate of gain of radioactivity by the protein. There has never been a doubt in my mind that this proved that amino acyl-sRNA is the major source of radioactivity for the *in vitro* labeling of nondefined protein. This experiment also shows that essentially all of the radioactive sRNA is functional in labeling the protein and argues strongly against the possibility raised above that only a small fraction may possess all of the biological activity. GTP was essential for this *in vitro* transfer of radioactivity from sRNA to protein and a large excess of C^{12}-leucine had a minimal effect on the transfer process. In another experiment it was shown that microsomal protein was much more efficiently labeled with pH 5 enzymes prelabeled with C^{14}-leucine than when free C^{14}-leucine was the source of radioactivity in the presence of unlabeled pH 5 enzymes. When C^{14}-leucine-sRNA was isolated by the phenol method, it was found that the transfer of radioactivity to protein depended on the addition of ATP in addition to GTP.

Mager and Lipmann described analogous studies pointing to a labeling of nucleic acid in *Tetrahymena pyriformis* with C^{14}-leucine (44). No nucleic acid was actually isolated but radioactivity which was precipitated with cold TCA and lost by treatment with hot TCA was taken to be representative of amino acyl-RNA. It was found that an enzymatic incubation with pyrophosphate caused a loss of RNA-associated radioactivity. The interesting observation was made that for incorporation of amino acid into protein, the microsomal and supernatant fractions of rat liver and *Tetrahymena* could be interchanged. Lipmann also reported that all 20 amino acids could stimulate PP^{32}-ATP exchange in a pigeon pancreas extract although the range varied over approximately one hundredfold for the different amino acids (45).

In the same volume of the journal, Zachau, Acs, and Lipmann elucidated the manner of binding of leucine in sRNA (46). They treated liver C^{14}-leucyl-sRNA with RNAase and isolated a fragment by paper electrophoresis which contained practically all the radioactivity. By chemical tests and by the liberation of equimolar amounts of amino acid and adenosine on hydrolysis the fragment was identified as a leucyl ester formed with the 2'- or 3'-hydroxyl group of the ribose in the adenosine. This information, combined with a report from Hecht, Stephenson, and Zamecnik on the terminal configuration of sRNA, led to the following formulation for the end sequence of amino acyl-sRNA (47):

$$
\begin{array}{c}
\text{Ad} \quad \text{Cy} \quad \text{Cy} \\
\text{O} \quad \quad \quad \quad \quad \quad \text{etc.} \\
\| \\
\text{R—CH—C—O—}|\ \text{P} \diagup \text{P} \diagup \text{P} \diagup \\
| \quad \quad \quad \quad \quad | \\
\text{NH}_2 \quad \quad \quad | \\
\quad \quad \quad \text{RNAase} \\
\quad \quad \quad \text{Split}
\end{array}
$$

All future work has been consistent with this representation.

This is the state of the field at the time of the Fourth International Congress of Biochemistry held in Vienna in 1958. The session on protein synthesis was primarily devoted to a consideration of the work just discussed, in papers from the laboratories of Lipmann, Hoagland, and Zamecnik. Lost in the deluge of data in support of the role of amino acid activation and amino acyl-sRNA in protein synthesis were the following remarks by Ochoa (who later on became a staunch supporter for the view of protein synthesis involving amino acid activation and amino acyl-sRNA).

"I must cast some doubt on the current belief that the amino acid activating enzymes that catalyze an amino acid-dependent exchange of PP^{32} with ATP (PP enzymes) are involved in protein biosynthesis. This view is based on the observation that crude enzyme preparations (pH 5 enzymes) which contain PP-exchange enzymes, stimulate the incorporation of amino acids into microsomal protein but there are no indications that purified PP-exchange enzymes have such an effect.

"Experiments in our laboratory, reported by M. Beljanski at these meetings, led to the discovery of an enzyme which stimulates incorporation of amino acids into protein in the apparent absence of PP enzymes. This enzyme, referred to as the amino acid-incorporation enzyme, was isolated in highly purified form from extracts of *Alcaligenes faecalis*. The purified preparations, which are completely free of PP-enzymes, stimulate the incorporation of all amino acids thus far tried (glycine, alanine, threonine, valine, leucine, lysine, proline, phenylalanine, tyrosine, tryptophan) into protein of a particulate fraction derived from the same microorganism which gives a negative assay for PP enzymes. Amino acid-incorporation by the *Alcaligenes* particles is inhibited by chloramphenicol, eliminating the possibility that one might be dealing with synthesis of cell wall protein.

"It was possible that the incorporation enzyme might act only in bacterial systems. However, we have recently obtained indications for the presence of this enzyme in rat liver supernatant and have also found that the purified enzyme (10 μg of protein) completely replaces the crude pH 5 enzymes (1.5 mg of protein) in stimulating the incorporation of C^{14}-leucine into protein of rat liver microsomes. This is convincing proof that the incorporation enzyme is involved in protein synthesis both in bacterial and mammalian cells.

"We have further found that highly purified preparations of the incorporation enzyme catalyze a Mg^{2+}-dependent exchange of radioactive ADP with ATP which appears to be related to their amino acid-incorporation activity. Although the mode of action of the incorporation enzyme is as yet unknown, the catalysis of the ADP-ATP exchange reaction may be of significance in view of the fact that glutathione synthetase, an enzyme which catalyzes the synthesis of a typical peptide, brings about a similar exchange (48, 49)."

Another summation of the state of the field as of 1959 is presented by Zamecnik in The Harvey Lectures (50). Most of the literature covered by Zamecnik has already been discussed in this historical survey. During the lecture, however, Zamecnik addressed himself again to the question of rapidity of labeling of sRNA with amino acid in an intact cell compared to the rate of labeling of the protein. He referred to his earlier measurements (discussed on page 60). Although the earlier measurements were believed to be adequate, he felt that the newer data were better because they were taken at 20° which sufficiently slowed the initial labeling of the sRNA in order to reduce the zero time blank. In the newer data (see Figure 2) the sRNA was labeled with C^{14}-valine in intact ascites cells at the rate of 140 cpm/(min)(mg)RNA.

The ribosomal protein was labeled at a rate of 315 cpm/(min)(mg) from the second to the eighth minute. Even including the initial delay, the rate of labeling of the protein during the first 4 min was 170 cpm/(min)(mg). Since there is only 1/3.5 times as much sRNA as ribosomal protein, if all of the amino acid must pass through the sRNA to get to proteins, the sRNA has to be labeled 3.5 times faster than the protein. We find that the rate of labeling of sRNA was from four

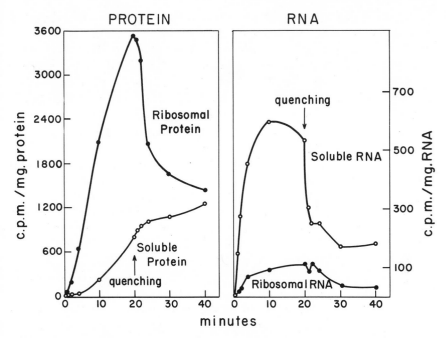

Figure 2 C^{14}-valine incorporation into intact ascites cells. Intact ascites cells were incubated in their own fluid with C^{14}-L-valine and with addition of 1 M glucose and 1 M tris buffer (pH 7.6) to final concentrations of 0.04 M glucose and 0.02 M tris buffer. The C^{14}-valine was 0.02 mM in the medium and had a specific activity of 1.3 × 10^7 cpm/μmole. The incubation was carried out at 20°C for the time interval indicated and immediately thereafter the samples were diluted with ice-cold wash medium. The cells were washed and lysed. Ribonucleoprotein particles (ribosomes) prepared by the sodium chloride method and the soluble cytoplasmic fractions were isolated and assayed for radioactivity in both the RNA and protein of the fractions. At 20 min, C^{12}-L-valine was added to a final concentration in the medium of 0.002 M. This represented a dilution of the intracellular C^{14}-valine concentration (endogenous plus added) to the order of a third of its previous specific activity. From P. C. Zamecnik, *Historical and Current Aspects of the Problem of Protein Synthesis, The Harvard Lectures 1958–1959*, Academic Press, New York.

to eight times too slow. It is really not valid to accept the delay in labeling of the protein during the first minute or two in the calculation because, regardless of the identity of the initial precursor, it would account for the initial delay and, on saturation of this precursor, the rate of entry of radioactivity into the protein (from the second to eighth minutes) would measure the true rate of entry of radioactive amino acid. This, then, constitutes the second of six published tests for the rate of labeling of sRNA in intact cells.

Another interesting observation was discussed by Zamecnik. He gave data which showed that when radioactive amino acid was transferred from sRNA to proteins, the terminal labeled AMP of the sRNA was also transferred to the microsome with the amino acid. Since enzymes exist in the supernatant fraction for adding back the terminal nucleotides to sRNA (discussed in the following), this removal and reformation of the terminal configuration of sRNA looked like a cyclic mechanism which accompanied amino acid-incorporation into proteins and this idea was incorporated into Zamecnik's over-all scheme.

In another study Hecht, Zamecnik, Stephenson, and Scott provided convincing evidence for reactions which added a cytosine nucleotide and an adenine nucleotide terminally to sRNA, with the latter nucleotide predominantly in the exterior position (51). Preincubation of sRNA with the ascites cell supernatant pH 5 fraction at 37° resulted in a loss of this terminal sequence which could then be regenerated from CTP and ATP in the medium. Although sRNA from rat liver and yeast served as acceptor for the adenine nucleotide end group, ascites nuclear and microsomal RNA were almost completely inactive. The reversibility of the reaction was demonstrated by the incorporation of PP[32] into CTP in the presence of sRNA. These results were in agreement with earlier reports by Canellakis (52), Heidelberger et al. (53), and Patterson and LePage (54), which showed that added adenine nucleotides were located adjacent to cytosine nucleotides. Hecht, Stephenson, and Zamecnik followed this report with another that showed that preincubation of sRNA with pH 5 enzyme fraction destroyed its ability to accept amino acids. This ability was completely recovered, however, when cytosine and adenine nucleotides were added back to the sRNA by incubation with CTP and ATP.

Karasek et al. helped to fill in a gap in the understanding of the amino acid activation reaction by showing the positive existence of the suspected amino acyl adenylate and the existence of an exchange mechanism between free tryptophan and free tryptophanyl adenylate catalyzed by the tryptophan activation enzyme (56). The formation of ATP from inorganic pyrophosphate and a variety of other alpha amino acyl adenylates catalyzed by the tryptophan activation enzyme was also shown so that the specificity of the reaction did not appear to hold for the reverse reaction.

For a short time Castelfranco, Meister, and Moldave were separated from the strong tide of acceptance for the concept of the involvement of amino acid activation and sRNA in normal protein synthesis. These authors showed that after boiling the liver microsomal supernatant preparation of Zamecnik et al., the incorporation of radioactive glycine from labeled glycyl adenylate was increased 12 times over the rate with the unboiled preparation and 1200 times over the rate for C^{14}-glycine plus adenylate in the unboiled preparation (57). Similar results were obtained with tryptophanyl adenylate. The "incorporated" radioactive amino acid was held by bonds as strong as those holding the rest of the protein together and no $C^{14}O_2$ was released from its radioactive carboxyl group by treatment with ninhydrin. More than 70% of the label, however, was held in N-terminal positions. It was also found that C^{14}-glycyl adenylate would label purified rat liver sRNA in an alkali-labile fashion. When such labeled sRNA was subsequently incubated with a heated or unheated protein preparation, significant quantities of radioactivity were transferred to the protein. Thus, the possibility was raised that some of the acellular labeling observed might be due to the chemical reactivity of this anhydride rather than to enzymatic reactions. These authors pointed out, however, that at the much lower concentration which these amino acyl adenylates might naturally occur, enzymatic catalysis may provide the major route of reaction. An analogous situation was described by Zioudrou et al. (58). A later study by Wong, Meister, and Moldave, (59) showed that the transfer of amino acid from amino acyl-sRNA to protein could be speeded up tenfold in the presence of the appropriate amino acid activating enzyme; thus, the prominent reaction in any enzymatic situation would be enzyme mediated.

As we approach the current state of research affairs in this field, the number of significant papers grows beyond the scope of this history. Therefore, although we shall continue to trace the mainstream in the development of our ideas and knowledge, it is impossible to cite all the additional corroborative studies.

In the interest of continuity the development of sRNA was followed to the neglect of some other pertinent investigations. At this time we shall leave the sRNA path briefly and drop back a few years in time to pick up the threads of various other routes of investigation.

In 1958 Monod reviewed the field of enzyme induction (60). The picture that he drew for the process of enzyme induction would be almost completely unrecognizable to the workers in this area today. Briefly restated, Monod proposed that amino acids are carried by sRNA to a stable template and then released to the template. He speculated that the sRNA may be a precursor for the formation of cellular RNA. Although Monod pointed out that the absence of a significant lag in enzyme induction might indicate unstable and short-lived templates, he argued against this idea in favor of stable templates. A preenzyme would be formed on the template, which could, when released, assume different configurations. These configurations, however, would not be associated with enzymatic activity unless the inducer, or more likely a metabolite of the inducer, brought about a specific type of folding. Furthermore, the difference between a constitutive and induced formation of the enzyme would be that, in the constitutive case, the normal inducer would be present endogenously.

At the Third International Congress of Biochemistry (pages 42–44) Gale summarized his findings which showed that disrupted Staphylococcal cells would incorporate amino acids into protein but that this ability was almost completely lost if the cells were depleted of nucleic acid. The incorporation ability could be restored by DNA or RNA, DNA being the more effective. Digestion of DNA with DNAase reduced the stimulatory effect, whereas digestion of RNA with RNAase enhanced its effect. Different fragments of the RNA digest were isolated by paper electrophoresis and these possessed amino acid-incorporation stimulating effects that exhibited a degree of specificity towards different amino acids. These factors cochromatographed with apparently unique di- or trinucleotides, and were thought of in terms of fragments of the protein-forming template. In the years that followed (this material is reviewed by Gale (61)) Gale found that activity persisted on hydrolysis completely to the mononucleotide stage and furthermore that when all of the nucleotides were moved off of the origin by electrophoresis, the activity could be recovered in the material remaining at the origin which was nucleotide free. Insufficient amounts of the isolated factor were available for chemical characterization at this time, but it was learned that the "incorporation factor" had no specific ultraviolet spectrum and was stable to boiling at 100° in 1 N sodium hydroxide for one hour but lost 60% of the activity on boiling one hour in 0.1 N HCl. This factor replaced nucleic acid completely for the incorporation of glycine, aspartic acid, phenylalanine, leucine, arginine, and glutamic acid and was much less effective for isoleucine, valine, proline, and tyrosine. Since in earlier studies with the RNA digest, a valine incorporation factor was described, it would seem that specific incorporation factors might still exist. It was also learned that the incorporation factor was effective for allowing nucleic acid-depleted cells to incorporate C^{14}-adenine into nucleic acid, but that the total repair of this defect required the addition of DNA with the factor. Gale speculated that perhaps the effect of the incorporation factor on amino acid incorporation may be mediated through its effect on nucleic acid synthesis.

This story was picked up once more in 1960 and has since fallen into the background unfinished and unclarified (62). Kuehl, Demain, and Rickes identified glycerol as the only component present in a highly purified but aged preparation of "incorporation factor" prepared by Gale. Gale reported that glycerol had comparatively little activity when introduced into an amino acid-incorporating system from disrupted Staphylococcal cells or intact coliform organisms, but after a lag of 60 to 90 min, activity increased to that found for fresh "incorporation factor." The situation at the moment is still unresolved and if Gale is correct it would appear that glycerol enters into a labile association with some other factor and then becomes bound by the cellular RNA from whence it exerts its stimulatory effect on amino acid-incorporation. This may or may not be the final stage of development in this curious and interesting situation but no further enlightenment has appeared as of 1967.

In my own work I pursued the possibility that the easily sedimentable fraction (cell debris), which contained the elements of the endoplasmic reticulum represented an intermediary stage between free amino acid and soluble ovalbumin. This was later established by finding that intact prelabeled oviduct, when subsequently homogenized, yielded a cell debris fraction which could release highly radioactive normal soluble proteins into an unlabeled supernatant fraction (63). During this time it was most surprising to see that the total amount of radioactivity contained in the protein increased. This was unexpected because the amount of free C^{14}-amino acid present in these washed broken-cell incubations was far too low to account for the observed extent of labeling of the protein. It therefore appeared that some labeled material present in the prelabeled cell debris fraction, other than free amino acid, was serving as a source of radioactivity for the proteins in the homogenate incubation (64). The incubation consisted only of a prelabeled cell debris and unlabeled supernatant fraction both obtained by homogenization and fractionation of intact oviduct.

Whatever the nature of the radioactive chemical entity that was supplying the proteins in the incubation, it had to come from the cell debris and might be identified by monitoring fractions of the cell debris for losses of radioactivity during this incubation. When this was done, it was found, entirely unexpectedly, that only the lipid-soluble fraction showed a decrease in radioactivity during the phase of increase of labeling of the proteins; the nucleic acid-associated radioactivity (which contained the lowest amount of radioactivity of all the fractions) showed a slow steady increase of radioactivity (65). Therefore the only fraction exhibiting the qualitative behavior predicted for the suspected intermediate was the lipid fraction. At that time I reviewed some of the experimental observations that served to confirm our knowledge about the involvement of RNA in protein synthesis and decided to apply the same criteria to test the possible involvement of lipids (66). First, it was pointed out that the microsomes, which are the major sites of protein synthesis, contained the major fraction of cellular RNA. (It is an interesting paradox that of all the cellular RNA shown to function in protein synthesis, absolutely no evidence has accrued so far to implicate a particular role for this microsomal RNA.) The microsomes also contain a major fraction of cellular lipid which, as a matter of fact, accounts for about five times as much of its total weight as the RNA. It was shown that by dissolving cellular RNA with 1 M sodium chloride, protein synthesis was impaired. I tested the effects of lipid-dissolving agents like lysolecithin and deoxycholate and found them to be potent inhibitors of protein synthesis.

A better test for the involvement of RNA was the inhibitory effects of small amounts of ribonuclease. I tested lysolecithinase A and found that on a weight basis it was four to five times more potent as an inhibitor than ribonuclease. These observations stimulated my interest in the possibility that protein synthesis may occur at a nonpolar site in the cell (a membrane), and my thinking about the problem has involved the importance of cytological considerations ever since.

In order to determine whether the rate of entry and exit of radioactive amino acid in the lipid fraction was consistent with the possibility that it could be considered *en route* to protein, I studied these rates as well as the rates of labeling and unlabeling of sRNA (65). This study marks the third attempt to measure the rate of labeling of sRNA in an intact cell in comparison to the rate of labeling of the protein (the other two are discussed on pages 60 and 64). The initial rates of labeling of the total nucleic acid fractions (sedimentable and soluble) with radioactive amino acid were manyfold lower than the labeling of the protein even when the measurement was made at five minutes of incubation at which time the rate of labeling of the protein was still in a lag phase (i.e., the extrapolation of the early linear part of the curve depicting the quantity of radioactivity in the protein intersected the abscissa at 4 min, showing a delay of appearance of radioactivity in the protein). Once labeled, the nucleic acid fraction showed little or no tendency to lose the radioactivity upon further incubation in a medium depleted of radioactive amino acids. It was true that the amount of radioactivity in 1 mg of RNA was somewhat higher than the amount in 1 mg of protein, thus agreeing with the observations of Zamecnik et al. (pages 61 and 65). The apparent quantity of radioactive material transferred by the nucleic acid fraction, however, was too small to provide support for the idea that a major fraction of the radioactive amino acids was passing through the fraction *en route* to protein. The passage of radioactivity through the lipid fraction is considered again in Chapter 8.

The fourth published test of the kinetics of labeling of sRNA with radioactive amino acid in comparison to the rate of labeling of proteins in an intact cell was published by Lacks and Gros who measured these parameters for *E. coli* (67). The results shown in Figure 3 show that radioactive methionine entered protein at least three times faster than the maximal rate of labeling of sRNA. As can be seen, the tangent to the beginning of the sRNA curve does not actually go through any of the experimental points. If a straight line were drawn through the origin and the first two points, the maximal rate of labeling of the sRNA would appear to be even less than indicated by the tangent drawn. To explain these results Lacks and Gros considered the possibility that perhaps only a portion of amino acid incorporated into protein may be proceeding via the amino acid-activation, sRNA route and the rest by a route not using these intermediates.

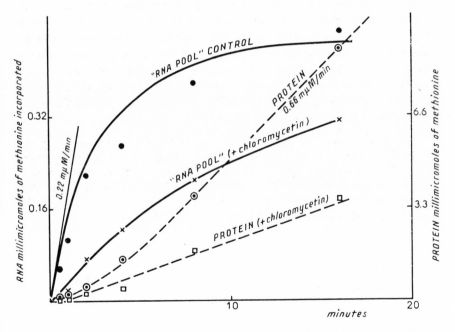

Figure 3 Comparative rate of incorporation of methionine into the RNA-methionine complex and into protein. From Lacks and F. Gros, "A Metabolic Study of the RNA-Amino Acid Complexes in *Escherichia coli*." *J. Mol. Biol.*, **1**, 301 (1959).

The same conclusion was reached independently by Rendi and Campbell for different reasons (68). These investigators found that amino acid-incorporation into rat liver microsomes could be stimulated by a preparation of Sachs' "S-protein" (69) isolated from rat supernatant fraction by precipitation with protamine, as well as by the conventional pH 5 precipitate. The protamine removes virtually all of the RNA from the S-protein. Incorporation supported by dialyzed S-protein was markedly stimulated by glutathione whereas incorporation by dialyzed or nondialyzed pH 5 fraction was either unaffected or inhibited by the addition of glutathione. Therefore a microsomal-bound form of sRNA does not seem to offer an explanation for the phenomenon. These authors concluded that amino acid-incorporation into rat liver microsomes proceeded by two routes: one employing sRNA and the conventional pH 5 fraction enzymes, the other by different non-sRNA dependent reactions supported by S-protein and glutathione.

Although the tissue system of choice up to 1959 had been rat liver, *E. coli* soon became the prestige organism for "molecular biologists." Tissières, Watson, Schlessinger, and Hollingworth carefully studied the properties of ribosomes liberated from these cells by alumina grinding (70). The subunit-composition and the equilibria among these units as influenced by magnesium concentration were reported. These studies served as a basis for later work employing *E. coli* ribosomes.

The fifth study of the kinetics of labeling of sRNA in an intact cell (*E. coli*) was described by Bolton and Roberts of the Carnegie Institution of Washington (71). They stated,

"The radioactivity of the sRNA fraction continued to increase after the time when the full rate of incorporation into protein was established. Furthermore, radioactivity persisted in the sRNA fraction after incorporation into protein had ceased. Thus the S^{35} bound to the sRNA cannot be in a form that is an intermediate precursor to protein. Instead it seems to equilibrate with the pool as indicated in B of [Figure 4]."

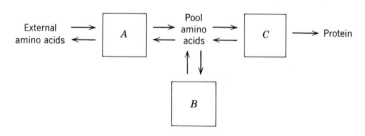

Figure 4

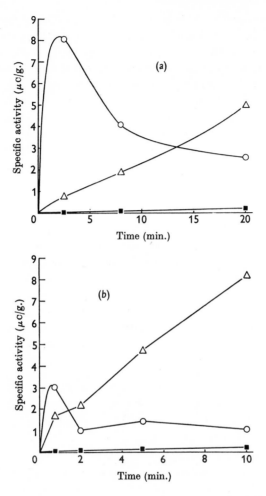

Figure 5   Uptake of amino acids into the lipid (◯),   protein (△) and RNA (■)
fractions by photoplasts: (*a*) protoplasts, (1.5 g) incubated with 30 μc of L-[*C*¹⁴]
arginine; (*b*) 1.4 *g*. of protoplasts incubated with 20 μc of L-[C¹⁴] threonine. Samples
were taken at the indicated times. From G. D. Hunter and R. A. Goodsall,
Biochem. J., **78**, 564 (1961).

A final published test of the rate of labeling of sRNA in an intact
cytoplasm appeared in 1961 and I shall skip ahead to that year just
to complete the picture with respect to this point (72). Hunter and
Goodsall incubated intact protoplasts of *B. megaterium* with L-C¹⁴-argi-
nine or L-C¹⁴-threonine. The results are shown in Figure 5.

The nucleic acid-bound radioactive amino acid is seen to be in a stable state as opposed to a dynamic state. This agrees with other intact cell results. The initial rate of labeling of the RNA fraction was small compared to that of the protein. These studies also pointed to the lipid fraction in terms of rapidity of labeling of this fraction with amino acids. This problem is considered further in Chapter 8.

A paper was published at about this time by Heller, Szafranski, and Sulkowski (73) in which they pointed out that in the silk glands of the silk worm two well-defined proteins are produced. "Fibroin" which contains 42% glycine and 28% alanine is formed by the posterior glands, and "sericin" which contains 30% serine is produced by the middle gland. When the pH 5 enzymes of the glands were obtained and tested for catalysis of hydroxamate formation it was found that the two amino acids which showed the highest rate of activation were tryptophan and tyrosine even though these amino acids occur only in traces in the proteins of silk. Glycine showed the lowest rate of activation. Alanine was similarly quite inert. To test the possibility that glycine may be activated by transacylation, its activation was studied in the presence of tryptophan or tyrosine, but only tryptophan or tyrosine hydroxamates could be formed from the mixtures—no glycine hydroxamate was formed. To evaluate the possibility that glycine may be activated outside of the gland, the haemolymph was tested but the activation was still negligible.

Further Clarification of the Nature of sRNA (1960)—
End Groups and Secondary Structure

Hecht, Stephenson, and Zamecnik found that by using a pH 5 enzyme fraction from mouse Ehrlich ascites cells, sRNA from either rat liver, calf liver or yeast liver was active as an aminoacyl acceptor (74). This was in contrast to the early studies of Berg and Ofengand (page 59) who reported a species specificity. Following incubation of pH 5 fraction at 37° in the absence of nucleoside triphosphate and amino acids, the ability of the endogenous sRNA to bind leucine or valine dropped about 95%. The leucine and valine-catalyzed ATP-PP exchangeability, however, dropped only slightly. Incubation of this damaged sRNA fraction with CTP plus ATP restored its ability to accept valine and leucine. The other nucleoside triphosphates could not replace CTP. The concentration of CTP required for optimal incorporation of valine into preincubated sRNA and for the incorporation of CTP and ATP into this sRNA was the same; CTP and ATP were incorporated in the ratio of 2:1. Splitting sRNA labeled with these nucleotides by means of snake venom diesterase quantitatively liberated the radioactivity as 5'AMP-C^{14}—only 5% of the sRNA was degraded. These facts suggest an end grouping of -pCpCpA in sRNA (see page 63). It was shown that the only function of CTP in restoring the ability of preincubated sRNA to accept amino acids was in reestablishing the terminal sequence. When amino acid was held by the sRNA, the terminal-CCA was protected against loss or replacement, suggesting that the amino acid was held by the terminal adenine nucleotide. The stability of the bond holding the amino acid to conditions of heating at 100° at pH 4 through 5 tends to eliminate a carboxyl phosphoanhydride linkage. The fact that borate ions inhibit the binding of amino acids by sRNA suggested the involvement of the 2' or 3'-hydroxyl groups of ribose. It was also found that bound amino acid protected the sRNA from reaction with periodate, a reagent that attacks a pair of hydroxyls held on adjacent carbons. This also points to amino acid held by the 2'- or 3'-hydroxyl groups of the ribose of the terminal adenine nucleotide.

The revelation of the manner of linkage of the amino acid to sRNA was further accomplished by the precise studies of Preiss, Berg, Ofengand, Bergmann, and Dieckmann (75). These investigators purified amino acid-activating enzymes for L-valine, L-methionine and L-leucine from *E. coli*. Treatment of sRNA with periodate destroyed its ability to accept valine or leucine. When sRNA was bound singly to either valine, leucine, or methionine, the whole fraction was treated with periodate. The removal of the amino acid through treatment with dilute alkali followed, and the resultant sRNA was tested for ability to accept amino acid; it was found that only the amino acid that was bound during periodate treatment could be re-accepted. This meant that the presence of a particular amino acid protected a particular sRNA from destruction of its amino acid accepting ability. All of the other noncharged sRNA's were destroyed. This experiment not only confirmed the binding to the 2'- or 3'-hydroxyl group but was also a dramatic demonstration of specific sRNA's. Via a route similar to the one used by Zachau et al. (46) with liver sRNA, Preiss et al. used pancreatic ribonuclease to liberate the terminal leucyl adenosine from leucyl sRNA and isolated and identified the product, which further established the universality of the -CCA-amino acid terminal configuration.

On the basis of hyperchromicity,[2] sedimentation diffusion properties and X-ray diffraction analysis, it was deduced that the most likely structure for sRNA from *E. coli* was a single chain of 70 to 80 nucleotides folded back on itself in the form of a double helix (76). This precise secondary structure, however, could be destroyed by heating and rapid

[2] Hyperchromicity—the increase in ultraviolet absorption that accompanies the splitting apart of nucleic acid chains that are held together by hydrogen bonds between complementary bases (i.e. guanine-cytosine and adenenine-uracil)

cooling without affecting the amino acid-acceptor ability of the molecule. Singer and Cantoni further contributed the findings that with rat liver-soluble RNA, all units possessed a terminal guanosine residue at the 5' end (77).

Later evidence from Spencer, Fuller, Wilkins, and Brown provided additional support for a structure of sRNA by X-ray diffraction techniques whereby a double helix is formed when the chain bends back on itself (78). This double helical "hairpin" consisting of approximately 80 bases is 100 A long and contains roughly 3.5 helix turns.[3]

Partial Fractionation of sRNA Specific for Particular Amino Acids

Smith, Cordes, and Schweet introduced the name transfer RNA (tRNA) to replace soluble RNA (sRNA) and to emphasize the functionally active species. By use of a starch-derived cation exchanger they obtained fractions showing various states of enrichment for tyrosine-specific and leucine-specific s- or tRNA's (79). Similar enrichments were obtained by ammonium sulfate fractionation. Holley, Doctor, Merrill and Saad attained almost complete separation of the sRNA's from rat liver for alanine and tyrosine by 250 transfers in a countercurrent distribution apparatus (80). Brown described a method to obtain purified tyrosine and histidine specific sRNA from *E. coli* (81). The method was based on the ability of the sRNA charged with these two amino acids to complex with polydiazostyrene and form precipitates. The subsequently released sRNA's were charged this time only with tyrosine and subjected to another complexing sequence in order to separate the tyrosine-specific from histidine-specific sRNA. Monier, Stephenson, and Zamecnik showed that sRNA could be isolated directly from whole yeast by an adaptation of Kirby's phenol extraction procedure, thus circumventing the longer procedure based on ultracentrifugation (82). Zamecnik, Stephenson, and Scott then obtained a partial purification of valyl-sRNA extracted from whole yeast by charging the sRNA only with valine (83). The uncharged sRNA's were then reacted with periodate and subsequently with the dye 2-hydroxy-3-naphthoic acid hydrazide to yield colored hydrazones which could be removed from the valyl sRNA on the basis of differing solubilities in aqueous n-propanol.

Further Clarification of the Nature of Ribosomes

Huxley and Zubay, in a fine combined ultracentrifugal and electron microscopic study of *E. coli* ribosomes, showed that 30s ribosomes ap-

---

[3] More recent concepts of sRNA structure can be found in *Science,* **153**, 531 (1966).

peared as crescent-shaped caps and the 50s ribosomes appeared somewhat "dome-shaped" (84). The 70s particles were composed of the 30s particles fitting like a cap on the flattened face of the 50s particle, with a cleft clearly shown between the two. The 100s particles were shown to consist of two 70s units with the smaller 30s particles apposed to each other.

Kurland further revealed that the 30s ribosome of *E. coli* contained only a single 16s RNA component of 5.6×10^5 molecular weight that could be removed by detergent or phenol extraction (85). The 50s ribosomes yielded both 16s and 23s RNA units. Since the molecular weight of 23s RNA is 1.1×10^6, just twice that of the 16s RNA, and since the 50s ribosomes have a molecular weight of 1.8×10^6, it appears that each 50s unit contains either one 23s or two 16s RNA pieces.

Characteristics of an Acellular Incorporating System from E. coli

Tissières, Schlessinger, and Gros described an *in vitro* system for incorporation of amino acids into *E. coli* ribosomes (86). The important observation was made that less than 10% of the total ribosome population accounted for almost all of the incorporation and that these "active 70s" ribosomes were more resistant than the inactive ribosomes to being split into subunits in a medium containing 5×10^{-4} M Mg^{2+}. The authors calculated that at maximum efficiency in the first minute of incorporation the ribosomes accounted for only 1% of the intact cells protein synthetic abilities. In later studies (87–89) Tani and I examined intact *E. coli* for the location of the active ribosomes and found evidence that they occurred in association with the cell membrane. In the *in vitro* approach of Tissières et al. the cell membrane fraction was found to be relatively inert. They used alumina grinding to break the cells and our work showed that this procedure destroyed membrane-associated ribosomes both by mechanical disruption and by liberation of active ribonuclease (88, 89). An acellular *E. coli* system similar to that of Tissières et al. was described by Lamborg and Zamecnik (90).

THE EMERGENCE OF A NEW LOOK AND A NEW VOCABULARY FOR PROTEIN SYNTHESIS

Toward the close of the 1950's some other developments were taking place. Studies with induced enzyme formation, particularly the β-galactosidase of *E. coli*, began to yield important results. Our concept of the nature of gene action comes principally from these studies and extensions in theory proposed by the Pasteur Institute groups working with Jacob and Monod.

Regulation of protein synthesis at the genetic level, repressors, inducers, operators, operons, messengers and "a-y-z-o-i." Monod and his colleagues had shown that in *E. coli* the uptake of β-galactosides into the cell and their subsequent metabolism were mediated by two inducible systems which were themselves under genetic control. The gene which contained the structural information for the synthesis of β-galactosidase was designated z and when this gene was present and able to direct the formation of functional enzyme, it was written as z^+, the absence of complete structural information was designated z^-. Similarly, although no such enzyme has actually been identified, the properties of the substrate-accumulation system appeared to justify the symbol y for the structural information of an inducible permease protein, y^+ when information is complete, y^- when it is functionally absent. These enzymes could be produced constitutively (in the absence of added inducer) in certain strains and be inducible in others. This property was also under genetic control and is designated by the symbol i^+ for inducibility and i^- for constitutivity. Although inducible systems seemed to hold a key for understanding the nature of genetic control for at least some microbial proteins, the question of what constitutes inducibility withstood all attacks for many years of intensive study.

The revolutionary breakthrough was published by Pardee, Jacob, and Monod (91). The chromosome of *E. coli* is haploid and is shown by mapping techniques to be representable by a circle (92). The genetic determinants (a, y, i, z) are located very close together in this mapping and constitute a region known as "lac." Pardee et al. used the technique of sexual mating to introduce small segments of the chromosome of male *E. coli* (Hfr strains) into the receptor cells of female *E. coli* (F^- strains) so that a small segment of the chromosome in the female ($\female$) cell became diploid. It was shown in control experiments that very little cytoplasm from the male ($\male$) strain accompanied the genetic material into the female strain. The following mating was carried out in the presence of inducer and streptomycin:

$$\male z^+y^+i^+Sm^s \times \female z^-y^+i^+Sm^r$$

Therefore, no enzyme could be formed in the donor cells because they were streptomycin sensitive (Sm^s). No enzyme could be formed in the original acceptor cells because, although they were streptomycin resistant (Sm^r), they lacked the structural information (z^-). The mated cells, however, since they were from the Sm^r strain, could make the enzyme once the z^+ was injected. It was found that enzyme synthesis proceeded only a few minutes after the first z^+ gene entered the zygotes.

When the mating ($\male z^+i^+ \times \female z^-i^-$) was carried out in the *absence* of inducer, enzyme synthesis proceeded almost immediately, but began to slow down after 30 min. Since genetic recombination would not occur until 60 to 90 min after injection, the immediate expression of the z^+ gene occurs in the *trans* $(z^+i^+/(z^-i^-)$ not *cis* $(z^+i^-/(z^-i^+)$ configuration (with respect to z^+ and i^-). Therefore the control that i exerts over z does not require physical continuity in the chromosome. Also the i^+ or inducibility influence seemed to take time to be expressed, whereas the production of enzyme or z^+ expression was immediate.

The mating $(\male z^-i^- \times \female z^+i^+)$, when performed in the *absence* of inducer, yielded no trace of enzyme at any time after mixing, although independent controls showed that conjugation and chromosome injection occurred normally. This result, taken with that of the previous mating, shows that the i^+ effect is exerted through the cytoplasm and is dominant over the i^- or constitutive influence.

Once more, the mating was performed in the *absence* of inducer:

$$\male z^+i^+Sm^sT_6{}^s \times z^-i^-Sm^rT_6{}^r$$

In order to block induction of the donor, a mixture of streptomycin and T_6 phage was used. It was observed that in the presence of streptomycin and T_6 phage, enzyme synthesis began immediately upon mating but stopped again in 2 hr. At that time the cells were entirely inducible and resumed synthesis in the presence of inducer, streptomycin and T_6 phage (i.e. the cytoplasm of the female cells which were constitutive allowed enzyme production initially but in a short while the i^+ influence became dominant).

These studies showed that the expression of the genetic information (z) is under the influence of another gene (i). This regulator gene seems to produce a product which is cytoplasmic in *E. coli* and which represses the expression of the z gene. This repressor is highly specific for the lac region enzymes governed by z and y. This negative control might work at the genetic level or at the cytoplasmic level, perhaps with the ribosomes. The repressor, in theory, could be a substance such as RNA or protein, or an enzyme which inactivates the inducer. On the basis of the known facts and exercises in logic, the favored view was that the repressor is an RNA which reacts at the genetic level.[4] Other systems of control of genetic activity are amenable to the same genetic picture. This is true of such examples as enzyme synthesis repression, whereby the final product, synthesized by a series of enzyme reactions, can inhibit the synthesis of one of the early enzymes in the biosynthetic

---

[4] The view in 1966 favored a protein repressor instead of an RNA repressor. See *Proc. Nat. Acad. Sci.*, **56**, 1891 (1966).

sequence. Another example is lysogeny, as expressed by the immunity to cell destruction of *E. coli* K^{12} which contain phage in the resting or prophage state. In the phenomenon of repression, the metabolite might react with the product of the *i* gene which normally could not block the related structural gene and the combination would cause repression for a particular enzyme.

This line of experimentation and inspired deductive reasoning is fully developed in a classic paper of Jacob and Monod in the *Journal of Molecular Biology* (93). These authors pointed out that with several other groups of related enzymes which worked in a common path such as histidine biosynthesis (94), mapping techniques revealed that the structural information for these related enzymes are located close together. Furthermore, repression of synthesis seems to affect all of the related enzymes together and in a proportional manner. Therefore, a repressor can act at a site which controls the synthesis of several related enzymes. The site is highly specific for a particular repressor. This hypothetical site is called the operator (*o*). The operator plus the group of structural genes whose activity it controls is called the operon. The lac operon consists of *a, y, z, o* (i.e., an enzyme designated *a* called thiogalactoside transacetylase that uses β-galactosides, galactoside permease, β-galactosidase and operator). These activities are influenced by the cytoplasmic product of the *i* gene which maps alongside of the operator in the lac region.

Mutants have been isolated which, although they contain the *i*$^+$, are constitutive because the operator cannot accommodate the *i*$^+$ product. These are designated *o*^{c}, and *o*^{c} is dominant over *o*$^+$ (i.e., constitutivity of the type *i*$^-$*o*$^+$ is recessive to *i*$^+$*o*$^+$ but constitutivity of the type *o*^{c} is dominant over the inducibility of the type *i*$^+$*o*$^+$). Some mutants that contain *z*$^+$ and *y*$^+$ are not able to express these activities either in the *i*$^-$ configuration or *i*$^+$ in the presence of inducer. These are designated *o*^{0}. In summary, the gene *i* and the gene *o* exert coordinated regulatory influence over the expression of groups of genes concerned with producing enzymes required to accomplish a particular function in the cytoplasm.[5]

Messenger RNA

A careful study of the kinetics of enzyme formation after genetic information had been introduced into a cell showed that enzyme synthesis reached a steady rate within 2 min (95). The interpretation was that either a short-lived message which turns on enzyme production

---

[5] More recent studies have split the operator region into an operator *per se* which recognizes and accepts the repressor substance and a promotor which initiates and controls the level of transcription of genetic information.

was rapidly made and kept at a dynamic steady state level, or that a limited number of stable structural messages were rapidly made and the gene then turned off. If the latter alternative is correct, then destruction of the gene should not prevent synthesis from continuing. Gene destruction was accomplished by the incorporation into the male cell of high levels of P^{32} before mating. The mated cells were allowed 25 min to express the z^+ gene by β-galactosidase formation. The zygotes were then frozen for various lengths of time during which the P^{32} decay destroyed the integrity of the male z^+. When the zygotes were warmed and β-galactosidase activity tested, it was found that the activity decreased sharply as a function of P^{32} decay. These studies (if we neglect any other possible destructive activity caused by the decaying P^{32} atoms) show that the genetic message is not stable and that the continued presence of the gene is necessary, presumably for constant rereading. The postulated unstable product of the structural genes which is apparently responsible for cytoplasmic enzyme production was designated as the messenger and assumed to be RNA in nature.

Strong support for the messenger hypothesis was amassed in two papers by Brenner, Jacob, and Meselson in *Nature* (96), and by Gros, Hiatt, Gilbert, Kurland, Risebrough and Watson (97). In 1963 I critically examined these and other pertinent papers which established the basis for early acceptance of the messenger hypothesis, but I shall delay discussing this data until we reach the year 1963 in this historical summary (see page 99).

Templates and Early Codes

A specific template model which employed protein itself as the template and made use of the forces of crystallization for attracting the required amino acids was discussed in detail by Haurowitz (98). Dounce proposed a rather ingenious model using RNA as the template (99). He postulated a phosphotransferase which used ATP to attach another phosphate to each phosphate group of the RNA. He then called for a series of amino acid-specific enzymes which would recognize a triplet of bases formed by a particular base attached to a diphosphoribose and each of its neighboring bases. These enzymes would specifically attach amino acids through their amino groups to the phosphates at the expense of the energy of the diphosphate bond which would be split.

Finally, another enzyme joined adjacent amino acids while peeling the nascent peptide from the template. The same diphospho-RNA-template could be used to synthesize itself by performing this process with nucleotides and four specific enzymes instead of with amino acids. Lipmann proposed that a template (of no particular kind) would possess nondefined amino acid-specific activation spots (100). Pyrophosphate would be transferred from ATP to nonspecified acceptors at these spots and in turn these would be displaced by amino acids binding through their carboxyl groups. Finally, neighboring bound amino acids would react together to form the peptide which would then exit from the template. Borsook modified Dounce's template mechanism by proposing that instead of charging the template with amino acids via the amino groups, the carboxyl-activated amino acids would react with the pyrophosphate groups of the template to form anhydrides (101). The concept of templates at the molecular level must inevitably lead to thinking about information storage and codes.

A Brief Review of Different Codes in Which Certain Base-Sequences of Nucleic Acids Specify Particular Amino Acids[6]

1. Diamond code using double-stranded RNA (102). Two points of a diamond are formed by a base pair, either A-T or G-C. The top point is formed by a separate base as well as the bottom. If it is assumed that a form such as G-(A-T)-C is the same as the bilaterally symmetrical forms C-(A-T)-G, then 20 such diamonds are formed. The code is overlapping in that two of the four points are shared in common for adjacent diamonds.

2. Triangular codes using one strand of nucleic acid. This code in essence represents a linear overlapping sequential code so that by using two bases in common (the compact triangular code) five amino acids would be specified in the example shown:

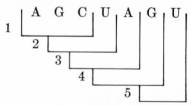

By using one base in common three amino acids would be specified:

3. Major-minor code (proposed by L. Orgel). The central nucleotide of a triplet is the major determinant for an amino acid, but the selection

[6] Some of these codes are further discussed in (102).

of the amino acid is influenced by the nature of the adjoining nucleotides. Thus it would be possible to have degenerate triplets for a given amino acid if the central triplet remained the same (degenerate means different particular triplets may specify the same amino acid).

4. Sequential code (E. Teller). An amino acid is specified by two bases and the preceding amino acid in the peptide chain which is being built in juxtaposition to an adjacent nucleic acid template.

5. Combination code (103). All permutations of a given triplet stand for the same amino acid. Thus GUU = UGU = UUG.

6. Transposable triplet codes. Use is made of the base pairing rules of a Watson-Crick double helix, so that for one triplet UCG, the related triplet would be AGC in a parallel reading or CGA in an anti-parallel reading. The related or transposed triplets may be equivalent to the original triplet or they may signify nonsense.

7. Two-letter codes. Either only two bases specify an amino acid, or two out of a triplet are functional.

8. Partly overlapping codes with codons longer than three but with a coding ratio of three (104). If the number of bases in a codon is designated as t and the number of bases which overlap for adjacent codons is s, the coding ratio will be equal to $(t-s)$. Therefore a code with four bases in the codon and one of these overlapping the adjacent codon will give a coding ratio of three and be compatible with many of the sequence data (as of 1962).

9. Logical codes designed to fit certain parameters (105). Woese assumed a degenerate nonoverlapping triplet code that would be compatible with the large range of differences in DNA base ratios for organisms having a similarity in over-all amino acid composition.

10. Commaless code (106). There are 64 possible triplets but only 20 are needed. Consider a string of adjacent nucleotide bases: only certain triplets would convey sense, others nonsense. Of the 64 possible triplets only 20 need make sense. Whenever two sensible triplets are adjacent, the intervening or overlapping triplets are not meaningful. The authors showed that such restrictions could yield 20 sensible triplets which, when written in any sequence, would only yield nonsense for overlapping triplets. However, certain triplets would have to be relegated to nonsense. For example, if any amino acid were specified by a repeating base such as UUU, then for the amino acid to occur next to itself, the template would have to be UUUUUU, but in this case the overlapping triplets would also be UUU and this cannot be allowed. Therefore, according to this code, repeating bases must signify nonsense. The results of Nirenberg and Matthaei destroyed this possibility (see page 88).

Most of these ingenious codes could be eliminated on the basis of restrictions imposed by overlapping since once an amino acid is specified, the overlapping portion of the next amino acid is also specified. Brenner showed that most overlapping codes could be eliminated by the wide range of differences already reported in the literature, for possible neighboring amino acids. Furthermore, by mutational alteration in overlapping codes, more than one amino acid should be replaced, whereas the literature already showed ample illustrations of mutational replacements of single amino acids. The ingenious speculations about possible codes was brought to an abrupt standstill by the acellular approach of Nirenberg and Ochoa and their groups.

Summary Statement on the Role of sRNA as of 1961

In 1961 there no longer seemed to be any question that the cell's machinery for making protein could employ sRNA in a key role. However, I had not considered this as obligatory for the following reasons, which were published by me the following year (107).

1. In the six published studies of the kinetics of labeling of sRNA in an intact cell compared to the kinetics of labeling for the protein, the rate of labeling the sRNA was manyfold too low.

2. Several systems that are capable of amino acid-incorporation into protein function with little or no sRNA present and with no ATP-PP^{32} exchange ability.

3. The most convincing evidence accrued in support of a quantitatively significant role of sRNA in protein synthesis was obtained with acellular systems where approximately 99% of the quantitative ability to make protein was destroyed. The residual fractional activity represented incorporation into noncharacterizable and unnatural products.

4. The amino acid activation activity of a tissue does not necessarily correspond to the amino acid-incorporating ability (see Fraser and Gutfreund (108) and also pages 51 and 74).

5. Several studies pointed to a direct route for external amino acid into proteins without equilibration with the internal pool which should be immediately accessible to free sRNA.

6. The wide occurrence and fast kinetics of labeling lipid compounds which contained amino acids suggested a possible role for these substances in protein synthesis.

7. Until 1961 there was no evidence that amino acid could be passed from sRNA to an authentic protein synthesized by a tissue system.

One other point could have been included in this critique but I was not aware of it at the time. The known and recognized perspicacity of Crick has created a great deal of acceptance for his intuitive analyses of the various complexities in the processes under evaluation. One of these analyses, however, has been neglected. Crick stated, "The sRNA is too long to join onto template RNA (in the microsomal particles) by base-pairing since it would take too great a time for a piece of RNA 25 nucleotides long, say, to diffuse to the correct place in the correct particle (109)." Actually, the sRNA is now known to be several times larger than the 25 nucleotides referred to by Crick.

Many of the above enumerated points still stand today (1967) as unanswered and disturbing features for the over-all and unique role considered for *free* sRNA in intact cells.[7] Point No. 7, however, was brilliantly resolved in a study published by Von Ehrenstein and Lipmann in 1961 (110). These authors recognized the predicament posed by the fact that all proteins previously labeled through the intermediacy of sRNA in acellular systems were "rather ill-defined." They turned to the acellular reticulocyte system of Schweet et al. (111) which was capable of producing labeled hemoglobin. *E. coli* sRNA was isolated and charged with C^{14}-L-leucine. Ribosomes were isolated from rabbit reticulocytes and incubated with C^{14}-leucyl sRNA. After the incubation reticulocyte lysate was added to provide carrier hemoglobin. The supernatant fraction was treated with carbon monoxide to convert the hemoglobin to carbon monoxyhemoglobin which was collected by precipitation with ammonium sulfate. The precipitate was redissolved, dialyzed, and chromatographed on CM-cellulose. The main hemoglobin peak comprising 75% of the applied radioactivity and 44% of the incorporated leucine was digested with trypsin and the resulting peptides fractionated in two dimensions on paper by electrophoresis and butanol acetic acid chromatography. The resultant "fingerprint" verified that authentic hemoglobin was indeed formed in this incubation with C^{14}-leucyl-sRNA. The fact that C^{12}-leucine was also present in the incubation medium in relatively large amounts would seem to exclude the possibility that the C^{14}-leucyl-sRNA was hydrolyzed before incorporation. This study was the first proof that sRNA can participate in the formation of a characteristic cellular protein. It was the proof that I was waiting for

[7] See Chapter 8 for the discussion of a possible resolution of this dilemma.

before accepting the idea that sRNA participates in the cells' normal protein synthesizing machinery. The question of the quantitative significance of this participation in terms of the intact cells' total protein synthesizing activities, however, as well as the question of the operation of alternate pathways, was still very much at issue.

The Sequential Nature of Peptide Synthesis

Bishop, Leahy, and Schweet presented evidence that hemoglobin was synthesized sequentially from the N-terminal valine onward (112). The evidence was obtained by briefly labeling the intact reticulocytes with C^{14}-valine and isolating the ribosomes. The ribosomes were then incubated *in vitro* with either unlabeled valine or C^{14}-valine. The ratio of N-terminal C^{14}-valine to total incorporated valine in the isolated hemoglobin supported the concept that the ribosomes obtained from the intact cell incubation had started the formation of hemoglobin with the N-terminal valine. Dintzis gave convincing proof that the synthesis of hemoglobin in rabbit reticulocytes proceeded linearly from the N-terminal end to the carboxyl end (113). Working at 15° in order to slow the reactions to more measurable rates, he incubated the cells with either H^3- or C^{14}-leucine for different periods of time. The hemoglobin was isolated, separated into α and β chains, hydrolyzed with trypsin, and the resulting peptides fractionated by paper electrophoresis and partition chromatography. The peptides were arbitrarily numbered 1 to 31 and plotted according to increasing content of radioactivity. It was found that the curve was steepest for the shorter periods of incubation and became more shallow with increasing time of incubations, approaching a slope of 0°; *but* that the relative order of the peptides did not change. This pattern of labeling was consistent with the idea that several nascent chains of different sizes were present when the isotope was introduced and that all of these chains could accept at least some isotopic amino acid at the end being completed. The first chain to be released from the template would be unlabeled except for the recently added terminal radioactive amino acids. Chains that were less nearly complete at the time of incubation with the label would, when released, contain radioactivity further along the chain until eventually chains would be started, completed, and released entirely labeled by the radioactive amino acid present in the medium. In order to determine whether the first peptide to be labeled came from the amino or carboxyl end, the following experiment was performed. Hemoglobin was labeled uniformly with C^{14}-leucine and then digested with carboxypeptidase to remove its carboxyl terminus. Uniformly H^3-leucine-labeled hemoglobin was prepared and not treated with carboxypeptidase. The two preparations were then mixed, treated with trypsin, and fractionated to identify the c-terminal peptides. The combined data indicated that the peptides that were labeled first for both the α and β chains were the peptides that contained the carboxyl end groups; therefore the concept of a linear process of labeling proceeding from the N-terminal to c-terminal ends appears to be well supported.

Subsequently Naughton and Dintzis made use of the known amino acid sequence of human hemoglobin to demonstrate that the order of peptides in rabbit hemoglobin deduced from the time-course experiments corresponded with the known order of the peptides in relation to the chemically characterized sequence, for the human hemoglobin (114). Thus the sequential growth of the chain from the amino terminal end seems firmly established. A similar situation has been indicated for the proteins of *E. coli* (Goldstein and Brown, 115) for a bacterial amylase (Yoshida and Tobita, 116) and egg white lysozyme (Canfield and Anfinsen, 117).

Early Indications of a Role for Large Molecular Weight
Nucleic Acid in the Amino Acid-Incorporation Process

Tissières and Hopkins showed that DNAase partially inhibited amino acid-incorporation into an *E. coli* ribosomal system (118). Added DNA or nucleoside triphosphates stimulated incorporation. Under conditions of amino acid-incorporation, it was also found that C^{14}-labeled nucleoside 5' monophosphates were incorporated into RNA. The newly labeled RNA sedimented both as 70s and 15 to 16s material in the presence of 10^{-2} M magnesium. In 10^{-4} M magnesium the radioactivity was found only in the 15 to 16s region. The authors arbitrarily identified this RNA as messenger RNA because it sedimented in the same region as material similarly called messenger RNA by Gros et al. (97). The interesting observation was made that a preparation of similar RNA, when added to the ribosomal incorporating system stimulated the incorporation of C^{14}-alanine into protein. Added sRNA did not produce such stimulation. Matthaei and Nirenberg also described experiments which showed that RNA derived from ribosomes stimulated the incorporation of C^{14}-L-valine into an acellular *E. coli* ribosomal amino acid-incorporating system (119). This stimulation occurred at saturation concentrations for sRNA and the system was inhibited by RNAase, chloramphenicol and puromycin. The system was dependent on an ATP generating system and was unaffected by DNAase. The active RNA sedimented at around 15s. As controls, equivalent concentrations of polyanions such as poly-A, salmon sperm DNA, and a polymer of glucose carboxylic acid were added. These compounds elicited no response.

The Breakthrough on the Code

One day in the Spring of 1961, Marshall Nirenberg dropped into my laboratory and asked if I could let him have some high specific activity U-C^{14}-phenylalanine that I had synthesized from $C^{14}O_2$ using algae. He found that by adding TMV-RNA to *E. coli* ribosomes, a pronounced stimulation of amino acid-incorporation occurred. He had just made arrangements to go out to California to H. Fraenkel-Conrat's laboratory to see if the *E. coli* ribosomes were actually making TMV protein. Since I was cynical about certain phases of the developing protein synthesis story that would be much easier to accept than this, I certainly thought that this was a farfetched possibility. Succeeding developments in the field during the next five years completely justified Nirenberg's boldness of thought even though, as it turned out, the product produced in this particular system was not authentic TMV protein.

At the Fifth International Congress of Biochemistry in Moscow, Nirenberg reported that polyuridylic acid (poly-U), when added to *E. coli* ribosomes, caused the formation of a product which had all the characteristics of polyphenylalanine and that polycytidylic acid appeared to cause the formation of polyproline. In other words, to borrow more recent terminology, he programmed the ribosomes by supplying the code for a specific polypeptide. The feverish activity that followed in the next few years in Nirenberg's laboratory and that of Ochoa who soon entered the field, is told in the many published accounts in and out of scientific journals and in the published lexicons of code words that rapidly grew larger. Both of these groups performed a thorough exploration of this important aspect in the development of our current picture (1967) of protein synthesis.

Matthaei and Nirenberg described a stable acellular ribosomal amino acid-incorporating system derived from *E. coli* that was DNAase-sensitive (120). Similar DNAase-sensitivities for acellular systems had also recently been described (121, 122). The system was strongly inhibited by puromycin, chloramphenicol, and RNAase. Nirenberg and Matthaei, in an accompanying paper, described the stimulatory effect of added RNA (123). This stimulation by added *E. coli* ribosomal RNA occurred with saturating concentrations of sRNA already present and was proportional to the amount of ribosomal RNA added. Incorporation was further enhanced if a mixture of all 20 amino acids was present. To demonstrate this effect of added ribosomal RNA, the use of washed ribosomes and 105,000 $\times$ *g* supernatant solution was required. It was found that yeast ribosomal RNA was much more effective than *E. coli* ribosomal RNA in stimulating C^{14}-valine incorporation, and that TMV

RNA was twice as effective as the yeast preparation. The remarkable fact was revealed that 10 μg of poly-U caused one thousandfold stimulation of phenylalanine incorporation. Poly-A, poly-C, and poly-I (inosinic acid) were ineffective in this respect and poly-A destroyed the effectiveness of poly-U. This latter observation indicated that single-strandedness was a requisite for activity since, as a result of Watson-Crick hydrogen-bonding affinities, poly-A binds with poly-U to form multiple strands. DNAase did not inhibit the incorporation, and a mixture of the other 19 amino acids did not help. The stimulatory effect of poly-U seemed to be quite specific for phenylalanine. The radioactive product in the incubation possessed properties of the suspected polyphenylalanine. In a note added in proof to the paper it was reported that the ratio of phenylalanine incorporated to uridylic acid units added approached 1:1 and that poly-C specifically mediated the incorporation of proline into a TCA-precipitable product.

A race to complete the codon dictionary rapidly developed between Ochoa's laboratory and that of Nirenberg. The Proceedings of the National Academy of Sciences serves as the historical archives for preserving the records of the fast-moving developments at that time. In the same issue that contains the report of Nirenberg and Matthaei; Lengyel, Speyer, and Ochoa made the first of many progress reports on their efforts to break the genetic code (124). These authors confirmed the effect of poly-U on phenylalanine incorporation and of high C-content polymers on proline incorporation in *E. coli*. In addition to testing poly-U and poly-C, they tested poly-UC (U:C = 5:1), poly UA (U:A = 5:1), and poly-CU (C:U = 5:1), finding enhancements of incorporation for phenylalanine, serine, leucine, and isoleucine by poly-UC; phenylalanine tyrosine, leucine, and isoleucine by poly-UA; and phenylalanine, isoleucine, and proline by poly-CU. It was suggested that code assignments might be deduced by correlating the relative incorporation for another stimulated amino acid to that of phenylalanine for a given mixed polynucleotide, to that of the frequency of all possible triplets in a randomly ordered nucleotide chain; for example, in a random polymer of poly-UC (U:C = 5:1), the relative chances of encountering a U is $\frac{5}{6}$; for C the chance is $\frac{1}{6}$. The chance for finding UUU is $\frac{5}{6} \times \frac{5}{6} \times \frac{5}{6}$; for finding a triplet containing two U's and one C it is $\frac{5}{6} \times \frac{5}{6} \times \frac{1}{6}$. The relative frequency of UUU triplets to those of U_2C would be $(\frac{5}{6} \times \frac{5}{6} \times \frac{5}{6}) \div (\frac{5}{6} \times \frac{5}{6} \times \frac{1}{6}) = 5:1$. The relative incorporation of phenylalanine to that of serine was 4.4 and that of phenylalanine to leucine was 4.7. Serine and/or leucine, on this basis, might be coded for by a triplet containing two U's and one C. This system was used in succeeding papers by both groups to make code word assignments.

Hall and Spiegelman made another startling observation (125). Much of the excitement at this time was based on the finding by Volkin and Astrachan that in T_2-phage infected $E.$ $coli$, a rapidly labeled RNA having a DNA-like base composition is formed (126). This RNA was automatically accepted as being the messenger RNA copy of the DNA. That this idea does not necessarily follow is discussed later on pages 101 and 120. Nonetheless, Hall and Spiegelman, in order to test whether the T_2-phage-induced DNA-like RNA really possessed a complementary base sequence to that of phage T_2-DNA, sought to make use of the reformation of a double-stranded helix from single strands melted out of a double helical configuration, as recently described by Marmur and Doty and colleagues (127, 128). Hall and Spiegelman looked for hybrid double strands between T_2-phage-induced RNA labeled with P^{32}, and heat denatured T_2-phage DNA labeled with H^3. Density gradient centrifugation in a $CsCl_2$ gradient was used and the co-occurrence of P^{32} and H^3 in a single peak was taken as a criterion for hybrid RNA-DNA formation. The system worked beautifully and hybrids were demonstrated for heat denatured T_2-DNA and T_2-induced messenger RNA (m-RNA), but not with native T_2-DNA or heat denatured $E.$ $coli$ DNA or $Ps.$ $aeruginosa$ DNA or T_5-DNA and T_2-induced RNA. This study added considerable support to the idea of complementarity between the RNA formed in $E.$ $coli$ as a result of T_2-phage infection and T_2-DNA.

Chamberlin and Berg described the isolation and characteristics of a DNA-dependent RNA polymerase from $E.$ $coli$, which, in the presence of primer DNA, the four nucleoside triphosphates and a divalent cation synthesized an RNA complementary in base sequence to the primer (129). When this system was introduced into a ribosomal system from $E.$ $coli$ which was capable of amino acid-incorporation, there was a marked stimulation of both the rate and extent of amino acid-incorporation. Although the basic system was insensitive to DNAase, the DNA-primed synthesis was markedly sensitive. Enzymatically synthesized RNA, in the absence of RNA polymerase and DNA, also stimulated amino acid-incorporation and this incorporation was not affected by DNAase. Thus, the results also pointed to the ability of RNA (other than sRNA), which could be copied from DNA, to cause a marked stimulation in the incorporation of amino acid by a ribosomal system from $E.$ $coli$.

Another important step was taken by Nirenberg, Matthaei, and Jones when they showed that phenylalanine was transferred directly to a peptide linkage from sRNA without going through the free amino acid stage when the incorporation was under poly-U mediation of the ribosomal system (130).

The Recognition of Codons and Amino Acids by sRNA

It became apparent that degeneracy exists in the codon assignments for the amino acids. Thus it was found that both poly-UC and poly-UG stimulated leucine incorporation. It was also known that more than one chromatographic species of sRNA could be found for a given amino acid. By isolating two such different *E. coli* sRNA's specific for leucine, Weisblum, Benzer, and Holley demonstrated that one of these responded to poly-UC in the medium but not to poly-UG, whereas the other responded to poly-UG but not to poly-UC, as measured by stimulation of incorporation of leucine into protein (131). This, then, provides one explanation for coding degeneracy. Particular codons would require particular sRNA's for a given amino acid. This type of degeneracy was further revealed when Bennett, Goldstein and Lipmann showed that yeast activating enzyme would attach leucine to only two thirds of the sRNA that could be so united with leucine in the presence of *E. coli* activating enzymes (132). Although fully charged leucyl sRNA from *E. coli* will respond both to poly-UC and poly-UG for incorporation into protein, that charged with the yeast enzymes responded preferentially to poly-UC while the other fraction of leucyl-sRNA responded preferentially to poly-UG. In fact, von Ehrenstein and Dais showed that *E. coli* leucyl acceptor sRNA could be separated into three peaks by counter current distribution (133). The first responded preferentially to poly-UC, the second to high U-content polymers, including poly-U, and the third preferentially to poly-UG.

A remarkable demonstration of the ability of sRNA to recognize the incorporation site for a particular amino acid was reported by Chapeville, Lipmann, von Ehrenstein, Weisblum, Ray, and Benzer (134). They found that they could chemically convert cysteine to alanine by reductive desulfhydration with Raney Nickel *while the cysteine remained attached to its specific sRNA*. The product then was alanine attached to sRNA which was specific for cysteine. The work of Nirenberg and colleagues, as well as Ochoa and colleagues, showed that poly-UG would stimulate the incorporation of cysteine into protein from either the free or sRNA-bound state, but would not stimulate alanine from either state. The observation was made that when the alanine was attached to the cysteinyl sRNA, poly-UG stimulated its incorporation just as though it were cysteine. Thus the adaptor hypothesis was neatly confirmed and it was conclusively demonstrated that the amino acid-incorporating site recognizes the sRNA and not the amino acid.

This finding makes an earlier report of R. Loftfield difficult to understand. Loftfield observed that isoleucine is favored over alloisoleucine

20:1 in labeling sRNA but 2000:1 in incorporation into protein (135). After the alloisoleucine is already in the sRNA, what would exercise this additional 100:1 recognition factor?

An interesting situation developed in the year 1962. The code word dictionary had been expanded to include all 20 amino acids. All of the code words, however, contained uracil, and these included only 10 out of 24 possible triplets (if the order of the bases is unimportant).

Several workers decided to test the authenticity of the code by use of data on mutational replacements found for amino acids. The criterion of a fit was that the two amino acids involved in the mutation should be represented by code word triplets which had two of the three bases in common. Emil Smith published a paper which showed a high degree of correlation for amino acid replacements involving human hemoglobin with the code word designations for the amino acids involved in the mutations (136). Similar analyses were cited by Matthaei et al. (137) and by Speyer et al. (138). At this time, it was also felt that the other triplets (not containing U) undoubtedly also functioned in coding, but that for some reason our techniques were not yet adequate to study their coding function. Therefore, it did not seem right that, knowing only a part of the dictionary, we should find such good agreement when natural processes would have available a dictionary with a larger vocabulary. Consequently I asked the question, "How good would such a correlation be if we picked any two amino acids at random"? It turned out that due to the degeneracy of the code, and also to the fact that order within triplets was not a factor, the random correlation was as good as that which was being found in the mutation data (139). This type of analysis also showed that data derived from particular mutations induced by the use of nitrous acid, which resulted in specific nucleotide changes (i.e., A→G and C→U), were also inadequate. The usefulness of this episode lies in the illustration that it is very easy to find evidence for an idea that one wants to believe. In other words, the best way to find a weakness in any data is to have them examined by someone who is unconvinced.

sRNA Association with Ribosomes and
the Mode of Action of Puromycin

Takanami clearly demonstrated that sRNA itself was also incorporated into ribosomes in the amino acid-incorporation system from rat liver (140). The incorporation of sRNA occurred with equal facility whether it was stripped of, or charged with, amino acids. In fact, by use of charged sRNA, 10 times more sRNA than amino acid on a molar basis was incorporated. Removal of the -CCA terminal configuration abolished the incorporation, but the loss was restored by enzymatically adding back cytidilic and adenylic acids. Similar incorporations of sRNA into microsomal RNA were also described by Hultin and van der Decken (141), by Bloemendal, Bosch, and Sluyser (142), and by Hoagland and Comply (143).

Several years earlier, Yarmolinsky and de la Haba observed the structural similarity between the antibiotic puromycin and the terminal adenyl acyl amino acid sequence of sRNA and investigated the possibility that puromycin might owe its activity to this analogy in structure (144). They found that the antibiotic did not inhibit the activation of leucine or transfer of activated leucine to sRNA by a rat liver pH 5 system. The transfer of leucine from sRNA to microsomal protein, however, was inhibited to an extent sufficient to account for the observed inhibition by puromycin of leucine incorporation into protein. Morris, Favelukes, Arlinghaus, and Schweet provided data which indicated that puromycin caused the release of incompleted globin chains from ribosomes of rabbit reticulocytes (145). This process seemed to require neither the soluble enzymes used for amino acid-incorporation, nor an energy source.

Allen and Zamecnik also studied the nature of puromycin action in a rabbit reticulocyte system (146). They found that in a complete amino acid-incorporating system at 37°, puromycin released a portion of the protein from the ribosomes to a nearly maximal extent (about 20 to 30%) in only 1 min. This release phenomenon required neither an energy source nor the soluble enzymes. Puromycin was also very effective in releasing incorporated radioactivity from partially heat denatured ribosomes. Ribosomes, after treatment with puromycin, seemed to be undamaged by the criterion of ultracentrifugal analysis. Although the soluble protein released from the ribosomes by incubation in the absence of puromycin chromatographed like authentic hemoglobin, that released in its presence was easily separable from authentic hemoglobin. The N-terminal amino acid of this puromycin-released material, however, was mostly valine as one would expect if the peptide was an intermediate in hemoglobin synthesis. By use of C^{14}-puromycin, it was determined that the analogue was bound to the released peptide approximately in a 1:1 ratio and that the average length of the peptide was such that five or six valine residues were contained out of 11 found in the complete chain. Although the nature of the peptide-puromycin link was not elucidated, the absence of dinitrophenyl p-methoxy phenylalanine (which is a derivative of the amino group-containing part of the puromycin molecule) after treatment with dinitroflurobenzene indicated that its amino group may have been bound to the carboxyl group of the growing peptide chain. A major portion of puromycin remained associated with the ribosomes even after the soluble peptides were released. These data would be compatible with the puromycin successfully masquerading as an sRNA so that the growing polypeptide chain is passed from the ribosome to the free puromycin rather than to the next amino acyl-sRNA on the ribosome.

A little later, Gilbert prelabeled a ribosome suspension with C^{14}-phenylalanine in the presence of poly-U (147). This labeled preparation was subjected to dialysis against 10^{-4} M magnesium and then fractionated by ultracentrifugation in a sucrose gradient. The radioactivity (of the nascent polypeptide) stayed with the 50s ribosome, not the 30s unit. When the subunits were further treated with 0.5% sodium dodecylsulfate, the radioactive component was released and then sedimented in the 4s region, in which sRNA is normally encountered. RNAase treatment caused the radioactive material to sediment at 2.6s. Treatment of the labeled ribosome with puromycin similarly stripped away the nascent peptide causing it to sediment at about 2.6s. Using a double isotope technique with C^{14}-phenylalanine and P^{32}-sRNA, nascent peptide, upon chromatography on Sephadex G-200, moved in coincidence with sRNA even though this column resolves the two substances when they are applied as separate compounds. Although mild alkalinity and hydroxylamine split the peptide-sRNA bond, it was manifold more stable than a single amino acyl-sRNA bond. As a result of these studies, Gilbert proposed a model that is largely retained through the present time (1967). The growing peptide is attached to sRNA which occupies a specific site on the 50s ribosome. The incoming amino acyl-sRNA, after finding its correct triplet on the messenger, breaks the peptide-sRNA bond by accepting the peptide on the free amino end of the amino acyl group, it shifts into the site previously occupied by the peptidyl sRNA and in so doing ejects the old sRNA and moves the messenger forward one codon to the next reading frame. The process is then set to continue in the same fashion with new ribosomes being added as enough of the messenger is moved along to provide a new starting point. This conclusion was also supported by the work of Takanami[1] (148).

The binding of sRNA by *E. coli* ribosomes was further studied by Cannon, Krug, and Gilbert, who found that *E. coli* ribosomes can reversibly bind sRNA molecules on a one-sRNA per one-ribosome basis (149). This binding is not enzymatic, is sensitive to magnesium concentration, can be localized principally in the 50s ribosomes, and occurs

---

[1] Although the 50s subunit binds charged and uncharged sRNA, the specific binding of sRNA to its proper position on the mRNA is mediated by the 30s subunit. See *J. Mol. Biol.*, **25**, 407 (1967).

with equal facility for amino acid-charged and uncharged sRNA. This binding was *not* affected by puromycin, which fact confirms the expectation that puromycin itself does not become bound. Yet puromycin, even though it is unbound, is capable of mimicking normal sRNA and removing the growing peptide chain from its association with a bound sRNA. This raises the question of what prevents other soluble sRNA's from causing puromycin-like abortive transfer. Also, if either amino acyl-charged or uncharged sRNA can bind to the ribosome, how does this affect the delay factor which is to be expected from diffusion processes that were anticipated by Crick? (see page 85). During protein synthesis, a species of tightly bound sRNA appears. This sRNA is resistant to washing in low magnesium concentration buffers, it is resistant to RNAase, and remains with the 5 to 10% of active "stuck" ribosomes which resist dissociation in low concentration magnesium solutions. This tightly bound sRNA can be removed by washing in 0.5% sodium dodecyl sulfate, a procedure which also removes nascent peptides. The active ribosomes appear to contain the firmly bound sRNA also on a one molecule per ribosome basis and these results confirm the hypothesis that the growing peptide chain may be linked through its carboxyl group to an sRNA held on the 50s component.

Gilbert supports the concept of Allen and Zamecnik that puromycin which resembles an amino acyl-sRNA works by accepting the growing peptide chain from the sRNA through which it is held to the ribosomes (147). Since puromycin lacks the rest of the sRNA chain, it is not held by the ribosomes and so the peptide chain is removed. If the condensation reaction of peptide plus amino acyl analogue of puromycin occurs without benefit of enzyme, according to Allen and Zamecnik as well as in these studies by Gilbert, why does an enzyme seem to be required for the natural condensation? (The enzyme requirement is discussed on pages 107–109.)

Morris, Arlinghaus, Favelukes, and Schweet described the results of a rather thorough study of the effects of puromycin on a reticulocyte ribosome system (150). Puromycin inhibited the rate and extent of amino acid-incorporation. The extent of this inhibition was greatly enhanced by preincubating the ribosomes with puromycin *in the presence of an energy source and the soluble enzymes required for amino acid-incorporation.* The inhibited ribosomes could be largely restored to activity by a simple washing procedure. Prelabeled ribosomes, upon subsequent incubation in a system containing soluble enzymes and a source of energy, released 75% of their radioactivity as TCA-precipitable material. By omitting the energy source, this release reduces to 8%. In the complete system puromycin caused a 55% release (counting both TCA-soluble and insoluble radioactivity) which in effect represents an inhibition of release by 20% compared to the complete system alone. In the system lacking energy puromycin caused the release of about 50% of the radioactivity with or without the enzyme being present. Therefore, although puromycin caused a release phenomenon in the system lacking an energy source, it inhibited release in the total system. This point, unfortunately, was not discussed in the paper of Morris et al., which concerned itself mainly with the release phenomenon. Incubation of ribosomes in the presence of puromycin did not alter the subsequent sedimentation behavior in an ultracentrifuge.

Another interesting finding was that in the presence of the soluble enzymes normally present, the released acid insoluble radioactive material was partly converted to acid soluble material. The puromycin-released material was of high specific radioactivity and although it was not hemoglobin it resembled the molecule in terms of the peptides released by tryptic digestion and by its N-terminal C^{14}-valine content. The latter parameter, by its increased content, indicated that the average chain length of the released material was shorter than that of fully formed hemoglobin.

Heavy Ribosomes—Then Polysomes

Risebrough, Tissières, and Watson in 1961 thought that the attachment of messenger RNA (mRNA) occurred with monomers or dimers of ribosomes (151). They reasoned that with a suspected molecular weight of at least 2.5×10^5 for mRNA its attachment should cause the active ribosomes to sediment a little more rapidly than inactive ones. Being conscious of the necessity to virtually eliminate ribonuclease activity during their fractionations, Risebrough et al. demonstrated that this factor could be controlled by working at a low temperature. They found that the active ribosomes, as measured by mRNA-binding and nascent protein formation, did actually sediment 5 to 10s faster than free ribosomes. A smaller amount of activity was also found in heavier regions. It is interesting that when the theory predicted active ribosomes to sediment only slightly faster than free ribosomes, the experimental data beautifully supported this model. Later the theory predicted that active ribosomes should be in much heavier aggregates of ribosomes (polysomes). Since that time, they have been found principally as the heavier aggregate. The active ribosome, which contained a piece of mRNA and sedimented 5 to 10s faster than the free ribosome, has faded into the background with its theoretical raison d'etre.

Gilbert showed that in E. coli the presence of both 30s and 50s ribosomal components are required in order to obtain a response to poly-U for polyphenylalanine synthesis (152). When the ribosomes were preincubated with poly-U, fractionated by ultracentrifugation in a sucrose gradient, and assayed for C^{14}-phenylalanine incorporation, it was found that the incorporating activity occurred in a broad band from 140 to 200s. After brief treatment with RNAase, the activity sedimented coincident with the 70s ribosomes. After labeling with C^{14}-phenylalanine followed by a "chase" with C^{12}-phenylalanine, it was shown that the C^{12} incubation seemed to cause radioactivity to shift to the 70s region and indicated a release of radioactive monosomes (single ribosomes) from the polysomes (aggregates of ribosomes). Using C^{14}-alanine and "natural messenger," radioactivity was also associated with particles in the 100s to 200s region.

Gierer presented the first convincing evidence for functional ribosome aggregates (later called polysomes) in reticulocytes (153). By first labeling intact reticulocytes and then isolating ribosomes, or by labeling ribosomes directly, he found that the major incorporated radioactivity sedimented in regions of up to 220s, although free ribosomes sedimented at 80s. By brief treatment with RNAase, both 260 mμ-absorbing material and radioactivity left the heavier regions and sedimented at 80s. Gierer calculated that an aggregate of 6 ribosomes would sediment at 220s. The ribosomes most responsive to poly-U sedimented at 80s, but after treatment with poly-U and radioactive phenylalanine, appreciable radioactivity was found to sediment in the heavier regions. Gierer proposed that one molecule of mRNA was simultaneously read by several attached ribosomes and that the messenger moved along the ribosomes. The same picture is retained today (1967) although newer terms such as "tape theory" and "polysomes" have been added.

The December 27, 1962, issue of *Science* had on its cover the first published picture of ribosomal clumps—newly named polyribosomes or simply polysomes (154). Clumped ribosomes were not new. At first, it had been considered as an artifact of preparation, and indeed, Mary Petermann has discussed the fact that ribosomes bind many substances including themselves (155). What was new to both this group and to Gierer was the concept that these clumps represented the cells' cytological machinery for protein synthesis and that they were held together by a strand of mRNA that was being simultaneously read by the ribosomes which traversed the strand from one end to the other. Warner et al. showed that after 45 seconds-labeling of reticulocytes with C^{14}-algal amino acids, followed by osmotic lysis, practically all of the radioactivity was distributed in a sucrose gradient ultracentrifugation with 260 mμ-absorbing material in a broad peak around 170s. The monosome or 76s region had much 260 mμ-absorbing material, but practically no radioactivity. By electron microscopy it was shown that the 170s region contained mostly pentamers of ribosomes or simply polysomes. One electron micrograph showed a slender thread (10 to 15 A thick) connecting three ribosomes. It was suggested that the thread is mRNA. Essentially the same data were again discussed in a later paper by these workers (156).

How Should Messenger RNA Resemble DNA?

Hayashi, Hayashi, and Spiegelman pointed out that the base composition of unfractionated T_2-phage-complementary RNA and the material isolated by hybridization to DNA on columns, all showed a significant inequality of guanine to cytosine (157). Furthermore, both ribosomal and sRNA are complementary to certain areas of DNA although it is well known that in these molecular species G does not equal C nor does A equal U; such equalities are characteristic of DNA. Therefore, mRNA may be a copy of only one complementary strand of DNA—not both. To answer this question these workers turned to the single-stranded DNA virus ϕX 174 of Sinsheimer. This virus, during replication in the host, E. coli c, forms a circular double-stranded DNA structure known as RF-DNA or replicating form-DNA. By separately pulse-labeling an uninfected host with H^3-uridine and an infected host with C^{14}-uridine, followed by isolation, mixing, and co-chromatography of the radioactive RNA's, they found a species of RNA characteristic of the infected system. This RNA would form an RNAase-resistant mixed hybrid only with heat denatured RF-DNA but not with vegetative state single-stranded ϕX 174 DNA in either the natural or heated state. It therefore seems that the RNA formed as a result of ϕX 174 infection is complementary to the complement of vegetative state DNA and not to the single strand contained in the virus. The base composition of the pulse-labeled, viral-induced RNA was very similar to that of the single-stranded ϕX 174 DNA. The A content did not equal U and the G content did not equal C in this pulse-labeled RNA. A note added in proof told of similar conclusions reached with studies involving normally double-stranded systems.

In view of these studies and the theoretical considerations discussed on page 105 mRNA would most likely not have a DNA-like base composition, in the sense of guanine content equal to that of cytosine, and adenine content equal to that of uracil. On the other hand, as a consequence of Watson-Crick base pairing the $A + T/G + C$ ratio for double-stranded DNA will be the same for either strand as for the total molecule. Therefore the $A + U/G + C$ ratio can be used to compare a particular single-stranded RNA to any particular variety of suspected

DNA template. DNA-likeness for mRNA would be expected for the percent of G + C content but not for the equivalence of complementary bases.

Some of the Weaknesses in the Over-all Scheme

By the year 1963, a unanimity in approach and interpretation had pervaded the entire field. Inconsistencies were not dwelled on; everything had to fit. I felt that it was important, however, to have a serious discussion about the actual strength of the foundations for these strong convictions and therefore published a discussion of some of these issues (158). This discussion, entitled "Protein Biosynthesis: Some Alternative Considerations on the Current Hypothesis," is reproduced below because it outlined, in a concise fashion, some of the principal weaknesses of the theory.

Several recent articles give the impression that the problem of protein biosynthesis is largely solved and that the remaining unsettled questions concern specific details (1, 2).* The purpose of this discussion is to emphasize that the current formulation still represents a working hypothesis. New findings which do not readily fit in the current visualization should be used to help modify the hypothesis. It is very important at this stage of our understanding to be careful not to select only those findings and interpretations which are consistent.

The essential features of the most generally accepted mode of protein synthesis are as follows:

1. All of the information for making specific proteins is contained in a linear sequence of nucleotides.
2. The information is written in the form of a code which uses three (or a multiple of three) nucleotides to specify a single amino acid. The information may be carried predominantly in two of three nucleotides (2, 3).
3. An exact replica of the DNA sequence is made in the form of RNA with uracil substituting for thymine.
4. This replica, called messenger RNA, attaches itself to ribosomes already existing in the cytoplasm and from this location directs the formation of protein.
5. Amino acid is activated for protein synthesis, and in chemical association with a specific molecule of soluble ribonucleic acid it is brought to a particular location on the messenger. This location recognizes the soluble ribonucleic acid and not the amino acid.
6. The messenger has a short life (2 to 3 min) and, after making several

* Reference numbers in this article refer to the list of readings at the end of this reprint on page 106.

proteins, it is degraded. More recently, in order to retain the concept of messenger in those cases where evidence for a rapidly renewed message was not obtained, the possibility of a more stable message is considered.

In regard to point 1, the evidence which demonstrates that nucleic acid is the carrier of genetic information is firmly supported by experiments on bacterial transformation, infection of hosts by virus nucleic acid, and induced and natural mutations which affect the nucleic acids and result in specific substitutions of amino acids. That nucleic acid is the only participant in the storage and transfer of information is a satisfying but not unique interpretation of these experiments. Another tenable explanation could be that genetic information is written in the form of a polynucleotide in juxtaposition to a non-nucleotide polymer existing in the host cell. Reservations of this kind have recently been stated (4).

In regard to point 2 it is worth remembering that the origin of the triplet concept was the result of a logical process and was not based on the interpretation of experimental data. If all of the information is carried in the linear sequence, then it is obvious that if only one nucleotide acted as a code for one amino acid, the four nucleotides would be insufficient for 20 amino acids. By similar reasoning, since two nucleotides alone would act as a code to $4 \times 4 = 16$ amino acids (an insufficient number), at least three nucleotides would be needed, since $4 \times 4 \times 4 = 64$ (2, 3).

In an analysis of acridine-induced mutations in *E. coli* infected with T_4-phage Crick et al. (5) provided evidence obtained in a study of changes in the B cistron of the R_{II} region of the phage which lends support to the triplet hypothesis.

Point 5 concerns the role of sRNA in protein synthesis. This subject has been treated in some detail (6) and will not be considered further. It should be noted, however, that according to the scheme as now conceived, the role of sRNA is directly linked to the function of messenger RNA. The fact that some question exists on the universality of the requirement for sRNA in processes of protein synthesis means that the same question would have to apply to the role of messenger in this scheme.

Points 3 to 6 constitute the messenger RNA concept, and the discussion that follows is concerned mainly with this subject. The modern idea of the messenger was emphatically stated with supporting experimental data by Brenner, Jacob, and Meselson (7), although the idea was originally discussed by Jacob and Monod (8). Since this statement launched the existing messenger concept and is often referred to, it is considered here in some detail. The support for the messenger concept

UNIVERSITY OF PITTSBURGH
BRADFORD CAMPUS LIBR

from Brenner et al. was of three kinds: namely, completely theoretical, logical, and experimental. The theoretical considerations were discussed in the first paragraph. The points raised are not particularly compelling, and alternative explanations could easily be proposed. The rest of the argument suffers from two very serious weaknesses.

First, three models are considered to represent the process of normal protein synthesis and the changes that take place during phage infection. Experimental data were interpreted as being incompatible with two of these models. Since, however, the three models in no way exhaust the possibilities of representing the events under study, the fact that data are incompatible with two of the three models does not strengthen the case for the third, even though these data may be compatible with the third model.

Second, an integral feature built into all three models is the assumption that the Volkin and Astrachan type of RNA* (9) participates in the transfer of information from DNA to the protein-synthesizing system. That this assumption does not uniquely follow from the experimental observations is demonstrated by the fact that a completely different explanation was considered when Astrachan and Volkin discussed their own original findings (10). Furthermore, newer observations with uninfected cells would suggest that much of this type of RNA functions as an early precursor stage in the formation of stable ribosomes (11, 12).

The problems associated with evaluating the data in terms of only three models can best be demonstrated by interpreting the experimental observations in terms of a fourth model. In this model the active site of protein synthesis is not considered to be free ribosome but a membrane-associated ribosome. This concept actually accounts for many published observations (6). More recent experiments suggest that in *E. coli* such a model might apply (13). In this model the Volkin and Astrachan type RNA is considered as an early stage of ribosome formation. The experiments performed by Brenner et al. showed the following:

1. The new C^{14}-uracil-labeled RNA formed after phage infection was associated with the ribosome band assumed to be of the 70s class.

2. This RNA had been metabolically turning over because after 16 min growth in C^{12}-labeled medium diluted two hundredfold, only 25% of the radioactivity was still associated with the 70s ribosomes.

3. If cells were grown in N^{15}-C^{13}-labeled medium, infected with T_4-bacteriophage, transferred to N^{14}-C^{12}-labeled medium, and fed P^{32}-labeled phosphate from the second to the seventh minute, the P^{32} would

* That is, RNA that is rapidly turning over and which has a base composition close to that found in DNA.

be associated mainly with the N^{15}-C^{13}-labeled 70s ribosomes, although there would be some associated with the lighter ribosomes.

Their interpretation was that the new RNA was associated only with already existing ribosomes and that no wholly new ribosomes were synthesized after phage infection. A completely different interpretation, however, could readily be offered on the basis of the fourth model; that is, ribosomes are assembled on the surface of the cell membrane. Therefore the host cells grown in N^{15}-C^{13}-labeled medium contain a whole chain of N^{15}-C^{13}-labeled ribosome precursors. The new RNA is incorporated into the new ribosomes being synthesized at the cell membrane and released during the process of cell disruption. The N^{14}-C^{12}-labeled components enter the ribosome-synthesizing sequence from the beginning and in the five minutes of growth in P^{32}-labeled medium, a small percentage of N^{14}-C^{12}-P^{32}-labeled ribosomes would be the result of a continuous production of new ribosomes containing the new RNA which serves as ribosome precursor. It should be noted that this interpretation is compatible with the recent findings cited above (11,12).

In an experiment to determine whether new ribosomes or those already existing make protein a similar approach was used by Brenner et al. (7). Cells were grown in N^{15}-labeled medium, infected with phage, transferred to N^{14}-labeled medium, and fed S^{35}-labeled sulfate for the first 2 min of growth. The result was that only N^{15}-labeled ribosomes contained S^{35}. Their interpretation was that protein was synthesized only on already existing ribosomes. According to model four, however, protein was made on the ribosomes newly synthesized from the N^{15}-labeled precursors or already existing on or in the cell membrane.

An important facet in the theoretical development of the messenger concept regarding synthesis of protein in bacteria is the high rate of metabolic turnover of the messenger molecule. This is dictated because of the fact that the kinetics of enzyme induction and repression are quite rapid; therefore the message must be quickly renewed or destroyed. Furthermore, as pointed out by Gros et al. (14), the amount of bacterial RNA, which could be messenger, is only a small percentage at most. For pulse-labeling of RNA to reflect base compositions of DNA, an appreciable percentage of the newly synthesized RNA must be messenger. Since the steady-state amount of messenger RNA would be only a small percentage of the total, it follows that the DNA-like RNA must be turning over very much faster than the bulk RNA. Indeed the experimental results of Brenner et al. (7) and Gros et al. (14) seem to confirm this property of rapidly turning over messenger RNA which transiently was associated with fully formed 70s or 100s ribosomes in *E. coli*.

Support for this interpretation was also drawn from the fact that a similar DNA-like RNA (D-RNA) in yeast appeared to be rapidly turning over. More recent information, however, with both the *E. coli* D-RNA (11) and yeast D-RNA (12) give a somewhat different picture. Ycas and Vincent, who originally described the D-RNA of yeast, in later experiments with Kitazume (12), reported that there was no obvious relationship between the synthesis of D-RNA and that of protein. They further reported that D-RNA was an obligatory precursor to the bulk of cellular RNA and therefore that it was not being extensively degraded such as is thought to be the case with messenger. These findings confirm those of McCarthy, Britten, and Roberts (11) who showed that the RNA fraction in *E. coli* which has the properties of the messenger RNA functions largely as a precursor to the ribosomes and that no evidence was obtained to support the contention of Gros et al. and Brenner et al. that this fraction of messenger RNA was being rapidly and extensively degraded to smaller stages. Tissières and Watson (15) studied the attachment of "messenger RNA" *in vivo* to ribosomes and its concomitant breakdown linked to amino acid-incorporation. They calculated that about 17 μg of messenger RNA was broken down for each 1 μg of amino acid incorporated (i.e., 4 to 5 moles of nucleotides per mole of amino acid). On the other hand, using synthetic messenger RNA, Barondes and Nirenberg (16) showed that *in vitro* about 80% of the polynucleotide was rapidly broken down before C^{14}-phenylalanine incorporation reached its maximal rate and that during the phase of active amino acid-incorporation, the nucleotide breakdown was at a minimal rate (i.e., approximately 2 μmole of amino acid were incorporated for each μmole of nucleotide released). It is important to mention that in both of these studies, the net disappearance of messenger was followed. A concomitant resynthesis, if it occurred, would complicate the interpretation.

Two very interesting papers have recently appeared which are concerned with the same problem of the dependence of protein synthesis on the presence of a metabolically unstable RNA messenger (17, 18). Both of the reports concerned the effects of actinomycin D with *B. subtilis in vivo* on the inhibition of RNA synthesis and protein synthesis. Hurwitz et al. (17) found that actinomycin (0.2 μM), inhibited DNA synthesis only 25% but inhibited RNA synthesis 90 to 95%. The small amount of RNA synthesized (2 to 8%), when examined after pulse-labeling with P^{32}, did not resemble in base composition either DNA or RNA of *B. subtilis*. Under these conditions, with messenger RNA synthesis completely blocked, protein synthesis continued at 25 to 50% of the normal rate. Hurwitz et al. concluded that, obviously, protein synthesis independent of unstable messenger RNA must proceed in this organism.

Levinthal et al. (18), using actinomycin D at a concentration 40 times greater than that of Hurwitz et al., found that synthesis of protein, as well as of RNA, was completely blocked. Since, however, at lower concentration a separation of the two inhibitory effects was apparently achieved, the double effect at the higher level cannot be interpreted in terms of a link between the two processes. The foregoing studies do not fit into a single clear picture of messenger activity. This fact is of further concern since the conflicting findings were made on very similar systems.

An essential type of evidence for the firm establishment of the messenger concept, as we now know it, would be the demonstration that a protein-synthesizing system of one species could be induced to make a specific protein characteristic of a completely different species, by the addition of RNA characteristic of the second species.* One attempt in this direction was recently described (19). Ribonucleic acid of tobacco mosaic virus was added to the *E. coli* system of Nirenberg, Matthaei, and Jones (20), and an appreciable stimulation of amino acid-incorporation was noted. An evaluation of the findings in relation to the controls shows that of the induced incorporation of amino acid, 9% at most, could possibly be tobacco mosaic virus (TMV) protein. This amount of protein would, however, be sufficient to establish the point, if it were demonstrated unequivocally to be TMV protein. Although there were some definite similarities between a part of the induced protein and TMV coat protein, no identity was established. The data were obtained by fractionation on Dowex 1 of peptides obtained from a trypsin digest of the purified induced TMV-like protein. The doubt about this kind of proof is reflected by the fact that after incubation with either C^{14}-labeled phenylalanine or tyrosine, peptide peaks not supposed to contain either of these amino acids contained appreciable radioactivity. It is true, of course, that other TMV-induced proteins could be blurring the picture, but this point must be resolved. A peptide "fingerprint" (chromatographic and electrophoretic analysis) would provide a more powerful demonstration of the formation of the TMV-protein.†

---

* This important control was successfully carried out in 1965 by Schwartz, Eisenstadt, Brawerman, and Zinder who showed that extracts of *Euglena gracilis* could make *E. coli* phage f-2 coat protein when supplied with phage f-2 RNA (269).

† This issue was subsequently resolved when G. Aach et al. reported that the radioactive proteins produced in this system in response to TMV-RNA were not representative of TMV proteins (268).

The beautiful demonstration by Nathans et al. (21) of the induced formation of f-2 coat proteins in an *E. coli* system does not serve to resolve this point. This phage is an RNA-phage of *E. coli*. There is a valid question as to how much information for the synthesis of f-2 coat protein could be stored in *E. coli*. Nathans et al described control experiments with RNA from TMV in this system. With the techniques used to study the formation of f-2 coat protein, these authors possess the facilities to provide the much-needed information concerning the possible formation of TMV-coat protein. A more recent attempt to identify a polynucleotide-stimulated formation of protein was reported as negative (22).

There are other unsettled questions which are not treated in the present scheme of protein synthesis. What happens in the case of synthesis of protein and specific enzymes in enucleated cells of algae and amebae (23)?* In the model system of Nirenberg and his associates, single-strandedness appears to be an essential feature for messenger activity (i.e., poly-U (polyuridylic acid) is active, but when poly-A (polyadenylic acid) is added along with it, the poly-U is ineffective (24)). From the fact that the base ratio of messenger RNA is supposed to be similar to that of DNA, it would appear that both complementary strands of the DNA are copied and are present. If so, how then does the accepted scheme solve the problem of removing the unwanted strand?

An alternative possibility would be that each single strand of the DNA had a base pair relation of adenine = thymine and guanine = cytosine. In the absence of any reason for such a postulation (such as base-pairing in the case of double strands) this possibility is not attractive at the present time. It should be noted that the situation does not appear to apply for the DNA which is supposed to serve as a template for ribosomal RNA (25). Another discussion of the limitations of the current theory in terms of experimental data derived with HeLa cells has recently appeared (26).

In conclusion, certain points should be clearly stated. This paper is not an attempt to disprove the validity of the messenger concept. The messenger concept is a very powerful tool with which to view new data and plan further experimental approaches. The messenger concept is, at the same time, only a working hypothesis and one which no doubt will have to be altered as more information is acquired. A potentially undesirable situation is developing where the distinction between working

---

* Even though more recent work showed the presence of DNA in mitochondria and chloroplasts, it has not been established in 1967 that this DNA codes for any particular protein or enzymes.

hypothesis and fact is becoming diffuse. The process of protein synthesis with respect to its direction control, utilization of energy, and integration with the rest of the cell's activities is not likely to be completely understood until the various implicated reactions are viewed in their relation to cellular organization. The problem should be approached in many different ways. Simple schemes representing our current thinking should not be treated as established patterns.

REFERENCES (FOR PAGES 99–106)

1. P. C. Zamecnik, *Biochem. J.,* **85,** 257 (1962); M. Calvin, *Am. Inst. Biol. Sci. Bull.,* **13,** 2 (1962); J. D. Watson, *Science,* **140,** 17 (1963).
2. F. H. C. Crick, in *Progress in Nucleic Acid Research,* Academic Press, New York, 1963.
3. R. B. Roberts, *Proc. Natl. Acad. Sci. U.S.,* **48,** 1245 (1962); R. W. Eck, *Science,* **140,** 477 (1963).
4. N. W. Pirie, *Nature,* **197,** 568 (1963); J. R. S. Fincham, *Ann. Rev. Biochem.,* **38,** 343 (1959).
5. F. H. C. Crick, L. Barnett, S. Brenner, and R. J. Watts-Tobin, *Nature,* **192,** 1227 (1962).
6. R. W. Hendler, *ibid.,* **193,** 821 (1962).
7. S. Brenner, F. Jacob, and M. Meselson, *ibid.,* **190,** 576 (1961).
8. F. Jacob and J. Monod. *J. Mol. Biol.,* **3,** 318 (1961).
9. E. Volkin and L. Astrachan, in *The Chemical Basis of Heredity,* Johns Hopkins Press, Baltimore, 1957, p. 686.
10. L. Astrachan and E. Volkin, *Biochim. Biophys. Acta,* **29,** 536 (1958).
11. J. McCarthy, R. J. Britten, and R. B. Roberts, *Biophys. J.,* **2,** 57 (1962).
12. Y. Kitazume, M. Yčas, and W. S. Vincent, *Proc. Natl. Acad. Sci. U.S.,* **48,** 265 (1962).
13. R. W. Hendler and J. Tani, in preparation.
14. F. Gros, H. Hiatt, W. Gilbert, C. G. Kurland, R. W. Risebrough, and J. D. Watson, *Nature,* **190,** 581 (1962).
15. A. Tissières and J. D. Watson, *Proc. Natl. Acad. Sci. U.S.,* **48,** 1061 (1962).
16. S. H. Barondes and M. W. Nirenberg, *Science,* **138,** 810 (1962).
17. J. Hurwitz, J. J. Furth, M. Malamy, and M. Alexander, *Proc. Natl. Acad. Sci. U.S.,* **48,** 1222 (1962).
18. C. Levinthal, A. Keynanand, and A. Higa, *ibid.,* p. 1631.
19. A. Tsugita, H. Fraenkel-Conrat, M. W. Nirenberg, and J. H. Matthaei, *ibid.,* p. 816.
20. M. W. Nirenberg, J. H. Matthaei, and O. W. Jones, *ibid.,* p. 104.
21. D. Nathans, G. Notani, J. H. Schwartz, and N. D. Zinder, *ibid.,* p. 1424.
22. B. R. Chatterjee and R. P. Williams, *Biochem. Biophys. Res. Commun.,* **9,** 72 (1962).
23. H. Chantrenne, in *The Biosynthesis of Proteins,* Pergamon, New York, 1962.
24. M. W. Nirenberg and J. H. Matthaei, *Proc. Natl. Acad. Sci. U.S.,* **47,** 1588 (1962); M. F. Singer, O. W. Jones, and M. W. Nirenberg, *ibid.,* **49,** 392 (1693).
25. S. A. Yankofsky and S. Spiegelman, *ibid.,* **48,** 1466 (1962).
26. H. Harris, H. W. Fisher, A. Rodgers, T. Spencer, and J. W. Watts, *Proc. Roy. Soc. London, Ser. B,* **154,** 177 (1963).

The Location of the Amino Acid in Amino Acyl-sRNA

By two independent methods, one employing tosylation[8] of free hydroxyl groups (159) and the other, nuclear magnetic resonance spectroscopy (160), the site of esterification for amino acyl-sRNA was fixed at the 3' rather than the 2' hydroxyl of the ribose of the terminal adenosine. Although fairly convincing, the results have an element of uncertainty due to the possible contribution of amino acyl migration which would lead to an equilibrium mixture containing 70% of the 3' ester. In both of the above cited studies however, more than 90% of the amino acid was found in the 3' position, and so this seems to be the preferred location.

Enzymes for sRNA Binding and Peptide Bond Formation

Bishop and Schweet obtained several fractions from reticulocyte supernatant fluid by ammonium sulfate precipitation and chromatography on DEAE-cellulose (161). Three of these fractions possessed different relative abilities for catalyzing the transfer of different radioactive amino acids from sRNA to ribosomes. Combining two of the fractions gave 8 to 10 times more activity than the sum of the two tested separately for C[14]-leucine transfer. The authors speculated that there may exist a separate transfer factor for each amino acid and the limitation of transfer found by testing one fraction may be due to an insufficiency of the transfer factors for other amino acids which may be provided in the second fraction.

Nathans and Lipmann isolated a transfer factor from *E. coli* supernatant fraction by ammonium sulfate precipitation and chromatography on DEAE cellulose (162). Their factor was nondialyzable and heat labile. The same peak of activity showed up when assayed with five different amino acids and so they concluded that the factor was not specific for the amino acid. The factor was specific to the tissue, however, and no cross reactions were found for ribosomes and factors from *E. coli* and rat liver. It was also shown that the factor was different than the activating enzyme or aminoacyl synthetase. Transfer catalyzed by the factor was inhibited by puromycin and chloramphenicol. After participation in a cycle of transfer, sRNA was fully active for another cycle and so it functioned catalytically. In the presence of puromycin the same amount of amino acid was enzymatically transferred from

---

[8] p-Toluenesulfonylchloride (tosyl) is a reagent used to block free hydroxyl groups. The process of reacting this reagent with sugars is called tosylation.

the sRNA but instead of becoming incorporated into the ribosomes, it was simply released.

Grossi and Moldave obtained transfer of C^{14}-leucine from sRNA to rat liver microsomal protein with a dialyzable plus a nondialyzable factor obtained from rat liver soluble supernatant at pH 5.2 (163). In an extension of this work the nondialyzable component was further purified by an ammonium sulfate precipitation and charcoal adsorption and appeared to be relatively stable at 100° heating for 10 min (164). The dialyzable fraction was tentatively identified as GTP. With ribosomes in place of the microsomes, the nondialyzable factor would not catalyze transfer of amino acid from sRNA unless an additional factor isolated from microsomes was added. This latter factor was also nondialyzable and heat labile, and was obtained by deoxycholate extraction of the microsomes (165). Fessenden and Moldave subsequently isolated both factors from the supernatant fraction of rat liver homogenate (166). The one originally found in the soluble fractions was named amino acyl transferase I, and the one originally obtained by a deoxycholate extraction of microsomes but now isolated directly from the supernatant fraction by ammonium sulfate precipitation was called amino acyl transferase II. Amino acyl transfer to ribosomes was shown to require GTP and a sulfhydryl compound. The isolation of these factors was subsequently improved by use of gel filtration techniques (167). More recent studies from Moldave's laboratory indicate an interaction between amino acyl-sRNA and the transferase I and between a sulfhydryl compound, ribosomes and transferase II. The significance of these interactions is not known at this time (168).

The factor described by Nathans and Lipmann was resolved into two components by chromatography on DEAE Sephadex A-50 (169). Each factor alone had little activity, but when combined an appreciable transfer of radioactivity from C^{14}-phenylalanyl-sRNA to E. coli ribosomes was observed. Both factors (called A and B) were heat labile, nondialyzable, ammonium sulfate precipitable, and essentially free of nucleic acid. Both Conway (170) and Spyrides (171) implicated fairly high concentrations of ammonium or potassium as being involved in the preliminary steps involving the interaction of phenylalanyl-sRNA, poly-U, and E. coli ribosomes. The binding of phenylalanyl-sRNA to ribosomes in this reaction was not inhibited by puromycin and GTP was not required. Ammonium ions were superior to potassium in the binding reaction. The binding reaction required the appropriate polynucleotide to be effective (i.e., poly-U for phenylalanine-sRNA, poly-A for lysyl-sRNA). Conway and Lipmann demonstrated a GTPase which yielded GDP and organic phosphate concurrent with polyphenylalanine synthesis in the poly-U-stimulated system (172). Although sRNA was required for GTPase activity it was active with or without attached amino acid and even after periodate decomposition of the terminal adenosine. Therefore GTP splitting is not directly dependent on peptide bond formation. The distribution of GTPase activity followed that for phenylalanine incorporation for ribosomes distributed in a sucrose gradient by ultracentrifugation. Activity seemed to require both ribosomal subunits. Antibiotics which inhibited phenylalanine incorporation did not inhibit GTPase activity. GDP inhibited both reactions. Heating at 55° destroyed incorporation ability but enhanced GTPase activity.

Nishizuka and Lipmann improved the purification technique for the two factors A and B by use of standard enzyme isolation techniques involving protamine treatment, alumina gel, hydroxy-apatite and a heat treatment to denature the A factor when present with the B factor (173). The A factor was renamed T factor to indicate its transfer function and the B factor was renamed G factor to indicate its GTPase activity. However, since the G factor overlapped another factor with high GTPase activity upon DEAE Sephadex chromatography, it is not readily apparent that the G factor possesses its own GTPase activity. At any rate, both factors, when present together, caused a very marked stimulation of transfer of amino acid from amino acyl-sRNA to *E. coli* ribosomes primed with the appropriate synthetic template.

Arlinghaus, Shaeffer, and Schweet continued their studies with their two transfer factors since called TF-1 and TF-2 and in this work renamed binding and peptide synthetase enzymes, respectively (174). Poly-U-stimulated phenylalanine incorporation from phenylalanyl-sRNA into reticulocyte ribosomes was again studied. The binding step required TF-1 and GTP but no sulfhydryl compound. The radioactivity which was released from the ribosomes and split from the sRNA was 97% phenylalanine and 3% diphenylalanine, the latter product being the result of a small contamination of TF-1 with TF-2. The bound phenylalanyl-sRNA moved with 80s ribosomes in a sucrose gradient containing 10^{-4} M magnesium. When TF-2 was added to the ribosomal bound phenylalanyl-sRNA and the system incubated for 10 min, 25% of the radioactivity was found in diphenylalanyl-sRNA and 70% in phenylalanyl-sRNA. In other words, essentially one cycle of condensation occurred and GTP did not seem to be involved in this step. If more phenylalanine-RNA and GTP was added triphenylanyl-sRNA could be formed.[9]

---

[9] For more recent information on the function of the soluble factors and GTP see references cited in footnote on page 55.

*The Transfer of Ribonucleic Acid from
the Nucleus to the Cytoplasm*

One of the major tenets of the modern concept of protein synthesis states that messenger RNA molecules, after being made as copies of portions of the chromosome, migrate to the cytoplasm where they direct protein synthesis. More recently hybridization experiments have shown that sequences complementary to the order of nucleotides in cytoplasmic ribosomes are present in the DNA of the nucleus and therefore it is felt that ribosomal RNA or even ribosomes themselves are made in the nucleus (specifically the nucleolus) and then migrate to the cytoplasm. This idea receives support from the similarities in base composition between nucleolar RNA and ribosomal RNA. Other observations in general accord with these assumptions are that RNA does indeed occur in close association with chromosomal DNA and that after pulse-labeling with various precursors to RNA it is found that nuclear RNA is labeled very rapidly with essentially no lag period, whereas cytoplasmic RNA is labeled much more slowly and only after a substantial lag. Many workers have shown that when the initial radioactive pulse is spent radioactivity leaves the nucleus and builds up in the cytoplasm.

These observations and their implications are very satisfying and comforting but, according to Henry Harris et al. (175) and an impartial reading of Harris' papers, the situation is not so clear or settled as it seems. Experiments with HeLa cells growing exponentially in suspension culture were performed by Harris et al. After a 10-min labeling with C^{14}-adenine the nuclei were isolated. These nuclei contained 99% of the DNA of the intact cells and all of the rapidly labeled RNA (within the limits of measurement). When the nuclei were lysed and the DNA recovered, it was found that approximately 90% of the pulse-labeled RNA was associated with the DNA. The rapidly labeled nuclear RNA had the same sedimentation distribution as that of cytoplasmic microsomal RNA, namely 16s and 28s. Most of the radioactivity in the cytoplasm after the 10 minute pulse-labeling was in RNA of the 3s class; the 16s and 28s varieties being barely labeled. After washing the labeled cells at the end of the 10 min exposure to C^{14}-adenine and incubating in unlabeled medium for 10 hr, the cytoplasmic 28s and 16s varieties slowly became labeled, but if unlabeled nucleic acid precursors were added during the second phase of incubation, no radioactivity appeared in the cytoplasmic RNA under conditions in which radioactivity continued to be lost from the nuclear RNA (176). When cells were labeled for 90 minutes with C^{14}-adenine and the nuclei and cytoplasm separated and incubated separately, it was found that the nuclear

RNA was degraded ultimately to nucleoside 5'-monophosphates, whereas the cytoplasmic RNA was not degraded, indicating that a mechanism for RNA breakdown was present in the nucleus. This was subsequently shown to be due to an enzyme having the properties of a polynucleotide phosphorylase and apparently found in actual association with the nuclear DNA and the rapidly labeled RNA (see page 114 and reference 188).

Base compositions of the various types of RNA as well as that for the RNA rapidly labeled with P^{32} were determined. It was found that none of the RNA fractions had a base ratio similar to that of DNA and that the distribution of radioactivity in the nucleotides changed rapidly with time. Since the P^{32} distribution was determined after alkaline hydrolysis (the technique usually used) the possibility existed that a DNA-like RNA might have been present, but was not detected because the P^{32} may have labeled the precursor nucleotides unevenly.[10] It may also be true that only a part of the labeled RNA was DNA-like but that this was obscured by the presence of other types of RNA with different base compositions. In conclusion, Harris et al. cited the experiments of Hämmerling with Acetabularia (177) which indicate:

1. that synthesis of specific proteins can continue for many weeks after removal of the nucleus.

2. that when an enucleate cell of one species receives a nucleus from another species, hybrid caps can be formed which possess the characteristics of both species.

These last two observations, however, would be compatible with a situation in which some information may be provided independently by the nucleus and the cytoplasm. Such a situation would be consistent with more recent findings of DNA in the cytoplasm.

In support of his contention that RNA rapidly synthesized in the nucleus can be degraded in the nucleus Harris cited the work of others (185). Thus Levy showed that when HeLa cells previously labeled with H^3-uridine were subsequently incubated in the presence of actinomycin D, radioactivity left the nuclear RNA but did not appear in the cytoplasmic RNA (178). Levy interpreted his work in terms of the prevention by actinomycin D of the migration of RNA from nucleus to cytoplasm.

---

[10] Under conditions of alkaline hydrolysis the phosphate is cleaved from its 5' bridge and is isolated as the (2',3') nucleotide monophosphate of the adjacent nucleotide. Therefore, phosphate is found not with the nucleotide with which it entered the nucleic acid, but instead with the adjoining nucleotide.

Harris pointed out that the work also showed that nuclear RNA may be degraded in the nucleus.

Paul and Struthers studied the effect of actinomycin D in LS cells (a subline of the strain L fibroblast) (179). To obtain 95% inhibition of nuclear RNA synthesis in 15 min, it was necessary to use 100 μg/ml of actinomycin D, which is a tremendously high concentration (see pages 116–119). They similarly found that in the presence of actinomycin D a previously H³-uridine pulse-labeled cell rapidly lost radioactivity from the nucleus without a concomitant rise of radioactivity in ribosomal RNA. Furthermore, in the presence of a high level of actinomycin D, both the nucleus and the cytoplasmic ribosomal RNA continued to incorporate radioactivity by a slow process. Harris also cited this work in support of his argument for the degradation of nuclear pulse-labeled RNA in the nucleus. Similarly, Harris cited the work of Lieberman, Abrams, and Ove who showed, with cultures of trypsinized kidney cells, that nuclear RNA was by far the most rapidly labeled species when a 20 minute pulse of H³-cytidine was used (180). When the cells were then replaced into a medium containing unlabeled cytidine and actinomycin D at a concentration of 0.25 μg/ml, radioactivity rapidly left the nucleus but did not appear in the cytoplasmic RNA. It is also possible, however, in these three studies cited by Harris, that actinomycin may interfere with the process of transfer of rapidly labeled nuclear RNA to the cytoplasm but the studies do show, as Harris contends, that a mechanism exists for rapidly decreasing the level of rapidly labeled RNA in the nucleus by hydrolysis in the nucleus, without transfer to the cytoplasm.

Another example of the breakdown of nuclear RNA and reincorporation of nucleotides into cytoplasmic RNA is provided by an experiment with radioactive adenine (181). Adenine is readily converted to guanine in tissues; the reverse situation does not so readily apply. After HeLa cells growing exponentially were labeled with adenine for 18 hr, then washed and transferred to a medium containing unlabeled adenosine at 10^{-3} M and guanosine at 2×10^{-4} M for subsequent reincubation, the nuclear and cytoplasmic RNA's were isolated and analyzed. It was found that whereas the relative content of radioactive adenine in the nuclear RNA decreased, the relative content of radioactive guanine in the cytoplasmic RNA increased. This conversion of adenine to guanine required the release of adenine nucleotides from the nucleic acid. This experiment shows that such release of nucleotides from nuclear RNA and subsequent resynthesis to cytoplasmic RNA occurs extensively. It does not rule out, however, the concomitant passage of some RNA intact from nucleus to cytoplasm. The ability of the cytoplasm to reuse breakdown products from nuclear RNA would tend to obscure the fact of nucleic acid breakdown. Watts and Harris, in a study of mammalian macrophages, showed that when the cells were labeled with P^{32} and subsequently "chased" with unlabeled phosphate, no loss of radioactivity from the RNA occurred (182). When the cells were labeled with C^{14}-adenine and then chased with unlabeled adenine or adenosine, appreciable loss of radioactivity from the RNA occurred. These experiments show how much the results can be altered by chasing under conditions where the precursor pool may be more or less accessible to the labeled precursor. These considerations led Harris to challenge the work of Scherrer, Latham, and Darnell (183), and Girard, Penman, and Darnell (184). These workers claimed to demonstrate the transfer of RNA from the nucleus to the cytoplasm with HeLa cells. In the experiments of Harris and others (Levy, Paul, and Struthers and Lieberman et al.) the labeled cells were isolated, washed, and returned to nonradioactive medium containing unlabeled nucleosides and actinomycin D in order to dilute out subsequent reincorporation of nucleotides released from nucleic acids. Girard et al. simply added actinomycin D to a culture growing in the presence of radioactive precursor. Harris performed the experiment both ways with C^{14}-adenine and showed that when the washing and transfer to nonlabeled medium containing actinomycin D technique was used, there was a fall in radioactivity of nuclear RNA which was not accompanied by any gain in cytoplasmic RNA labeling (185). When actinomycin D was added to the labeled culture however, both nuclear and cytoplasmic RNA continued to increase in content of radioactive adenine.

These experiments of Harris, however, do not resolve the ambiguity of the situation. In the work of Darnell et al., the addition of actinomycin D caused an immediate cessation of incorporation into RNA. In the work of Harris incorporation proceeded for a few hours before stopping. Darnell et al. showed a greater increase of radioactivity in cytoplasmic ribosomal RNA than in general cytoplasmic RNA after the addition of the actinomycin D, Harris did not distinguish between these varieties in his early time points during which radioactivity content in both nuclear and cytoplasmic RNA continued upward. Darnell et al. actually separated the cytoplasm from the nucleus for isolation of RNA. Harris used a preferential extraction with phenol water. Harris showed the continued incorporation of radioactivity into the cytoplasmic RNA in the presence of radioactive precursor. It may be well to recall that it has been reported that a high proportion of the actinomycin-resistant incorporation in the cytoplasmic L cell fibroblasts was into a 4s variety as opposed to the heavier classes of RNA (186). This was also true for HeLa cells (187). Harris pulse-labeled for 10 and 20 minutes; Darnell et al. pulse-labeled for 30 minutes. A recent study employing Ehrlich ascites cells comes to the conclusion that little, if any, of the rapidly labeled RNA of the nucleus was transported into the cytoplasm. However, after a 2 hr incubation with P^{32} followed by incubation in the presence of actinomycin D for 2.5 hr the content of radioactivity in the 28s and 18s components of the cytoplasm increased by a factor of 1.2 to 2. This may represent a slow transfer of stable RNA which, under normal conditions, may even be larger. Such a possibility is also compatible with Harris' results.

A point raised by Harris is that the enzyme for breakdown of nuclear RNA exists in the nucleus and appears to have ready access to the rapidly labeled RNA (188). Nuclei isolated from HeLa cells contain an active enzyme system which degrades rapidly labeled RNA to 5′ nucleotides (175). The nuclei also contain an enzyme which converts nucleoside diphosphates to 5′ monophosphates. DNA is not attacked by this enzyme. From the facts that ADP inhibited breakdown of nuclear RNA, phosphate stimulated the breakdown, and P^{32} exchanged with ADP during breakdown Harris concluded that the enzyme controlling breakdown is of the polynucleotide phosphorylase type. This enzyme was not detected in the cytoplasmic fraction but when the nuclear contents were dissolved in 1 M sodium chloride and the DNA precipitated out by dilution with water, the enzyme together with the rapidly labeled RNA came down apparently with the DNA fibers. It was found that the enzyme exerted a selective preference for degrading the rapidly labeled RNA, and it exhibited the same temperature sensitivity found for the loss of nuclear RNA from intact cells. As to the presence of a rapidly labeled, rapidly degraded RNA in bacteria, Harris pointed out that there was no convincing evidence on this point (189).

Harris et al. cited several examples of cases where cytoplasmic protein synthesis appears to occur in the absence of formation of new RNA (190). Thus Goldstein et al. have shown that synthesis of protein, measured by the incorporation of labeled amino acids, can take place at approximately normal rate in amnion cells in culture for many hours after removal of the cell nucleus (191) and essentially similar results have been obtained by Prescott in acanthamebae (192). Reich et al. have shown that the synthesis of protein can continue in L cells (cells occurring without a cell wall) at an approximately normal rate for several hours after RNA synthesis has been inhibited by the administration of actinomycin D (193); Greengard et al. have shown that the tryptophan-induced accumulation of trytophan pyrrolase in rat liver cells is not abolished by high concentration of actinomycin D (194). Seed and Goldberg have shown that thyroglobulin continues to be synthesized by cells of the thyroid gland under the same conditions (195). Garren *et al.* have shown that a hormone-induced increase in the rate of synthesis of tryptophan pyrrolase and tyrosine α-ketoglutarate transaminase can take place in liver cells after virtually all RNA synthesis has been inhibited by this antibiotic (196). It is, moreover, well known that the giant unicellular algae, acetabularia, not only can synthesize protein, but also actually undergoes regeneration in the absence of the nucleus, and Spencer and Harris have recently shown that the synthesis of three phosphatase enzymes in the cytoplasm of acetabularia can be regulated in an apparently normal way for many weeks after removal of the nucleus (197).

This list may be enlarged to include many other observations which show that cells can continue to synthesize protein for many hours after actinomycin D has reduced the rate of synthesis of RNA to a small percentage of the normal (198). Hiatt reported that a rapidly labeled RNA that possessed properties attributable to messenger RNA, was found in the nuclei of cells of regenerating rat liver; however, there was no indication that this material was transferred to the cytoplasm (199). Eboué-Bonis, Chambaut, Volfin, and Clauser reported that when actinomycin (10 μg/ml) completely inhibited RNA labeling in isolated rat diaphragm, amino acid-incorporation into protein was not inhibited and the 100% stimulation of amino acid-incorporation induced by insulin was not impaired (200). Scott and Malt reported that when the incorporation of C^{14}-uridine into the RNA of turtle and chicken erythrocytes was stopped by actinomycin D at a level of 10 μg/ml, protein synthesis continued, but at a slowly deteriorating rate for at least 24 hr (in the absence of new messenger RNA production) (201). Dure and Waters reported that germinating cotton embryos treated with actinomycin D at a level of 20 μg/ml for 12 hr virtually stopped incorporating P^{32} into nuclear RNA, but protein synthesis was unimpaired (202). Gross, Malkin, and Moyer found that sea urchin embryos retain the capacity to synthesize protein in the absence of RNA synthesis (203). Kirk reported that actinomycin D at a level of 10 μg/ml added to a two-day culture of embryonic chick neural retinal cells did not prevent an increase in glutamine synthetase activity in the developing cells, although the incorporation of C^{14}-uridine into RNA in 30 min was inhibited by at least 95% (204). Revel and Hiatt showed that levels of actinomycin D which inhibited the labeling of rat liver RNA had no effect on cytoplasmic amino acid-incorporation either *in vivo* or *in vitro* (205).

The Mode of Action of Actinomycin D

The most widely used tool in the study of messenger RNA formation, turnover, location, transfer, and function is actinomycin D. These studies are based on the ability of this antibiotic to combine with DNA and prevent its use as a primer for the DNA-dependent RNA polymerase system. It has been widely assumed in these studies that the agent is specific in its action and that other manifestations of metabolic deterioration can be traced to a primary interference with DNA-dependent RNA formation. There are two crucial aspects of these important studies that I should like to subject to scrutiny. The first concerns the specificity of action of the antibiotic, and the second concerns the reliability of the criteria that have been used to recognize messenger RNA.

The first clue to the mode of action of actinomycin D came from a brief paper by Melnick, Crowther, and Barrera-Ora (206); L cells (cells that occur normally as protoplasts) were allowed to incorporate H^3-cytidine and were then examined by radioautography before and after treatment with DNAase or RNAase. It was found that actinomycin D treatment resulted in a total loss of ability to incorporate the tracer into RNAase-sensitive material without exerting any appreciable effect on the labeling of DNAase-sensitive material. In one quantitative study a cell culture was treated with actinomycin D (0.2 μg/ml) for 4 hr, and radioactive leucine, uridine, and thymidine were added after the first 30 min exposure. Incorporation of uridine into RNA was depressed 58%, but incorporation of precursors into DNA and protein were depressed only 7 and 2%, respectively. It is of some interest that although more than one-half of the synthesis of the rapidly labeled RNA was inhibited, there was virtually no depression of protein synthesis. Another important finding was that the growth of Vaccinia, a DNA virus, was sensitive to actinomycin, but Mengo virus, an RNA type, was quite insensitive. This again points to an interference in the transcription of information from DNA to RNA. Goldberg and Rabinowitz isolated from HeLa cell nuclei a DNA-dependent RNA-synthesizing enzyme system (207). It was shown to be very susceptible to inhibition by actinomycin D. Inhibition was lessened for a particular concentration of actinomycin D by increasing either the amount of enzyme complex or by the adding back of DNA.

Kirk showed that actinomycin D at 0.25 to 0.75 μg/ml prevented growth and stopped RNA synthesis immediately when added to exponentially growing S. aureus (208). Kirk also demonstrated the combination of DNA and actinomycin D by a spectral shift for the DNA. Hurwitz, Furth, Malamy, and Alexander showed that purified RNA-polymerase from E. coli was extremely sensitive to actinomycin D, 0.25 μg/ml depressing the activity by 76% (209). It was shown that calf thymus DNA could competitively reverse the inhibitory effect of actinomycin D. The purified DNA-polymerase was unaffected by concentrations of actinomycin D below 1 μg/ml but was 50% inhibited at 7.5 μg/ml concentration. Here, too, calf thymus DNA could reduce the extent of inhibition. Actinomycin D was a potent inhibitor of the RNA-polymerase with various types of native DNA primers and with the single-stranded DNA of phage ϕX 174, but when polydeoxythymidylate was the primer, no inhibition resulted. With *B. subtilis* it was found that 0.25 μg/ml completely blocked growth. RNA synthesis was inhibited 90 to 95% and the remaining small percentage of activity produced a product that did not resemble either the RNA or DNA of *B. subtilis*. DNA synthesis was inhibited 25%. Protein synthesis, however, continued at 25 to 50% of normal. The authors concluded that evidently some protein synthesis did not depend on short-lived messenger RNA.

Later studies, up to the present day, employ actinomycin D under the assumption that its only effect is on the DNA-dependent RNA polymerase. Thus, for example, Levinthal, Keynan, and Higa used actinomycin D at 40 times the concentration shown by Hurwitz et al. (209) to be capable of stopping messenger RNA production (210). Although Hurwitz et al. found an appreciable fraction of the protein synthetic ability intact, Levinthal et al., at the higher concentration, found protein synthesis to decay rapidly after 2 min and interpreted their results in terms of the loss of protein synthetic ability as a direct result of the loss of messenger RNA. In an extensive study of the effects of actinomycin D in *B. subtilis* on chloramphenicol-induced RNA formation, Acs, Reich, and Valanju presented evidence that actinomycin itself caused the breakdown of rapidly labeled RNA (211). That such breakdown was not simply the result of inhibiting the synthetic reactions in a dynamic state of synthesis and breakdown was shown by the fact that 6-azouracil, which effectively blocks the synthetic reactions, did not lead to breakdown. The authors cautioned that the assumption that a turnover of RNA exists, and that its existence might be demonstrated by actinomycin, appeared questionable at this time.

Revel, Hiatt, and Revel critically examined the evidence that actinomycin D exerts its effect by preventing the formation of mRNA (212). They pointed out that in fibroblast cells and in Ehrlich ascites tumor cells (as well as in the case of *B. subtilis*), actinomycin D causes a rapid loss of preexisting RNA. These authors previously demonstrated that livers from rats injected with actinomycin D at a level of 1.5 mg/kg of body weight incorporated amino acid into protein normally despite a reduction of over 90% in labeling of RNA of the microsomal fraction, which contains the bulk of cytoplasmic mRNA (213). On the other hand, Staehelin, Wettstein, and Noll reported (214), and Revel et al. confirmed, that injecting actinomycin D at a level of 5 mg/kg body weight resulted in profound impairment of amino acid-incorporation into ribosomes and microsomes. But inhibition of isotope incorporation into microsomal RNA is only slightly greater at the higher concentration. In fact, at 3 mg actinomycin D/kg body weight, microsomal RNA synthesis is 99% inhibited, but C^{14}-leucine incorporation by the microsomal fraction *in vitro* is not reduced. This situation is analogous to that of the study of Levinthal et al. compared to Hurwitz et al. in *B. subtilis* (see page 103).

It was further shown that although homogenates of liver from rats treated with 5 mg actinomycin D/kg body weight showed fewer polyribosomes, the intact liver cell examined by electron microscopy showed an abundance of polyribosomes. Similarly, the homogenates produced preparations with large numbers of free ribosomes, whereas the intact cells under electron microscopy showed very few free ribosomes. The authors contend that the alterations of structure were induced by homogenization of the actinomycin D treated livers rather than as a direct result of the actinomycin D treatment itself. Furthermore, if intact slices and microsomal suspensions were made from livers of animals receiving actinomycin D at the 5 mg/kg level, it was found that C^{14}-leucine incorporation was unimpaired in the slices but was markedly reduced in the microsomal preparation. In slices prepared 11 hr after injection when orotic acid-incorporation into RNA was about 80% inhibited, leucine incorporation into protein was virtually uninhibited, but when microsomes were made from the same tissue, leucine incorporation was inhibited 65%. In slices prepared 2.5 hr after actinomycin D injections there was virtually no inhibition of orotic acid or leucine incorporation, but when microsomes were prepared and tested there was about 40% inhibition.

Honig and Rabinovitz presented conclusive evidence that the defect in protein synthetic ability induced by actinomycin D in sarcoma 37 ascites cells was not due to a deficiency of template RNA (215). Actinomycin D at a concentration of 1.7 μg/ml inhibited incorporation of C^{14}-lysine into all protein fractions including cytoplasmic and nuclear components. It was found that glucose could prevent such inhibitions and if the inhibition was allowed to develop such that less than 20% of the protein synthetic ability remained after 90 min incubation, the simple addition of glucose immediately restored incorporating ability to a level of about 110% compared to the uninhibited control. Clearly, the inhibition caused by actinomycin was not due to a deficiency in content of RNA templates. As a matter of fact, RNA synthesis was concomitantly effected by actinomycin D in the presence or absence of glucose but the effect on protein synthesis was mediated by glucose and not by the level of RNA. A somewhat related situation was described by Laszlo, Miller, McCarty, and Hochstein (216). These authors showed that actinomycin D, in addition to inhibiting RNA and protein synthesis in human leukocytes, also inhibited respiration, anaerobic glycolysis, and the maintenance of ATP levels. That these effects on energy metabolism were not simply due to a loss of templates for the synthesis of key enzymes was shown by the fact that protein synthesis could be deliberately blocked with puromycin with no concomitant effect on respiration. Another indication that the mode of action of actinomycin D on protein synthesis need not be the direct result of its effect on template RNA synthesis is seen in a report by Korn, Protass, and Leive (217). These workers showed that actinomycin D can block the development of T_4-phage progeny in cells of *E. coli* made sensitive by treatment with EDTA,* but RNA synthesis was not measurably inhibited.

These last several cases are surprising only because from the early sixties so much of our thinking has been in terms of RNA. Actually, papers in the literature prior to that time described effects of actinomycin on cellular respiration. Porro and Cima showed that actinomycin D at a level of 7 μg/ml inhibited the oxidation of several Krebs cycle intermediates in tissue slices from kidney and spleen (218). A series of three papers on the mode of action of actinomycin by Horini demonstrated a profound effect of the antibiotic on succinoxidase activity in slices of various rat tissues and in human lymph nodes (219).

Actinomycin D, however, has been extensively used with the assumption that it acts specifically against DNA-dependent RNA-polymerases. Not only has this been true in the recent literature (since 1961) but it is also true in the current literature.

The "Informosome," a 45s particle and the Measurement of Messenger RNA

Several studies since 1964 have been concerned with a cytoplasmic subribosomal particle which, it has been claimed, contains messenger RNA. Unfortunately, mRNA has not been sufficiently well characterized that it can be unambiguously measured or even identified. The various investigators have taken one or more criteria that they believe are sufficiently characteristic of mRNA and measured the type of RNA that fits these criteria.

1. Rapidity of labeling with radioactive precursors.
2. A DNA-like base composition.

* Ethylene diamine tetraacetate, a metal binder.

3. The ability to form hybrids with DNA.

4. The ability to stimulate amino acid-incorporation in an acellular ribosomal system, usually from *E. coli.*

None of these criteria, however, may be considered to be sufficiently representative of mRNA. In *E. coli* and in yeast most of the rapidly labeled RNA (including a DNA-like component) appears to be ribosomal precursor. We have already discussed the fact that since mRNA is believed to copy only a single strand of DNA a similar base composition to native DNA in the sense of G = C and A = U would not be expected. Similarity to a particular DNA in terms of G + C composition implies only that a particular DNA served as a template for the RNA (see page 98).

Studies with inhibition of RNA synthesis by actinomycin D have led to the conclusion that virtually all RNA is synthesized on a DNA primer. It has been specifically shown that ribosomal RNA, which is a prime example of non-DNA-like RNA, is hybridizable with DNA. This is also true for sRNA. Therefore under the proper conditions of testing all RNA should be capable of hybrid formation with DNA and so this test for identification of mRNA would have to be somewhat ambiguous in a homologous system.

Recent studies have shown that in addition to synthetic template RNA, ribosomal RNA and sRNA, when heated in the presence of kanomycin, function as good templates for amino acid-incorporation into protein in a ribosomal system from *E. coli* (220). Therefore this test may be a measure of the secondary structure of RNA rather than of biological function. At any rate, it is not a specific test for mRNA. In the discussion which follows, the use of the term mRNA means that the substance discussed has met one or more of the above cited criteria.

Ribonucleoprotein (RNP) particles with a sedimentation rate of approximately 45s have recently been found in a variety of cell types. Thus Spirin, Belitsina, and Ajtkhozhin described such particles in fish embryos (221). Evidence of mRNA was by criterion No. 4. Spirin and Nemur reported on a group of similar particles from sea urchin embryos with values of 20 to 65s (222). The identification of mRNA was by criteria Nos. 3 and 1. It was found, however, that although the mRNA was located in the light RNP and light polysome classes, the heavy polysomes which were responsible for most of the incorporation of C^{14}-leucine possessed a markedly deficient content of this type of mRNA. Although the RNA they identified as messenger was not shown to function as a carrier of genetic information, the authors, on the basis of this assumption, used the term "informosome" to describe the light RNP particles. ("Informosome" was originally coined by Spirin et al., reference 221.) The inactivity of these particles in protein synthesis was taken as an indication of a means of possible cytoplasmic regulation of protein synthesis by suppressing the use of the information contained in these particles.

Girard, Latham, Penman, and Darnell found that after a 30-min labeling of HeLa cells with H^3-uridine the newly synthesized RNA was contained in a cytoplasmic particle which sedimented at 45s (223). On longer incubation radioactivity also appeared in a 60s particle. The isolated RNA from the 45s RNP unit was at least 80% in the form of the 16s variety, whereas the RNA from the 60s unit was both 28s and 16s. The normal 74s ribosomes could also produce 60s and 45s subunits by dissociation in a cytoplasmic extract of low magnesium concentration. By tracing the appearance of radioactivity from radioactive uridine into the 16s and 28s varieties of RNA of the 45s, 60s, and 74s RNP particles, the authors concluded that the smaller particles represented ribosomal precursor stages. It was also concluded that the new 74s ribosomes went directly to the polysome stage without equilibrating with the cell pool of free 74s ribosomes.

Joklik and Becker confirmed the presence of 74s, 60s, and 40s RNP particles in HeLa cytoplasm (224). The 74s unit possessed 28s and 16s RNA, whereas the 60s possessed only 28s and the 40s particle contained only 16s. The 74s ribosomes in the 1.5 mM magnesium concentration used in these studies did not dissociate into 60s and 40s subunits and the authors felt that the origin of the 60s and 40s units encountered in the HeLa cell cytoplasm was not from the breakdown products of normal ribosomes. By following the location of new RNA synthesized after a pulse of C^{14}-uridine it was found that the appearance of radioactivity in the 40s particles preceded that in the 60s and 74s particles and the 60s particles received radioactivity prior to the 74s particle. At all times after pulsing, however, 80% of the label in the cytoplasm was found in the heavy pellet at the bottom of the centrifuge tube. It was also found that newly formed 74s ribosomes seemed to appear in polysomes before they were found in the free pool of 74s ribosomes, as though the 60s and 40s units were assembled into 74s ribosomes on the mRNA. These latter observations confirmed those of Girard et al. just described. Another interesting finding was that puromycin caused newly made ribosomes to leave the polysomes, and when the puromycin was removed, the just-released ribosomes went back to the polysomes without pooling with the free ribosomes. A role of cell structure in the maintenance of separate pools may have been involved.

In an accompanying paper, Joklik and Becker studied polysome formation in HeLa cells infected with Vaccinia, a DNA-containing virus which leads to a rapid cytoplasmic synthesis of virus mRNA uncomplicated in early time points by host mRNA synthesis (225). At the earliest time points (1 min) the newly synthesized RNA was in the heavy pellet, 45s, and 16 to 20s regions. On slightly longer incubation the proportion of radioactivity in the 16 to 20s region decreased and the proportion in the 70 to 85s region increased. The authors believed that the label appeared first as free mRNA (16 to 20s), then attached to a subribosomal particle (40s), and then, in combination with a 60s particle, made a ribosome (or ribosomes) attached to mRNA. A brief labeling period followed by a chase with unlabeled uridine in the presence of actinomycin D (2 μg/ml) led to a loss of radioactivity from the 40 to 45s region and an appearance in the pellet (identified as polysomes). Therefore, it was interpreted that the 40 to 45s particle is a direct precursor to polysomes. In a pulse experiment with uninfected cells, most of the newly made RNA was found in the polysomes and the rest was spread throughout a sucrose gradient (not necessarily located with the 40s particles).

In a previous study of rat liver cytoplasmic RNA Revel and Hiatt concluded that the bulk appears to be stable (213). Later, Henshaw, Revel, and Hiatt reported on a study of a rapidly labeled variety of RNA in rat liver cytoplasm (226). Rats were injected with C^{14}-orotic acid 40 min before sacrifice and isolation of liver fractions. The major fraction of newly synthesized RNA in cytoplasm, either treated with deoxycholate or not, was in a particle with a 45s sedimentation value. This particle readily adsorbed added poly-U which may indicate an affinity for natural mRNA. Although most of the pulsed radioactivity was localized in the 45s particles and the specific ability to stimulate amino acid-incorporation in an *E. coli* ribosomal system was similar for the RNA from membrane bound ribosomes and 45s particles, the bulk of such stimulating activity resided in the membrane-bound fraction. The authors cautioned that rapid uptake of radioactive material may not be used as a parameter of total cytoplasmic mRNA. In effect, parameter No. 1 (page 119) is ambiguous. The authors cautioned further that whether the RNA in the 45s particle includes messenger cannot be answered unequivocally until a practicable, specific assay for mRNA becomes available. The 45s particle contained exclusively 18s RNA and this RNA qualifies as messenger by criteria Nos. 1 and 4 (page 119). Somewhat heavier particles were found in the nucleus and these contained highly labeled 18s RNA. The authors believed these particles to be the precursors to the cytoplasmic 45s particles, but there was no evidence to support this assumption. In conclusion, the authors stated, "The presence of rapidly labeled RNA with many characteristics of messenger in a tissue in which the bulk of the messenger fraction is believed to be stable warrants further comment." Unfortunately, no resolution of this quandary was suggested.

McConkey and Hopkins also studied rapidly labeled RNA in HeLa cells (227). They confirmed the existence of 18s RNA in the 45s particle and 18 to 28s RNA in the 62s particle in agreement with Girard *et al.*, but in conflict with Joklik and Becker. McConkey and Hopkins sought to identify mRNA by use of criterion No. 3. The authors believed that the 45s particle represents the small subunit of the 74s ribosome with attached mRNA and contains 18s ribosomal RNA plus mRNA of the 15 to 16s class. From the relative efficiencies of hybridization for the RNA's of the polysomes, 45s and rapidly labeled RNA, the authors concluded that only a small number of 45s particles are carrying mRNA at any one time. This leads to the dilemma of why the basic 45s particles with or without attached mRNA have the same sedimentation coefficient.

Shatkin, Sebring, and Salzman reported that in HeLa cells infected with Vaccinia virus newly synthesized RNA has a DNA-like (G + C content) base composition and is found both in polysomes and in a broad band from 30s to 74s in a sucrose density gradient (228). When such cells are exposed to actinomycin D at 5 μg/ml, most of the DNA-like RNA is degraded from the polysome region but that in the 34 to 74s region is unchanged. Protein synthesis is decreased by 90%. It seems that the lighter material was not being used for protein synthesis and was protected from degradation.

Spirin has presented a theory based on the new cytoplasmic RNP particle which he has called an informosome. Faced with the problem that in several systems, protein synthesis does not seem to be at all dependent on the formation of new mRNA, he proposes that the mRNA is not new and unstable, but rather old, stable, and masked with overlying protein. This mRNA protein mask unit he calls "informosome." It is once again interesting to trace the development process of such an idea. Before 1961 it was thought that the information for protein synthesis was contained in a stable form in the ribosomes. In order to satisfy the situation where rapid changes occurred in the cell's demand for particular proteins, such as in enzyme induction in *E. coli*, it was proposed that unstable templates are rapidly formed, attached to preexistent ribosomes, utilized, and destroyed. In a large number of cases it turned out that such a continuous flow of template information did not accompany protein synthesis. This situation could easily have been explained by the original ideas employing stable information built into ribosomes. This newer information could have been considered as an argument against the mRNA concept. On the other hand, the original mRNA concept, which had for its birth a situation in which a supply of new information was considered necessary, could be altered so that it now contains old mRNA held in a masked form in a RNP particle and is stable. Conceptually, this is different than the stable ribosome idea and future research may establish it as true. The point of interest, however, is how we can retain a popular theory by completely reworking it even to the extent of getting back to a situation that could be easily explained under an earlier theory. Spirin feels that in a nondifferentiating cell the 45s particle represents the transit form of mRNA from the nucleus to the cytoplasm, but that in a differentiating system a series of particles exists ("informosomes") which consists of a series of different mRNA's and masking proteins. An informosome may be "unmasked" as such or it may be incorporated into an inactive polysome, and when unmasking occurs the polysome will be activated. The possibility that the 45s particle represents the smaller subunit of the 74s ribosome associated with a piece of mRNA, as concluded by Darnell, Girard, Latham, and Penman and by Joklik and Becker, seems to be unlikely according to Spirin since he would expect such a combination to produce a particle with a greater sedimentation velocity than 45s.

RULES OF PUNCTUATION AND GRAMMAR
FOR READING THE GENETIC CODE

The Direction of Reading

Eikenberry and Rich treated reticulocyte polysomes with each of two specific phosphodiesterases, namely spleen phosphodiesterase which attacks the 5'-OH end of RNA, and venom phosphodiesterase, which destroys the 3'-OH end (230). Subsequent acellular incorporations of tyrosine revealed that destruction of the 5'-OH end, presumably of mRNA, led to a moderate decrease in the incorporation of H^3-tyrosine and a decreased production of c-terminal peptides. Venom diesterase, on the other hand, increased the ratio of c-terminal to n-terminal peptides. Since proteins seem to be built from the n-terminal to the c-terminal end (see pages 86–87), these results are compatible with the mRNA being read from the 3'-OH to the 5'-OH end.

This same question was met with a different answer in another approach. In an earlier study Salas, Smith, Stanley, Wahba, and Ochoa employed *E. coli* ribosomes and *L. arabinosus* supernatant of low nuclease activity (231). Poly-A nucleotides terminated by a single C produced peptides of lysine with asparagine at the c-terminal end. These results indicated that AAC is the codon for asparagine and that the message is read from the 5' to the 3' end of the polynucleotide chain since the terminal C was at the 3′ end. In a more recent paper, Smith, Salas, Stanley, Wahba, and Ochoa tried to check the theory by producing N-terminal asparagine polylysine peptides with a mRNA starting with AAC at the 5′ end (232). It was found that very little asparagine (or threonine with the codon ACA) was incorporated relative to lysine if an A_2CA_n type of polymer was used. With A_5CA_n polymers, much more asparagine was incorporated and there was some indication of its being in the N-terminal position. With A_4CA_n polymers, more threonine relative to asparagine was incorporated. These results indicated that the first three A nucleotides were used for "threading" the ribosome and were not read themselves. The authors felt that these results provide conclusive evidence that the message is read from the 5′ to the 3′ end. Support for this view comes from studies of Terzaghi, Okada, Streisenger, Tsugita, Inouye and Emrich (233), and Thach, Cecere, Sundararajan, and Doty (234), the latter authors showing that A_3U_3 is read as lysyl phenylalanine.

Colinearity

If the genetic material consists of a linear sequence of nucleotides specifying a series of proteins, and the order of amino acids within these proteins; one may ask if in the final translation of this information a linear sequence of amino acids and ultimately proteins is produced which corresponds to the order in which this information is stored in the gene structure. This is the problem of colinearity. For a number of years Yanofsky and coworkers at Stanford University had been studying the genetic control of the synthesis of tryptophan synthetase in *E. coli*. This enzyme activity is carried out by two separable protein components designated A and B, each under the control of its own gene. A series of 16 mutants of the A gene were carefully mapped by a combination of the techniques of recombination frequencies, deletion mapping, and three point crosses. Amino acid substitutions had been detected in each of the A proteins of the 16 mutants and the locations of these substitutions had been pretty well specified by conventional fingerprint techniques. It was found that the positions of the amino acid replacement in the A protein were in the same relative order as in the order of the mutationally altered sites of the corresponding mutants in the A gene. There was even a fair degree of agreement of relative distances between mutations in the genetic map and between residues in the peptides of the A protein. This study reported by Yanofsky, Carlton, Guest, Helinski, and Henning provides strong evidence in support of the concept of colinearity between gene structure and protein structure in this system (235).

In a similar but less exact, although no less elegant, approach to the problem, Sarabhai, Stretton, Brenner, and Bolle arrived at the same conclusion by analyzing a series of Amber (page 131) mutants of *E. coli* bacteriophage T_4 head protein (236). Although the actual amino acid sequence of the head protein is not known, it is possible to line up the mutants in terms of the increasing number of common peptides that can be isolated from either a tryptic or chymotryptic digest. There was general agreement between the order for 9 out of 11 mutants arranged on this basis and the order based on genetic mapping techniques.

Start

Marcker and Sanger demonstrated that S^{35}-methionine, when incorporated into sRNA in an *E. coli* supernatant fraction system, could be recovered from the sRNA in the form of its N-formyl derivative by mild alkaline hydrolysis (237). It was shown by its sensitivity to RNAase that the formylation required the intactness of RNA, and it seems that methionyl-sRNA rather than free methionine is the formyl group acceptor. The N-formylmethionyl-sRNA was detected in normally growing *E. coli* and yeast but not in rat liver or hen oviduct. In an *E. coli* ribosomal system it was shown that S^{35}-labeled N-formylmethionyl-sRNA was a better precursor for incorporation into protein than was S^{35}-methionyl-sRNA. Using purified transformylase isolated from *E. coli*, Marcker found that the formylation of methionyl-sRNA was dependent on the addition of 10-formyl tetrahydrofolic acid (238). Methionyl-sRNA synthetase was not involved in the formylation of methionyl-sRNA. It was also established that at least two species of methionine acceptor sRNA exists in *E. coli*. One allows formylation of its bound methionine and the other does not. It was firmly established that formylation occurs with the methionyl-sRNA derivative and not free methionine. It was suggested that formylmethionyl-sRNA might function as a chain initiator in protein synthesis. Clark and Marcker (239) showed that both methionyl-sRNA and N-formyl methionyl-sRNA responded to the triplets UUG and AUG in the ribosomal binding assay of Nirenberg and Leder (240). When incorporation into ribosomal protein was tested in a system containing methionyl-sRNA and N-formyl methionyl-sRNA, it was found that poly-UG (1.6/1) caused a marked stimulation. Analysis of the TCA-insoluble product showed that the formylated methionine accounted for 80 to 90% of the total incorporation. Therefore it was concluded that only methionine which can be formylated, responds to poly-UG, and a code word using U and G is believed to provide the signal for incorporation.

Adams and Capecchi demonstrated the transfer of the formyl group from N^{10}-formyl-tetrahydrofolic acid (formyl THFA) to methionyl-sRNA in an *E. coli* supernatant fraction system (241). No formylalanine could be detected in this system despite the fact that both alanine and methionine are prominent N-terminal amino acids found in *E. coli* proteins. By use of RNA from R-17 virus (which is related to f-2 virus) as messenger with an *E. coli* ribosomal system and H^3-formylmethionyl-sRNA or simply H^3-formyl-THFA, a rapid incorporation of formylmethionine, apparently into viral coat protein, resulted. This presented a paradox because R-17 coat protein has alanine in the N-terminal position. By labeling the coat protein with C^{14}-alanine and H^3-formyl-THFA or C^{14}-methionine and H^3-formyl-THFA, followed by pronase digestion and electrophoretic fractionation, it was found that alanine occurred next to N-formylmethionine in the protein. The authors postulated that an enzyme must exist which could cleave the N-formylmethionine from the viral proteins and perhaps also split the N-formyl group away from *E. coli* proteins.

Webster, Engelhardt, and Zinder presented evidence that in the *in vitro* synthesis of f-2 phage coat protein by *E. coli* ribosomes under the direction of f-2 RNA, the coat protein, which normally has a free amino terminal alanine, is masked by an N-formylmethionine residue (242). When the nonsense mutant (see page 130) SuS 4A RNA was used, an incomplete portion of the N-terminal sequence of coat protein appeared to be formed and this fragment too was terminated by the N-formylmethionylalanyl sequence. These observations correspond to those of Adams and Capecchi (241) regarding the initiation of phage protein by an N-formylmethionyl group which is destined for subsequent removal from the completed peptide chain.

Nakamoto and Kolakofsky, using an *E. coli* ribosomal system at a magnesium concentration of 0.008 M, which is far below optimal for the poly-AGU-stimulated uptake of phenylalanine, showed that N^5-formyltetrahydrofolic acid markedly enhanced the stimulatory effect of the polynucleotide for phenylalanine incorporation. When N-formyl-C^{14}-methionyl-sRNA was used in place of the formylating system, it was found that a stimulation of phenylalanine uptake occurred only at low magnesium concentration (0.008 M), whereas at higher magnesium concentration (0.016 M) the system was quite active without added formylmethionyl-sRNA. Similarly, the system was dependent on formylmethionyl-sRNA addition at high temperature (45°) but not at low temperature (25°). It was found that peptidyl-sRNA (diphenylalanyl-sRNA) was also an effective stimulator of C^{14}-phenylalanine incorporation and that its effectiveness was also more marked at 45° than at 37° (243). Two major conclusions were drawn from these studies, both of which are questionable:

1. The fact that low magnesium concentration and high temperature enhance the apparent effectiveness of the chain initiators means that the ribosomal messenger complex is unstable in the absence of a particular type of stabilizing or initiative codon. It is known, however, that changing magnesium concentration and temperature will modify the incorporation response for a large variety of amino acids believed to be incorporated into the interior of chains (see page 133) and so a unique interpretation in terms of chain initiation is not a straightforward conclusion.

2. Since the peptidyl-sRNA also behaved like N-formylmethionyl-sRNA, it was concluded that the effectiveness of the formylated initiator resided in its similarity to peptidyl-sRNA; but, according to current concepts, the peptidyl-sRNA does not act as an initiator by binding to a messenger ribosome complex. It is formed by a stepwise process starting with a free amino acyl-sRNA.

In a study involving the effect of synthetic block nucleotides[11] on the binding of specific amino acyl-sRNA's to ribosomes Thach and Sundararajan concluded that in their *in vitro* system chain initiation could occur with little specificity regarding the trinucleotide codon (244); that is, trinucleotides at the 5′ end, the 3′ end, and the interior seemed to be effective for causing the binding of amino acyl-sRNA to the ribosomes and various different sequences of U and/or A were capable of competition to create a binding site for amino acyl-sRNA. The authors suggested that perhaps some distinctive feature of the *in vivo* system which controls chain initiation was absent in this system. A general discussion of N-formylmethionine as a chain initiator can be found in a recent paper by Noll (245).

A recent paper by Yegian, Stent, and Martin has a distinct bearing on the question of the natural occurrence of masking groups of amino acyl residues which are attached to sRNA (246). Using auxotrophs of *E. coli* (that require some particular metabolites for growth), they determined the total sRNA amino acid acceptor ability for each amino acid. Then by use of periodate oxidation which destroys nonamino acylated-sRNA they measured the percentage of sRNA which was acylated for each amino acid. During exponential growth more than 70% of the individual sRNA's were found to be bound to their corresponding amino acids. When mild alkaline hydrolysis was used to liberate the amino acids and a quantitative determination of the amino acids was performed, it was found that in most cases much more acyl group was removed from the sRNA than could be accounted for as free amino

---

[11] A group of nucleotides of known sequence produced by synthetic procedures.

acid released. A subsequent strong acid hydrolysis released additional amounts of alanine, glutamic acid, leucine, methionine, threonine, and valine, and less of aspartic acid, glycine, phenylalanine, and proline, but very little, if any, isoleucine, serine and tyrosine. Dowex column chromatography was used to isolate amino acid derivatives with blocked amino groups. Such were found for glutamic acid and methionine and to a lesser extent for alanine, aspartic acid, serine, and threonine. Hydrazinolysis was used to release amino acids from N-acyl bonds. Glycine and methionine were the two main products of this reaction. The possibility that some of the glutamic acid derivative may be pyrolidone carboxylic acid was supported by the generation of additional glutamic acid through medium-strength acid hydrolysis. Very mild acid hydrolysis which would be capable of removing N-formyl groups, generated the free amino acid predominantly from methionine and to a much smaller extent from a few of the other amino acids.

The authors thus confirmed that an appreciable fraction of methionine exists as N-formylmethionine attached to sRNA. In addition, their results indicate the possible occurrence of N-formyl derivatives of glycine, serine, asparagine, and glutamine. Furthermore, the possible normal occurrence of pyrolidone carboxylic acid linked to sRNA raises the question of whether it too might function as a chain initiator. The possible explanation of the cryptic[12] amino acids being present as peptides linked to sRNA was discounted because far too much of the amino acids appeared in this bound form than could be expected from the total 2% of all existing RNA believed to be carrying nascent peptide chains. Not only do these findings question the concept of N-formylmethionine as the *sole* initiator of peptide chain synthesis, but they suggest that other acylated forms of the amino acids may be playing a role in protein synthesis[13] (see page 300 for lipoamino acids).

Pause (Polarity and Modulation)

It frequently happens that for a series of enzymes that catalyzes a related metabolic sequence the genetic material specifying its structure is aligned in a contiguous manner. A classic example is that of the enzymes involved in histidine biosynthesis (247). A group of mutants has been encountered that has lost one or another of these enzymes. In addition to the loss of one particular enzyme, all enzymes that map in the genes on the side of the missing enzyme away from the operator

[12] Amino acids present as some chemical derivative.
[13] Later developments in this field have shown the involvement of initiation factors or enzymes. See *Nature,* **214,** 759 (1967).

are present in decreased amounts. This situation is called polarity. It has been determined that all the enzymes that are decreased in amount are decreased by the same fraction, a situation known as coordinate repression. In a sense the translation of the genetic information has been modulated from one point in the sequence. Ames and Hartman proposed that one enormous string of mRNA (polycistronic message) contains the information for the whole series of enzymes and that ribosomes attach at one end and sequentially translate the information into the enzymes one-by-one. They further suggested that particular codons might be modulating triplets in the sense that they attract an inefficient sRNA that might either cause ribosomes to fall off the messenger or to slow down the rate of translation of the message. In this manner the basis of degeneracy in the code is to control the rates of synthesis of various enzymes through modulation. Ames and Hartman pointed out that this idea relating degeneracy of the code to differences in the rates of enzyme synthesis was first suggested by Itano (248).

More recent views on modulation discard the concept of modulating triplets and altered sRNA (249). It is felt that the existence of nonsense triplets combined with the idea that a "dislocated ribosome," after passing a nonsense triplet, will have to be properly realigned on the mRNA by an initiating triplet is a sufficient basis for explaining the phenomenon. Further modulation is thus provided by the efficiency of the starting codon. This idea assumes several different codons for initiation, whereas at this time (1967) one predominant view considers that all peptide chains are started by the codon UUG signifying N-formylmethionine.

Stop (Nonsense and Suppression)

A nonsense codon was recognized when Weigert and Garen studied several revertants of *E. coli* which yielded alkaline phosphatase positive mutants from phosphatase negative mutants (250). It was thought that the cause of the deficiency in enzyme formation was the presence of a nonsense coding triplet in the messenger for this enzyme. It had been previously shown that a mutation at another locus (a suppressor mutation—discussed below) could cause the formation of active enzyme and also the substitution of a serine for a tryptophan in the enzyme molecule. It was decidedly of interest that in all of the revertants the same tryptophan residue was replaced by one of six different amino acids. If the inactivity is due originally to a nonsense codon at the tryptophan location that is related to the codons of all its possible substitute amino acids by a single nucleotide change, the only triplet compatible with this restriction would be UAG, which must be a nonsense codon.

Brenner, Stretton, and Kaplan proposed that nonsense triplets would constitute the natural message for chain termination (251). In certain instances this triplet could occur in the middle of a peptide chain and abort the completion of the chain. If another mutation would alter the codon recognition site on an sRNA so that it could read the nonsense codon as meaningful, a particular amino acid would be inserted and the chain completed. This second type of correcting mutation is called a suppressor mutation (Su$^+$). By examination of a large number of cases of suppression of *E. coli* phage rII mutants by *E. coli* Hfr it was found that suppressors could be divided into two classes. One class would suppress a given group of mutants and have no effect on others. This group of suppressors was arbitrarily called Amber. Another class of suppressors was found, however, which could suppress the mutations of the group not affected by the Amber suppressors. This latter class of suppressors also suppressed various of the mutants corrected by the Amber suppressors and is referred to as the Ochre suppressors. It would seem that at least two different types of nonsense mutation were involved. Ochre suppressors in general are weak in contrast to Amber suppressors. From the discovery that Ochre mutants could be easily converted to Amber mutants either spontaneously or by treatment with the mutagen, 2-amino purine, the authors concluded that the Amber and Ochre codons are related by a single nucleotide change. By a somewhat involved use of various mutagens, amino acid substitution data, known codon assigments, and logic it was decided that UAG corresponds to the Amber triplet and UAA the Ochre triplet. The authors suggest that particular sRNA's not containing amino acids but rather some chain-ending and peptidyl-sRNA-cleaving substance recognize these triplets and release the peptide chain from the template. Suppressor mutants, on the other hand, allow the chain to be completed. It was stated that it would be most undesirable for a suppressor also to suppress the normal messages for chain termination. This complication was not resolved.

The conclusion that a triplet codon exists for chain termination and release of the growing peptide is also indicated by the results of Bretscher, Goodman, Menninger, and Smith (252). These workers separated sRNA peptides from free peptides by either sucrose gradient centrifugation or an adsorption technique. By use of a variety of synthetic polynucleotide templates and an acellular *E. coli system* it was found that the poly-UA template produced the highest percentage of sRNA-free peptides, and so it was concluded that a triplet codon to terminate peptide growth and cause its release both from the ribosome and sRNA must contain A and U.

The theory of nonsense mutations then includes the abortion of peptide synthesis due to the formation of a triplet codon which signals chain completion. The theory of suppression states that a second mutation effects the formation of the codon recognition site for a particular sRNA. The altered site was originally related to the nonsense codon by two out of three bases so that the affected sRNA could now recognize the nonsense triplet and make sense of it by substituting its amino acid for the one originally intended in the wild strain. That suppression really is exerted by the sRNA is supported by the work of Engelhardt, Webster, Wilhelm, and Zinder (253). They showed in an *in vitro* system consisting of an amino acid-incorporating system from either Su^+ or Su^- *E. coli*, RNA from a nonsense mutated f-2 phage, and sRNA from either the Su^+ or Su^- strain, that the addition of sRNA from the Su^+ strain could largely repair the difficulty of incorporation evidenced by the system of Su^- ribosomes and Su^- sRNA and that the amino acid inserted under these conditions was serine. The same general picture for the mechanism of chain termination and peptide release from sRNA is supported by the work of Ganoza and Nakamoto (254). They devised a rapid method for separating peptidyl-sRNA from free peptides by adsorption on ECTEOLA-cellulose. They found that with most synthetic polypeptide messengers most of the peptides produced were bound to sRNA when removed from the ribosomes. With natural messengers such as f-2 RNA and *E. coli* endogenous RNA and with synthetic messengers of the type AU (3.3:1), most of the nascent peptide was unattached to sRNA. These results indicate that a particular codon is required for peptide release and that this codon is present in endogenous messengers and in the synthetic messenger of the AU type.

Misreading

Davies, Gilbert, and Gorini found that streptomycin caused mistakes in the translation of synthetic polynucleotide messengers by drug sensitive *E. coli* ribosomes (255). Thus poly-U could be translated predominantly as isoleucine rather than phenylalanine. Such errors in translation could be further elicited by changes in the magnesium or ammonium ion concentrations. Streptomycin-resistant ribosomes were not so affected by streptomycin but were sensitive to similar misreadings by the presence of the antibiotics kanomycin and neomycin. These alterations of translation also occurred with messengers not containing U. Thus in the presence of poly-CA streptomycin depressed the incorporation of proline and enhanced that of histidine; kanomycin enhanced the incorporation of threonine, proline, and histidine; and neomycin B enhanced the incorporation of threonine more than did kanomycin but increased the incorporation of proline and histidine less than did kanomycin.

Because such changes in environmental conditions produce marked changes in code translation, the authors questioned the validity of code assignments arrived at by *in vitro* incorporations. They further suggested that ribosomes may participate in translation and that suppressor mutations may be explained by ribosomal alterations as well as by alterations in sRNA.

Misreadings caused by temperature changes and the presence of various cations and polycations including magnesium, spermine, spermidine, and dihydrostreptomycin were reported by Friedman and Weinstein (256). These conditions effected the translation of poly-U into phenylalanine, leucine, isoleucine, serine, valine, and tyrosine. Coding properties of poly-UG and poly-UC were also affected. Thus alterations of magnesium concentration and temperature allowed poly-UG either to code for arginine better than for glycine or to prefer glycine to arginine by 30 to 1. Once again doubts were expressed of the validity of the existing *in vitro* code assignments and ideas were expressed relating these findings to a role for ribosomes in suppression phenomena.

The pH and concentration of sRNA were shown by Grunberg-Manago and Dondon to be factors also capable of influencing the fidelity of translation of polynucleotide messengers in an *E. coli* ribosomal system (257). The authors felt that it would be difficult to explain the effects of magnesium and pH in terms of codon-anticodon binding affinities and therefore suggested that an influence on ribosome configuration is a more plausible explanation.

One explanation of miscoding has been that when two out of three bases in a codon are in common for different amino acids a slight alteration in ribosome configuration would allow sufficient binding for translation to occur. It is of particular interest that some cases of miscoding involve codons in which two out of three of the bases are dissimilar. Thus Friedman and Weinstein have shown that poly-UG and poly-UGA, in the presence of dihydrostreptomycin, can code for proline (258). The codon assignments for proline include CCC, CCU, CCA and CCG. Therefore it would seem that miscoding must be extended to cases in which only one out of the required three nucleotide bases are in common.

Szer and Ochoa, using an *E. coli* ribosomal system, showed that poly-U could be induced to code for leucine by lowering the temperature of incubation from 37 to 20° (259). Raising the temperature to 45° further enhanced the incorporation of phenylalanine relative to leucine. Similarly, altering the magnesium concentration from 0.014 M to 0.022 M greatly enhanced the translation of poly-U into leucine incorporation. These two variables could also be manipulated to favor incorporation of isoleucine, tryosine, and serine. Grunberg-Manago and Michelson, in a similar system, showed that by manipulating pH from 7.97 to 7.43 the incorporation of leucine relative to that of phenylalanine could be decreased by a factor of two (260).

So and Davie reported that organic solvents in low concentrations could significantly enhance the extent of amino acid-incorporation into poly-U or poly-C-directed *E. coli* ribosomes (261). This ability was markedly influenced by both magnesium concentration and polynucleotide concentration. By manipulating these two variables, extensive miscoding could be induced so that at a concentration of 100 μg/ml of poly-U and 0.011 M magnesium acetate, the incorporation response was predominant for phenylalanine with no added ethanol, predominant for leucine from 0.64 to 1.28 M ethanol, and predominant for isoleucine above 1.28 M ethanol. A similar type of misreading response, although of a lower order of magnitude, involved proline, leucine, and threonine incorporation directed by poly-C. In a later paper So, Bodley, and Davie added ammonium ion and urea to the list of environmental factors that perturb the coding mechanism (262). They found that the ability of both streptomycin and ethanol to enhance poly-U-directed isoleucine incorporation was additive and independent. As an example of the relevance of these findings to current code assignments, they showed that with poly-UA (5:1) the calculated triplet frequencies for phenylalanine, leucine, isoleucine, and tyrosine, respectively, were 100, 20, 24, and 20, which are in agreement with incorporation data obtained at low magnesium concentrations (see page 89). By raising the magnesium and polymer concentrations, however, the observed incorporation ratios change to 100, 200, 300, and 30, respectively. The authors warned that code assignments based upon *in vitro* experiments are valid only when environmental conditions reflect those found *in vivo*. In this connection, and in view of the major theme of this monograph, it is important to consider what influence could be exerted on the over-all system by attachment to membranes which would provide a nonpolar environment. Sarin and Zamecnik studied the charging of *E. coli* sRNA by amino acids catalyzed by *E. coli* synthetases in the presence and absence of various organic solvents (263). A given solvent would either raise or lower or not affect acceptor ability for individual amino acids; for example 5% propoylene glycol nearly doubled the uptake of leucine, cut the uptake of lysine in half, and did not affect the loading of valine, phenylalanine, or tyrosine; 0.1% toluene increased the uptake of aspartic acid by 22% but cut the uptake of lysine by 93%. Many other similar examples were given. When the ribosomal incorporating system was used, 5% ethylene glycol monomethyl ether caused a 50% stimulation of poly-U-directed phenylalanine incorporation. Although 10% ethylene glycol monomethyl ether stimulated the charging of sRNA with phenyl-alanine by 54%, this concentration of the solvent inhibited the incorpora-tion of phenylalanine into ribosomes by 40% in the presence of 40 μg/ml of poly-U. Doubling the concentration of poly-U, however, tended to reverse this inhibition. These authors examined the optical rotary disper-sion of poly-U, ribosomes, sRNA, valyl synthetase, and poly-U ribosomes in the presence and absence of 10% ethylene glycol monomethyl ether. It was found that the organic solvent influenced the magnitude of the Cotton effect in all cases. The Cotton effect is an optical phenomenon which is believed to be quantitatively influenced by the secondary struc-ture of polymers which contain chromophoric groups. The interpretation of these data was that the solvent probably affects the structures of the enzyme, the ribosomes and the sRNA. In addition, it appears that the combination of messenger and ribosomes results in an alteration in ribo-somal and/or messenger configuration perhaps leading to a more effective template. So and Davie further showed that coding responses are influ-enced by the concentration of sRNA present and also by the make-up of the free amino acid pools in the incubation medium (264); for exam-ple, supplementary sRNA added to a poly-U-stimulated *E. coli* ribosome system stimulated the incorporation of phenylalanine, inhibited the in-corporation of isoleucine, and had little or no effect on the incorporation of leucine. In the presence of ethanol, leucine reduced phenylalanine incorporation. Since some of the amino acid effects were observed with the incorporation system but not at the level of loading the sRNA, it was concluded that these effects are operating at the level of protein synthesis. It was considered that the effects of organic solvents might be explainable in terms of tautomeric shifts for some of the nucleotide bases which would result in different base-pairing behavior. In a detailed and systematic study of miscoding induced by amino glycoside anti-biotics, Davies, Gorini, and Davis reached the following conclusions. In poly-U and poly-C-directed peptide synthesis only one base at a time is misread, but in poly-A and poly-I (believed to act like poly-G) all three bases may be misread (265). Misreading in the case of the pyrimidine codons seems to involve only a nucleotide in the 5' or internal position of a triplet, but all positions are involved for the purine codons. The misreadings seem to be both the transition type (purine misread for another purine or pyrimidine for another pyrimidine) and the trans-version type (purine read for a pyridimidine and vise-versa).

The Code Dictionary

Söll, Ohtsuka, Jones, Lohrmann, Hayatsu, Nishimura, and Khorana (266) reviewed the fact that codon assignments have been made by

Table 1   Genetic code*
NUCLEOTIDE SEQUENCES OF RNA CODONS

UUU	PHE	UCU	SER	UAU	TYR	UGU	CYS
UUC		UCC		UAC		UGC	
UUA	LEU	UCA		UAA	OCHRE	UGA	?
UUG		UCG		UAG	AMBER	UGG	TRP
CUU	LEU	CCU	PRO	CAU	HIS	CGU	ARG
CUC		CCC		CAC		CGC	
CUA		CCA		CAA	GLN	CGA	
CUG		CCG		CAG		CGG	
AUU	ILE	ACU	THR	AAU	ASN	AGU	SER
AUC		ACC		AAC		AGC	
AUA		ACA		AAA	LYS	AGA	ARG
AUG	MET$_{M,F}$	ACG		AAG		AGG	
GUU	VAL	GCU	ALA	GAU	ASP	GGU	GLY
GUC		GCC		GAC		GGC	
GUA		GCA		GAA	GLU	GGA	
GUG	met$_F$	GCG		GAG		GGG	

* Nucleotide sequences of RNA codons were determined by stimulating binding of *E. coli* aminoacyl-tRNA to *E. coli* ribosomes with trinucleoside diphosphate templates. MET$_F$ corresponds to N-formyl-methionyl-tRNA$_F$, an initiator of protein synthesis. MET$_M$ corresponds to methionyl-tRNA$_M$, which responds to noninitiator methionyl-codons. UAA and UAG are terminator codons. The status of UGA is uncertain; possibly UGA corresponds to stop or terminate. Internal phosphates of trinucleoside diphosphates (X_pY_pZ) are 3',5'-phosphodiester linkages. The first position of each triplet is occupied by a nucleoside with a free 5'-terminal hydroxyl; the third position is occupied by a nucleotide with free 2'- and 3'-terminal hydroxyl groups. References: 1. Nirenberg, M. W., et al., *Cold Spr. Har. Symp. Quant. Biol.* **31**, 11 (1966). 2. Khorana, H. G., et al., *ibid.* **31**, 39 (1966). This table and legend were kindly provided by Dr. M. W. Nirenberg.

the techniques of the stimulation of amino acid-incorporation in ribosomal systems by specific and random polynucleotides and by the stimulation of binding of amino acyl-sRNA to ribosomes by specific trinucleotides (267). The latter technique, however, is not always clear cut because
of the following:

1. Some codons cause a stimulation of binding for amino acyl-sRNA
not corresponding to these codons by the incorporation technique.

2. Some codons which did stimulate binding of the corresponding
amino acyl-sRNA also stimulated the binding of other amino acyl-
sRNA's. The extent of this type of misbinding could be influenced in
some cases by altering temperature and magnesium concentrations but
in other cases this did not change the situation.

3. The codon AGG, not yet assigned, stimulated the binding of
both glycyl-sRNA and phenylalanyl-sRNA.

4. In at least five cases, codons which are assigned to particular
amino acids by incorporation studies failed to give any detectable stimulation of binding of the appropriate amino acyl-sRNA's under a variety
of conditions.

The current status of the codon dictionary was then assessed, giving
maximum confidence to those assignments that had been consistent with
both the amino acid-incorporation and amino acyl-sRNA binding techniques. Their compilation was accurate but the existing data were
incomplete.

A more recent edition of the codon dictionary, which is now virtually
complete is shown in Table 1.

REFERENCES

1. M. B. Hoagland, *Biochim. Biophys. Acta,* **16,** 288 (1955).
2. W. K. Maas and G. D. Novelli, *Arch. Biochem. Biophys.,* **43,** 236 (1953).
3. P. Berg, *J. Amer. Chem. Soc.,* **77,** 3163 (1955).
4. J. A. DeMoss and G. D. Novelli, *Biochem. Biophys. Acta,* **18,** 592 (1955).
5. M. B. Hoagland, E. B. Keller, and P. C. Zamecnik, *J. Biol. Chem.,* **218,** 345 (1956).
6. J. A. DeMoss, S. M. Genuth, and G. D. Novelli, *Fed. Proc.,* **15,** 241 (1956).
7. J. A. DeMoss, S. M. Genuth, and G. D. Novelli, *Proc. Nat. Acad. Sci.,* **42,** 325 (1956).
8. J. A. DeMoss and G. D. Novelli, *Biochim. Biophys. Acta,* **22,** 49 (1956).
9. E. W. Davie, V. V. Koningsberger, and F Lipmann, *Arch. Biochem. Biophys.,* **65,** 21 (1956).
10. H. Chantrenne, *Compt. Rend. Trav. Lab. Carlsberg,* **26,** 297 (1948).
11. P. Berg, *J. Biol. Chem.,* **222,** 1025 (1956).
12. R. S. Schweet, *Fed. Proc.,* **16,** 244 (1957).
13. G. D. Novelli and J. A. DeMoss, *J. Cell. Comp. Physiol. Suppl.,* **1,** 173 (1957).

14. M. B. Hoagland, P. C. Zamecnik, N. Sharon, F. Lipmann, M. P. Stulberg, and P. D. Boyer, *Biochim. Biophys. Acta,* **26,** 215 (1957).
15. N. Sharon and F. Lipmann, *Arch. Biochem. Biophys.,* **69,** 219 (1957).
16. A. B. Pardee, V. G. Shore, and L. S. Prestidge, *Biochim. Biophys. Acta,* **21,** 406 (1956).
17. R. D. Cole, J. Coote, and T. S. Work, *Nature,* **179,** 199 (1957).
18. B. Nismann, F. H. Bergmann, and P. Berg, *Biochim. Biophys. Acta,* **26,** 639 (1957).
19. R. W. Hendler, *J. Biol. Chem.,* **223,** 831 (1956).
20. E. F. Gale and J. P. Folkes, *Biochem. J.,* **59,** 661 (1955).
21. G. C. Webster and M P. Johnson, *J. Biol. Chem.,* **217,** 641 (1955).
22. J. W. Littlefield, E. B. Keller, J. Gros, and P. C. Zamecnik, *J. Biol. Chem.,* **217,** 111 (1955).
23. M. L. Petermann, N. A. Mizen, and M. G. Hamilton, *Cancer Res.,* **13,** 372 (1953).
24. G. E. Palade, *J. Biophys. Biochem. Cyt.,* **1,** 59 (1955).
25. G. E. Palade and P. Siekevitz, *Fed. Proc.,* **14,** 262 (1955).
26. G. E. Palade and P. Siekevitz, *J. Biophys. Biochem. Cyt.,* **2,** 171 (1956).
27. G. E. Palade and P. Siekevitz, *J. Biophys. Biochem. Cyt.,* **2,** 671 (1956).
28. E. B. Keller and P. C. Zamecnik, *J. Biol. Chem.,* **221,** 45 (1956).
29. G. E. Palade, *J. Biophys. Biochem. Cyt. Suppl.,* **2,** 85 (1956).
30. R. W. Hendler, A. J. Dalton, and G. G. Glenner, *J. Biophys. Biochem. Cyt.,* **3,** 325 (1957).
31. M. Rabinovitz and M. E. Olson, *Exp. Cell. Res.,* **10,** 747 (1956).
32. J. W. Littlefield and E. B. Keller, *J. Biol. Chem.,* **224,** 13 (1957).
33. M. B. Hoagland, P. C. Zamecnik, and M. L. Stephenson, *Biochim. Biophys. Acta,* **24,** 215 (1957).
34. K. S. Kirby, *Biochem. J.,* **64,** 405 (1956).
35. R. W. Holley, *J. Amer. Chem. Soc.* **79,** 658 (1957).
36. K. Ogata and Nohara, H., *Biochim. Biophys. Acta,* **25,** 659 (1957).
37. F. Crick, *Biochem. Soc. Symp. No. 14,* (1957), p. 25.
38. S. B. Weiss, G. Acs, and F. Lipmann, *Proc. Nat. Acad. Sci.,* **44,** 189 (1958).
39. P. Berg and E. J. Ofengand, *Proc. Nat. Acad. Sci.,* **44,** 78 (1958).
40. R. S. Schweet, F. C. Bovard, E. Allen, and E. Glassman, *Proc. Nat. Acad. Sci.,* **44,** 173 (1958).
41. P. C. Zamecnik, M. L. Stephenson, and L. I. Hecht, *Proc. Nat. Acad. Sci.,* **44,** 73 (1958).
42. M. B. Hoagland, M. L. Stephenson, J. F. Scott, L. I. Hecht, and P. C. Zamecnik, *J. Biol. Chem.,* **231,** 241 (1958).
43. J. W. Littlefield and E. B. Keller, *J. Biol. Chem.,* **224,** 13 (1957).
44. J. Mager and F. Lipmann, *Proc. Nat. Acad. Sci.,* **44,** 305 (1958).
45. F. Lipmann, *Proc. Nat. Acad. Sci.,* **44,** 67 (1958).
46. H. G. Zachau, G. Acs, and F. Lipmann, *Proc. Nat. Acad. Sci.,* **44,** 885 (1958).
47. L. I. Hecht, M. L. Stephenson, and P. C. Zamecnik, *Fed. Proc.,* **17,** 239 (1958).
48. M. Beljanski and S. Ochoa, *Proc. Nat. Acad. Sci.,* **44,** 494 (1958).
49. M. Beljanski and S. Ochoa, *Proc. Nat. Acad. Sci.,* **44,** 1157 (1958).
50. P. C. Zamecnik, The Harvey Lectures, (1958–1959), p. 256.
51. L. I. Hecht, P. C. Zamecnik, M. L. Stephenson, and J. F. Scott, *J. Biol. Chem.,* **233,** 454 (1958).
52. E. S. Canellakis, *Biochim. Biophys. Acta,* **23,** 217 (1957).

53. C. Heidelberger, E. Harbers, K. C. Leibman, Y. Takagi, and V. R. Potter, *Biochim. Biophys. Acta,* **20,** 445 (1956).
54. A. R. P. Patterson and G. A. LePage, *Cancer Res.,* **17,** 409 (1957).
55. L. I. Hecht, M. L. Stephenson, and P. C. Zamecnik, *Biochim. Biophys. Acta,* **29,** 460 (1958).
56. M. Karasek, P. Castelfranco, P. R. Krishnaswamy, and A. Meister, in *Microsomal Particles and Protein Synthesis,* Ed., R. B. Roberts, Wash. Acad. Sci., Washington, D.C., 1958, p. 109.
57. P. Castelfranco, A. Meister, and K. Moldave, in *Microsomal Particles and Protein Synthesis,* Ed., R. B. Roberts, Wash. Acad. Sci., Washington, D.C., 1958, p. 115.
58. C. Zioudrou, S. Fujii, and J. S. Fruton, *Proc. Nat. Acad. Sci.,* **44,** 439 (1958).
59. K. K. Wong, A. Meister, and K. Moldave, *Biochim. Biophys. Acta,* **36,** 531 (1960).
60. J. Monod, *Rec. des Trav. Chim. des Pays-Bas,* **77,** 569 (1958).
61. E. F. Gale, *Rec. des Trav. Chim. des Pays-Bas,* **77,** 602 (1958).
62. F. A. Kuehl, A. L. Demain, and E. L. Rickes, *J. Amer. Chem. Soc.,* **82,** 2079 (1960).
63. R. W. Hendler, *J. Biol. Chem.,* **229,** 553 (1957).
64. R. W. Hendler, *Biochim. Biophys. Acta,* **25,** 444 (1957).
65. R. W. Hendler, *J. Biol. Chem.,* **234,** 1466 (1959).
66. R. W. Hendler, *Science,* **128,** 143 (1958).
67. S. Lacks and F. Gros, *J. Mol. Biol.,* **1,** 301 (1959).
68. R. Rendi and P. N. Campbell, *Biochem. J.,* **72,** 435 (1959).
69. H. Sachs, *J. Biol. Chem.,* **228,** 23 (1957).
70. A. Tissières, J. D. Watson, D. Schlessinger, and B. R. Hollingworth, *J. Mol. Biol.,* **1,** 221 (1959).
71. E. T. Bolton and R. B. Roberts, *Carnegie Institution of Washington Year Book,* 59, 268 (1960).
72. G. D. Hunter and R. A. Goodsall, *Biochem. J.,* **78,** 564 (1961).
73. J. Heller, P. Szafranski, and E. Sulkowski, *Nature,* **183,** 397 (1960).
74. L. I. Hecht, M. L. Stephenson, and P. C. Zamecnik, *Proc. Nat. Acad. Sci.,* **45,** 505 (1959).
75. J. Preiss, P. Berg, E. J. Ofengand and F. H. Bergmann, and M. Dieckmann, *Proc. Nat. Acad. Sci.,* **45,** 319 (1959).
76. G. L. Brown and G. Zubay, *J. Mol. Biol.* **2,** 287 (1960).
77. M. Singer and G. Cantoni, *Biochim. Biophys. Acta,* **39,** 182 (1960).
78. M. Spencer, W. Fuller, M. H. F. Wilkins, and G. L. Brown, *Nature,* **194,** 1014 (1962).
79. K. Smith, E. Cordes, and R. S. Schweet, *Biochim. Biophys. Acta,* **33,** 286 (1959).
80. R. W. Holley, B. P. Doctor, S. H. Merrill, and F. M. Saad, *Biochim. Biophys. Acta,* **35,** 272 (1959).
81. G. L. Brown, in Microbial Genetics," *Symp. Soc. Gen. Microb.,* **10,** 208 (1960).
82. R. Monier, M. L. Stephenson, and P. C. Zamecnik, *Biochim. Biophys. Acta,* **43,** 1 (1960).
83. P. C. Zamecnik, M. L. Stephenson, and J. F. Scott, *Proc. Nat. Acad. Sci.,* **46,** 811 (1960).
84. H. E. Huxley and G. Zubay, *J. Mol. Biol.,* **2,** 10 (1960).
85. C. G. Kurland, *J. Mol. Biol.,* **2,** 83 (1960).
86. A. Tissières, D. Schlessinger, and F. Gros, *Proc. Nat. Acad. Sci.,* **46,** 1450 (1960).

87. J. Tani and R. W. Hendler, *Biochim. Biophys. Acta,* **80,** 279 (1964).
88. R. W. Hendler and J. Tani, *Biochim. Biophys. Acta,* **80,** 294 (1964).
89. R. W. Hendler, W. G. Banfield, J. Tani, and E. L. Kuff, *Biochim. Biophys. Acta,* **80,** 307 (1964).
90. M. R. Lamborg and P. C. Zamecnik, *Biochim. Biophys. Acta,* **42,** 206 (1960).
91. A. B. Pardee, F. Jacob, and J. Monod, *J. Mol. Biol.,* **1,** 165 (1959).
92. F. Jacob and E. L. Wollman, *Symp. Soc. Exp. Biol.,* **12,** 75 (1958).
93. F. Jacob and J. Monod, *J. Mol. Biol.* **3,** 318 (1961).
94. B. N. Ames and B. Garry, *Proc. Nat. Acad. Sci.,* **45,** 1453 (1959).
95. M. Riley, A. B. Pardee, F. Jacob, and J. Monod, *J. Mol. Biol.,* **2,** 216 (1960).
96. S. Brenner, F. Jacob, and M. Meselson, *Nature,* **190,** 516 (1961).
97. F. Gros, H. Hiatt, W. Gilbert, R. W. Kurland, R. W. Risebrough, and J. D. Watson, *Nature,* **190,** 581 (1961).
98. F. Haurowitz, in *The Chemistry and Biology of Proteins,* Academic Press, New York 1950, p. 340.
99. A. L. Dounce, *Enzymologia,* **15,** 251 (1952–1953).
100. F. Lipmann, in *The Mechanism of Enzyme Action,* Eds., W. D. McElroy and H. B. Glass, Johns Hopkins University Press, Baltimore, 1954, p. 599.
101. H. Borsook, *J. Cell. and Comp. Physiol.* Suppl. 1, **47,** 35 (1956).
102. G. Gamow, A. Rich, and M. Yčas, *Adv. Biol. and Med. Phys.,* IV, 23 (1956).
103. G. Gamow and M. Yčas, *Proc. Nat. Acad. Sci.,* **41,** 1011 (1955).
104. R. Wall, *Nature,* **193,** 1268 (1962).
105. C. R. Woese, *Nature,* **194,** 1114 (1962).
106. F. Crick, J. S. Griffith, and L. E. Orgel, *Proc. Nat. Acad. Sci.,* **43,** 416 (1957).
107. R. W. Hendler, *Nature,* **193,** 821 (1962).
108. M. J. Fraser and H. Gutfreund, *Proc. Roy. Soc. London,* **149B,** 392 (1958).
109. F. Crick, *Soc. Exp. Biol. Symp. No. 12,* 1958, p. 138.
110. G. von Ehrenstein and F. Lipmann, *Proc. Nat. Acad. Sci.,* **47,** 941 (1961).
111. R. Schweet, H. Lamfrom, and E. Allen, *Proc. Nat. Acad. Sci.,* **44,** 1029 (1958).
112. J. Bishop, J. Leahy, and R. Schweet, *Proc. Nat. Acad. Sci.,* **46,** 1030 (1960).
113. H. M. Dintzis, *Proc. Nat. Acad. Sci.,* **47,** 247 (1961).
114. M. A. Naughton and H. M. Dintzis, *Proc. Nat. Acad. Sci.,* **48,** 1822 (1962).
115. A. Goldstein and B. J. Brown, *Biochim. Biophys. Acta,* **53,** 438 (1960).
116. A. Yoshida and T. Tobita, *Biochim. Biophys. Acta,* **37,** 513 (1960).
117. R. E. Canfield and C. B. Anfinsen, *Biochem.,* **2,** 1073 (1963).
118. A. Tissières and J. W. Hopkins, *Proc. Nat. Acad. Sci.,* **47,** 2015 (1961).
119. H. Matthaei and M. W. Nirenberg, *Biochim. Biophys. Res. Comm.,* **4,** 404 (1961).
120. J. H. Matthaei and M. W. Nirenberg, *Proc. Nat. Acad. Sci.* **47,** 1580 (1961).
121. A. Tissières, D. Schlessinger, and F. Gros, *Proc. Nat. Acad. Sci.,* **46,** 1450 (1960).
122. T. Kameyama and G. D. Novelli, *Biochim. Biophys. Res. Comm.,* **2,** 393 (1960).
123. M. W. Nirenberg, and J. H. Matthaei, *Proc. Nat. Acad. Sci.,* **47,** 1588 (1961).
124. P. Lengyel, J. F. Speyer, and S. Ochoa, *Proc. Nat. Acad. Sci.,* **47,** 1936 (1961).
125. B. D. Hall and S. Spiegelman, *Proc. Nat. Acad. Sci.,* **47,** 137 (1961).
126. E. Volkin and L. Astrachan, *Virology,* **2,** 149 (1956)
127. P. Doty, J. Marmur, J. Eigner, and C. Schildkraut, *Proc. Nat. Acad. Sci.,* **46,** 461 (1960).
128. J. Marmur and D. Lane, *Proc. Nat. Acad. Sci.,* **46,** 453 (1960).
129. M. Chamberlin and P. Berg, *Proc. Nat. Acad. Sci.,* **48,** 81 (1962).

130. M. W. Nirenberg, J. H. Matthaei, and O. W. Jones, *Proc. Nat. Acad. Sci.,* **48,** 104 (1962)

131. B. Weisblum, S. Benzer, and R. W. Holley, *Proc. Nat. Acad. Sci.,* **48,** 1449 (1962).

132. T. P. Bennett, J. Goldstein, and F. Lipmann, *Proc. Nat. Acad. Sci.,* **49,** 850 (1963).

133. G. von Ehrenstein and D. Dais, *Proc. Nat. Acad. Sci.,* **50,** 81 (1963).

134. F. Chapeville, F. Lipmann, G. von Ehrenstein, B. Weisblum, W. J. Ray, and S. Benzer, *Proc. Nat. Acad. Sci.,* **48,** 1086 (1962).

135. R. B. Loftfield, *J. Cell. Comp. Physol.,* Suppl. 1, **54,** 1959, p. 83.

136. E. L. Smith, *Proc. Nat. Acad. Sci.,* **48,** 677 (1962).

137. J. H. Matthaei, O. W. Jones, R. G. Martin, and M. W. Nirenberg, *Proc. Nat. Acad. Sci.,* **48,** 666 (1962).

138. J. F. Speyer, P. Lengyel, C. Basilio and S. Ochoa, *Proc. Nat. Acad. Sci.,* **48,** 441 (1962).

139. R. W. Hendler, *Proc. Nat. Acad. Sci.,* **48,** 1402 (1962).

140. M. Takanami, *Biochim. Biophys. Acta,* **55,** 132 (1962).

141. T. Hultin and A. van der Decken, *Exptl. Cell. Res.,* **16,** 444 (1959).

142. H. Bloemendal, L. Bosch, and M. Sluyser, *Biochim. Biophys. Acta,* **41,** 454 (1960).

143. M. B. Hoagland and L. T. Comply, *Proc. Nat. Acad. Sci.,* **46,** 1554 (1960).

144. M. B. Yarmolinsky and G. L. de la Haba, *Proc. Nat. Acad. Sci.,* **45,** 1721 (1959).

145. A. Morris, S. Favelukes, R. Arlinghaus, and R. Schweet, *Biochim. Biophys. Res. Comm.,* **7,** 326 (1962).

146. D. W. Allen and P. C. Zamecnik, *Biochim. Biophys. Acta,* **55,** 865 (1962).

147. W. Gilbert, *J. Mol. Biol.,* **6,** 389 (1963).

148. M. Takanami, *Biochim. Biophys. Acta,* **61,** 432 (1962).

149. M. Cannon, R. Krug, and W. Gilbert, *J. Mol. Biol.,* **7,** 360 (1963).

150. A. Morris, R. Arlinghaus, S. Favelukes, and R. Schweet, *Biochem.* **2,** 1084 (1963).

151. R. W. Risebrough, A. Tissières, and J. D. Watson, *Proc. Nat. Acad. Sci.,* **48,** 430 (1962).

152. W. Gilbert, *J. Mol. Biol.* **6,** 374 (1963).

153. A. Gierer, *J. Mol. Biol.* **6,** 148 (1963).

154. J. R. Warner, A. Rich, and C. E. Hall, *Science,* **138,** 1399 (1962).

155. M. Petermann, in *The Physical and Chemical Properties of Ribosomes,* Elsevier, Amsterdam, 1964.

156. J. R. Warner, P. M. Knopf, and A. Rich, *Proc. Nat. Acad. Sci.,* **49,** 122 (1963).

157. M. Hayashi, M. N. Hayashi, and S. Spiegelman, *Proc. Nat. Acad. Sci.,* **50,** 664 (1963).

158. R. W. Hendler, *Science,* **142,** 402 (1963).

159. H. Feldman and H. G. Zachau, *Biochim. Biophys. Rev. Comm.,* **15,** 13 (1964).

160. J. Sonnenbichler, H. Feldman, and H. G. Zachau, *Z. Physiol. Chem.,* **334,** 283 (1963).

161. J. O. Bishop and R. Schweet, *Biochim. Biophys. Acta,* **54,** 617 (1961).

162. D. Nathans and F. Lipmann, *Proc. Nat. Acad. Sci.,* **47,** 497 (1961).

163. L. G. Grossi and K. Moldave, *Biochim. Biophys. Acta,* **35,** 275 (1959).

164. L. G. Grossi and K. Moldave, *J. Biol. Chem.,* **235,** 2370 (1960).

165. J. M. Fessenden and K. Moldave, *Biochem.,* **1,** 485 (1962).

166. J. M. Fessenden and K. Moldave, *J. Biol. Chem.,* **238,** 1479 (1963).

167. E. Gasior and K. Moldave, *J. Biol. Chem.,* **240,** 3346 (1965).
168. R. P. Stutter and K. Moldave, *J. Biol. Chem.,* **241,** 1698 (1966).
169. J. E. Allende, R. Monro and F. Lipmann, *Proc. Nat. Acad. Sci.,* **51,** 1211 (1964).
170. T. W. Conway, *Proc. Nat. Acad. Sci.,* **51,** 1216 (1964).
171. G. J. Spyrides, *Proc. Nat. Acad. Sci.,* **51,** 1220 (1964).
172. T. W. Conway and F. Lipmann, *Proc. Nat. Acad. Sci.,* **52,** 1462 (1964).
173. Y. Nishizuka and F. Lipmann, *Proc. Nat. Acad. Sci.,* **55,** 212 (1966).
174. R. Arlinghaus, J. Schaeffer, and R. Schweet, *Proc. Nat. Acad. Sci.,* **51,** 129 (1964).
175. H. Harris, H. W. Fisher, A. Rodgers, T. Spencer, and J. W. Watts, *Proc. Roy. Soc. London,* **157B,** 177 (1963).
176. H. Harris and J. W. Watts, *Proc. Roy. Soc. London,* **156B,** 109 (1963).
177. J. Hämmerling, *Int. Rev. Cytol.,* **2,** 475 (1953).
178. H. B. Levy, *Proc. Soc. Exp. Biol. Med.,* **113,** 886 (1963).
179. J. Paul and M. C. Struthers, *Biochem. Biophys. Res. Comm.* **11,** 135 (1963).
180. I. Lieberman, R. Abrams, and P. Ove, *J. Biol. Chem.,* **238,** 2141 (1963).
181. H. Harris, *Nature,* **202,** 249 (1964).
182. J. W. Watts and H. Harris, *Biochem. J.,* **72,** 147 (1959).
183. K. Scherrer, H. Latham, and J. E. Darnell, *Proc. Nat. Acad. Sci.,* **49,** 240 (1963).
184. M. Girard, S. Penman, and J. E. Darnell, *Proc. Nat. Acad. Sci.,* **51,** 205 (1964).
185. H. Harris, *Nature,* **202,** 1301 (1964).
186. R. P. Perry, *Proc. Nat. Acad. Sci.,* **48,** 2179 (1962).
187. T. Tamaocki and G. C. Mueller, *Biochem. Biophys. Res. Comm.,* **9,** 451 (1962).
188. H. Harris, *Proc. Roy. Soc. London,* **158B,** 79 (1963).
189. H. Harris, *Nature,* **201,** 863 (1964).
190. H. Harris and L. D. Sabath, *Nature,* **202,** 1078 (1964).
191. L. Goldstein, J. Micou, and T. T. Crocker, *Biochim. Biophys. Acta,* **45,** 82 (1960).
192. D. M. Prescott, *Exp. Cell. Res.,* **19,** 29 (1960).
193. E. Reich, R. M. Franklin, A. J. Shatkin, and E. L. Tatum, *Proc. Nat. Acad. Sci.,* **48,** 1238 (1962).
194. O. Greengard, M. A. Smith, and G. Acs, *J. Biol. Chem.,* **238,** 1548 (1963).
195. R. W. Seed and I. H. Goldberg, *Proc. Nat. Acad. Sci.,* **50,** 275 (1963).
196. L. D. Garren, R. R. Howell, G. M. Tompkins, and R. M Crocco, *Proc. Nat. Acad. Sci.,* **52,** 1121 (1964).
197. T. Spencer and H. Harris, *Biochem. J.,* **91,** 282 (1964).
198. R. B. Scott nd E. Bell, *Science,* **145,** 711 (1964).
199. H. H. Hiatt, *J. Mol. Biol.,* **5,** 217 (1962).
200. D. Eboué-Bonis, A. M. Chambaut, P. Volfin, and H. Clauser, *Nature,* **199,** 1183 (1963).
201. R. B. Scott and R. A. Malt, *Nature,* **208,** 497 (1965).
202. L. Dure and L. Waters, *Science,* **147,** 410 (1965).
203. P. R. Gross, L. I. Malkin, and W. A. Moyer, *Proc. Nat. Acad. Sci.,* **51,** 407 (1964).
204. D. L. Kirk, *Proc. Nat. Acad. Sci.,* **54,** 1345 (1965).
205. M. Revel and H. H. Hiatt, *Proc. Nat. Acad. Sci.,* **51,** 810 (1964).
206. J. L. Melnick, D. Crowther, and J. Barrera-Ora. *Science,* **134,** 554 (1961).
207. I. H. Goldberg and M. Rabinowitz, *Science,* **136,** 315 (1962).
208. J. M. Kirk, *Biochim. Biophys. Acta,* **42,** 167 (1960).

209. J. Hurwitz, J. J. Furth, M. Malamy, and M. Alexander, *Proc. Nat. Acad. Sci.,* **48,** 1222 (1962).
210. C. Levinthal, A. Keynan, and A. Higa, *Proc. Nat. Acad. Sci.,* **48,** 1631 (1962).
211. G. Acs, E. Reich, S. Valanju, *Biochim. Biophys. Acta,* **76,** 68 (1963).
212. M. Revel, H. H. Hiatt, and J. P. Revel, *Science,* **146,** 1311 (1964).
213. M. Revel and H. H. Hiatt, *Proc. Nat. Acad. Sci.,* **51,** 810 (1964).
214. T. Staehelin, F. O. Wettstein, and H. Noll, *Science,* **140,** 180 (1963).
215. G. Honig and M. Rabinovitz, *Science,* **149,** 1504 (1965).
216. J. Laszlo, D. S. Miller, K. S. McCarty, and P. Hochstein, *Science,* **151,** 1007 (1966).
217. D. Korn, J. J. Protass, and L. Leive, *Biochem. Biophys. Res. Comm.,* **19,** 473 (1965).
218. A. Porro and L. Cima, *Arch. Int. Pharmacodyn.,* **100,** 63 (1954).
219. K. Horini, *Osakashiritsu Daigaku Igaku Zashi,* **10,** 1305–1323 (1958).
220. J. J. Holland, C. K. Buck, and B. J. McCarthy, *Biochem.,* **5,** 358 (1966).
221. A. S. Spirin, N. V. Belitsina, and M. A. Ajtkhozhin, *ch. Obstich. Biol.,* **25,** 321 (1964).
222. A. S. Spirin and M. Nemur, *Science,* **150,** 214 (1965).
223. M. Girard, H. Latham, S. Penman, and J. E. Darnell, *J. Mol. Biol.,* **11,** 187 (1965).
224. W. K. Joklik and Y. Becker, *J. Mol. Biol.,* **13,** 496 (1965).
225. W. K. Joklik and Y. Becker, *J. Mol. Biol.,* **13,** 511 (1965).
226. E. C. Henshaw, M. Revel, and H. H. Hiatt, *J. Mol. Biol.,* **14,** 241 (1965).
227. E. H. McConkey and J. W. Hopkins, *J. Mol. Biol.,* **14,** 257 (1965).
228. A. J. Shatkin, E. D. Sebring, and N. P. Salzman, *Science,* **148,** 87 (1966).
229. A. S. Spirin, in *Current Topics in Developmental Biology,* **1,** 1 (1966).
230. E. F. Eikenberry and A. Rich, *Proc. Nat. Acad. Sci.,* **53,** 668 (1965).
231. M. Salas, M. A. Smith, W. M. Stanley, Jr., A. J. Wahba, and S. Ochoa, *J. Biol. Chem.,* **240,** 3988 (1965).
232. M. A. Smith, M. Salas, W. M. Stanley, Jr., A. J. Wahba, and S. Ochoa, *Proc. Nat. Acad. Sci.,* **55,** 141 (1966).
233. E. Terzaghi, Y. Okada, G. Streisenger, A. Tsugita, M. Inouye, and J. Emrich, *Science,* **150,** 387 (1965).
234. R. E. Thach, M. A. Cecere, T. A. Sundararajan, and P. Doty, *Proc. Nat. Acad. Sci.,* **54,** 1167 (1965).
235. C. Yanofsky, B. C. Carlton, J. R. Guest, D. R. Helinski, and U. Henning, *Proc. Nat. Acad. Sci.,* **51,** 266 (1964).
236. A. S. Sarabhai, A. O. W. Stretton, S. Brenner, and A. Bolle, *Nature,* **201,** 13 (1964).
237. K. Marcker and F. Sanger, *J. Mol. Biol.,* **8,** 835 (1964).
238. K. Marcker, *J. Mol. Biol.,* **14,** 63 (1965).
239. B. F. C. Clark and K. Marcker, *Nature,* **207,** 1038 (1965).
240. M. W. Nirenberg and P. Leder, *Science,* **145,** 1399 (1964).
241. J. M. Adams and M. R. Capecchi, *Proc. Nat. Acad. Sci.,* **55,** 147 (1966).
242. R. E. Webster, D. L. Engelhardt, and N. D. Zinder, *Proc. Nat. Acad. Sci.,* **55,** 155 (1966).
243. T. Nakamoto and D. Kolakofsky, *Proc. Nat. Acad. Sci.,* **55,** 606 (1966).
244. R. E. Thach and T. A. Sundararajan, *Proc. Nat. Acad. Sci.,* **53,** 1021 (1965).
245. H. Noll, *Science,* **151,** 1241 (1966).
246. C. D. Yegian, G. S. Stent, and E. M. Martin, *Proc. Nat. Acad. Sci.,* **55,** 839 (1966).

247. B. N. Ames and P. E. Hartman, *Cold Spring Harbor· Symposium*, **28**, 349 (1963).
248. H. Itano, in *Abnormal Haemoglobin in Africa (1963) a Symposium* Ed., J. H. P. Jonxis, Blackwell Scientific Pub., Oxford, 1965, p. 3.
249. R. G. Martin, D. F. Silbert, W. E. Smith, and H. J. Whitfield, Jr., *J. Mol. Biol.*, **21**, 357 (1966).
250. M. G. Weigert and A. Garen, *Nature*, **206**, 992 (1965).
251. S. Brenner, A. O. W. Stretton, and S. Kaplan, *Nature*, **206**, 994 (1965).
252. M. S. Bretscher, H. M. Goodman, J. R. Menninger, and J. D. Smith, *J. Mol. Biol.*, **14**, 634 (1965).
253. D. L. Engelhardt, R. E. Webster, R. C. Wilhelm, and N. D. Zinder, *Proc. Nat. Acad. Sci.*, **54**, 1791 (1965).
254. M. C. Ganoza and T. Nakamoto, *Proc. Nat. Acad. Sci.*, **55**, 162 (1966).
255. J. Davies, W. Gilbert, and L. Gorini, *Proc. Nat. Acad. Sci.*, **51**, 883 (1964).
256. S. M. Friedman and I. B. Weinstein, *Proc. Nat. Acad. Sci.*, **52**, 988 (1964).
257. M. Grunberg-Manago, and J. Dondon, *Biochem. Biophys. Res. Comm.*, **18**, 517 (1965).
258. S. M. Friedman and I. B. Weinstein, *Biochem. Biophys. Res. Comm.*, **21**, 339 (1965).
259. W. Szer and S. Ochoa, *J. Mol. Biol.*, **8**, 823 (1964).
260. M. Grunberg-Manago and A. M. Michelson, *Biochim. Biophys. Acta*, **80**, 431 (1964).
261. A. G. Soand E. W. Davie, *Biochem.*, **3**, 1165 (1964).
262. A. G. So, J. W. Bodley, and E. W. Davie, *Biochem.*, **3**, 1977 (1964).
263. P. S. Sarin and P. C. Zamecnik, *Biochem. Biophys. Res. Comm.*, **19**, 198 (1965).
264. A. G. So and E. W. Davie, *Biochem.*, **4**, 1973 (1965).
265. J. Davies, L. Gorini, and B. D. Davis, *Mol. Pharm.*, **1**, 93 (1965).
266. D. Söll, E. Ohtsuka, D. S. Jones, R. Lohrmann, H. Hayatsu, S. Nishimura, and H. G. Khorana, *Proc. Nat. Acad. Sci.*, **54**, 1378 (1965).
267. M. W. Nirenberg and P. Leder, *Science*, **145**, 1399 (1964).
268. H. G. Aach, G. Funatsu, M. W. Nirenberg, and H. Fraenkel-Conrat. *Biochem.*, **3**, 1362 (1964).
269. J. H. Schwartz, J. M. Eisenstadt, G. Brawerman, and N. D. Zinder, *Proc. Nat. Acad. Sci.*, **53**, 195 (1965).

Chapter 4

CELL STRUCTURE—THEN AND NOW[1]

A study of the history of protein synthesis during the last twenty years has shown that the contemporary view of the field always presents a plausible image supported by a wealth of experimental observations and intelligent interpretations of existing data. Under the glare of historical perspective it can be seen that this contemporary image is constantly changing its form so that the image of 1968 does not resemble those of earlier years. This situation is no less true in regard to cell structure.

It will be helpful to review briefly the knowledge of cell structure from the beginning of its discovery up to the time that we picked up the threads of the story of protein synthesis in 1947—the time corresponding to the advent of atomic piles and radioactive amino acid tracers. At about the same time many cytological laboratories were beginning to use a tool every bit as powerful as the radioisotopes of biochemical laboratories, namely, the electron microscope.

BACK TO THE BEGINNING

Robert Hooke, approximately three hundred years ago, was appointed the first curator of the English Royal Society. Soon thereafter he and Nehemiah Grew, who was later named as a curator, independently published their observations which were made with an early compound microscope belonging to the Society. In 1665 Hooke described microscopic pores in petrified wood in John Evelyn's book, *Sylva*. A year later in his famous "Micrografia" he went into considerable detail concerning

---

[1] In this chapter the main emphasis is placed on membrane bounded organelles of the cytoplasm and the historical development of knowledge concerning these structures.

145

"pores or cells" in cork, the pith of Elder and in the pulp or pith of various other vegetables. Grew published *The Anatomy of Plants* in 1682 in which similar descriptions of pores are found for a number of other plants.

Beginning in 1673, in a series of letters to the Royal Society, Mr. Antony van Leeuwenhoek, a citizen of Delft in Holland and a linen draper by trade, described "animalcules," bacteria, spermatozoa, and blood corpuscles. These observations were made with homemade microscopes possessing single lenses which he ground himself.

During the eighteenth century the use of microscopes spread throughout Europe and as various technological improvements were introduced which increased magnification and eliminated various types of aberrations, the literature describing cells in a variety of tissues from both plants and animals increased. Thus in 1737 Jan Swammerdam described the corpuscles in animal blood. The general theory of the cell structure of tissues is attributed jointly to Theodore Schwann, who published his work on microscopic investigations in 1839, and to M. J. Schleiden, whose work appeared about the same time.

The cell was considered as a chamber or globule and the aim of cytologists has since been to find out what is inside of this globule and in what manner the contents are arranged. The first unit found was the nucleus and although this structure was seen earlier by Leeuwenhoek and Felix Fontano (1781) the concept of the nucleus as a basic unit of cell structure was due to Robert Brown in 1833. The recognition of an important organelle within the nucleus, the nucleolus, seems to have been due to Felix Fontano.

The area between the nucleus and the cell wall was given various names, such as sarcode, parenchym, hyaloplasm, sap, and protoplasm. Various authors described its adhesive, contractile, and jelly-like properties and the existence of movements and currents and granules within it. In a sense, the nucleus is a chamber within a chamber and cytologists were much more successful in furnishing the smaller room than the larger compartment which we now designate as the cytoplasm.

The concept of the germ-cell as a container of molecular information for the reproduction of the adult organism is credited to August Weismann (1885). It was shown in 1876 by O. Hertwig and in 1879 by H. Fol that during fertilization of an egg the two nuclei fused, and so it seemed that germ plasm was to be found in the nucleus. When E. Strasburger showed in 1884 that the male element from flowering plants appeared to be essentially devoid of cytoplasm, the conclusion seemed clear that the nucleus alone carried heredity. Among the earliest descriptions of chromatin threads and spindles in cells undergoing division are those of W. Flemming in 1882 and E. Selenka in 1878.

In 1893 A. Weismann, in his work *The Germ Plasm, A Theory of Heredity,* gave convincing arguments on why the chromosomes (chromatin threads) should be considered as the seats of heredity. In fertilization it was known that equal numbers were contributed by each parent and that the complex mechanism for cell division exists practically for the sole purpose of dividing the chromatin. Weismann was wrong in many important details concerning the process but his general views were prophetically correct. Observations of individual chromosomes, their pairing during cell division and their distribution between daughter cells reawakened an interest in the unit character of inheritance described in 1865 by Gregor Mendel. The further linking of these concepts in heredity to the structural unit of the chromosome was accomplished through the efforts of the American school of cytologists in the early 1900's; members included W. A. Cannon, E. B. Wilson, and W. S. Sutton. The most decisive studies of chromosomal heredity and the linear arrangements for genes controlling particular characteristics were carried out by Thomas Hunt Morgan using the fruit fly *Drosophila melanogaster* during the second decade of this century (only 40 years ago). Work with the giant salivary chromosomes of the larval fruit fly carried out by T. S. Painter at the University of Texas in the 1930's allowed the direct visualization of bands along the chromosome and the correlation of alterations in such structures with specific mutations developed in the organism.

From the nuclei of pus cells Frederick Miescher in 1869 prepared a substance which he called nuclein. Subsequently he isolated a similar substance from Rhine Salmon sperm. Because of its strong acidic character he referred to the material as nucleic acid and proceeded to study its chemistry; for example, this material was dissolved by alkali but resisted the action of pepsin. E. Zacharias applied these tests to cells under the microscope and found such material localized in the nucleus. Fleming, in his work of 1882, described and named chromatin (because of its affinity for dyes) and suggested that it was the chromatin of the nucleus which possessed the nucleic acid characteristics described by Zacharias. Later work better characterized the chemical make-up of nucleic acid and established the dichotomy of DNA and RNA. Original ideas mistakenly restricted RNA to plants and DNA to animals and later DNA to the nucleus and RNA to the cytoplasm. More recent work has shown that both types of nucleic acid are found in both the nuclei and cytoplasm of plant and animal cells.

FIRST INDICATION OF CYTOPLASMIC ORGANELLES

As already mentioned, the knowledge of cytoplasmic structures has lagged considerably behind that of the nucleus. In fact, this lag has placed the gathering of knowledge of cytoplasmic structures right up to contemporary times (1967). In the late 1800's various cells were fixed and stained with chemicals and many reports of cytoplasmic filaments, fibrils, tubes, and networks appeared, though many were due to poorly understood artifacts. These postulated structures were thought of as a framework of solid material that held the cytoplasm together. Other groups of observers described the protoplasm as a collection of granules within an amorphous matrix.

Several cytologists in the second half of the nineteenth century described granular or threadlike components of the cytoplasm which we now recognize as mitochondria. In 1886 Richard Altmann published a method for staining such granules which he thought were analogous to a cellular type of bacteria. The term mitochondrion comes from two Greek words meaning, respectively, thread and granule and was introduced by C. Benda of Berlin in 1898, who studied these granules with the use of special staining procedures in various tissues. Toward the close of the century and the beginning of the 1900's mitochondria were shown to be present in most cells of plant and animal origin.

The presence of the centrosome as a permanent cell structure which plays a role in cell division was suggested in 1887 by von Beniden and Neyt and Boveri. Fine structural studies have not revealed any discrete membrane-bounded organelle to be identified with the centrosome seen in light microscopy. It may be that a particular type of gel formation in the cytoplasm as well as the presence of connected microtubules contributes to its appearance. Inside of the centrosphere region and within the centrosome are the centrioles. These are two hollow cylinders about 300 to 500 mμ in length and 150 mμ in diameter. They are usually situated at right angles to each other. The walls are composed of nine evenly spaced triplet hollow fibrils or tubules which are embedded in a dense amorphous matrix. One end appears closed and one end open. During cell division spindle fibers radiate from the chromosomes to the centrioles. In some undetermined manner these bodies are involved in the process of chromosome separation (see Figures 31 and 32 on pages 198 and 199).

In 1898 the Italian neurologist, C. Golgi, described a network of material within Purkinje cells of an owl's cerebellum which could be impregnated with silver. This structure, which has since become known as the Golgi apparatus, has gone through a stormy history where for the most part cytologists attributed its very existence to artifacts of fixation or staining. In 1949, Palade and Claude suggested that the conventional Golgi net was an artifact produced by the action of fixatives on the lipoid-containing neutral red-staining spheroids present in the Golgi region of the cell.

As late as 1954 a symposium on the Golgi apparatus held by the Royal Microscopical Society was divided between believers and skeptics of the structure's very existence. Its identity was firmly established, however, by Dalton, Felix, Schneider, and Kuff in the 1950's (see page 168).

By the technique of differential ultracentrifugation, Claude, in 1943, isolated minute granules with diameters of about 0.1 μ. These he named microsomes (1). Although both T. Caspersson and J. Brachet, in the early 1940's, demonstrated the localization of RNA in the cytoplasm, no indications of any structural components were available at the time. The elucidation of the nature of ribosomes, microsomes, and ergastoplasm or endoplasmic reticulum is an achievement of more recent times and will be considered later in this historical development.

Thus in a rather sketchy manner[2] we have acquired a picture of cell structure at the beginning of the modern study of protein synthesis (1947). Developments in these two fields followed entirely separate paths. It has been only in rather recent times that the two routes have been approaching a joining and an occasional traveler has crossed back and forth between these paths.

In 1947, when Tarver, Greenberg, Borsook, Zamecnik and their co-workers were finding that radioactive amino acids could be incorporated into the proteins of various cell systems, Frey-Wyssling, the Swiss cytologist, published an important monograph on the latest concepts of cell structure entitled, "Submicroscopic Morphology of Protoplasm and Its Derivatives." Professor Frey-Wyssling sought to demonstrate that, contrary to the views of many cytologists, the cytoplasm could not be considered as a true liquid but rather as a substance that possesses some flexible internal structure. An inner elasticity could be demonstrated by suspending iron filings in the liquid cytoplasm and moving these by means of a magnetic field. As soon as the field was switched off, the particles returned elastically to their original position. Furthermore, when plasmolysis caused the cytoplasm to separate from the cell wall, very fine fibers could be seen stretching between the two structures.

Frey-Wyssling concluded that the only component of cytoplasm

[2] A more thorough review of the early historical development of cytology can be found in *A History of Cytology*, by Arthur Hughes, published by Abelard-Schuman, London and New York, 1959.

capable of forming high polymeric structural units was protein. His picture of cytoplasmic structure was that of extensive anastomosing peptide networks providing a framework. The peptides were held together by a variety of bonds from the very labile hydrophobic attractions of nonpolar side chains to stable covalent ester and disulphide bridges. The liquid part of the cytoplasm, with its dissolved and suspended aqueous and lipoidal components, was contained in the interstitial spaces of the network.

This structure could account for most of the observed properties of the cytoplasm. Moreover, he anticipated the concept of allosterism[3] by suggesting that certain configurations of the polypeptides were essential for metabolic activity and that these crucial configurations could be affected by small molecules such as salts or hormones. After considering the Danielli and Harvey and Davson-Danielli lipid bimolecular leaflet-globular protein models for the plasma membrane (see Chapter 5), Frey-Wyssling argued that it is more reasonable to think of the cytoplasmic boundary as an extension of the same polypeptide network running throughout the cytoplasm with a modification towards larger interstices filled with higher lipid content. Vacuoles or tonoplasts, on the other hand, were believed to be bounded by simple bi- or polymolecular lipid leaflets.

The other structural elements of the cytoplasm were considered under the heading of "biosomes." Since the Golgi apparatus was not mentioned in this section or any other part of his book, Frey-Wyssling must be numbered among the prominent cytologists who, in 1947, gave no credulity to its existence. Chondriosomes or mitochondria were defined as microscopic particles representative of a special system of the cell designated as *chondriome*. They consist of lipid, 35 to 44%, and two proteins of different isoelectric points comprising 56 to 65%. Recent electronmicrograms (1945) showed fixed chondriosomes possessing a lipid cortex and a watery, less dense, central zone (2). In sea urchin and Tubifex eggs is found a framework of coarse plasma fibrils with microscopic dimensions, possessing lipid and carrying small bodies of ribonucleic acid of almost submicroscopic dimensions. These bodies were called *chromidia*. It was believed that both mitochondria and chromidia possessed the capability of self-duplication. Frey-Wyssling's concept of nuclear structure was quite similar in that he visualized a comparable network of peptide chains with a nuclear sap held in the interstices.

In 1948 an authoritative textbook entitled, *Cytology and Cell Physi-*

---

[3] The ability of a small molecule, by combining with a protein, to change the configuration of the protein from an inactive to an active form or vice versa.

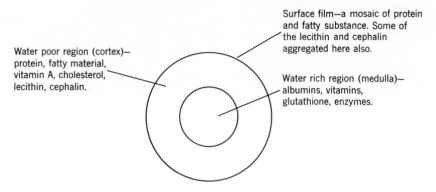

Surface film—a mosaic of protein and fatty substance. Some of the lecithin and cephalin aggregated here also.

Water poor region (cortex)— protein, fatty material, vitamin A, cholesterol, lecithin, cephalin.

Water rich region (medulla)— albumins, vitamins, glutathione, enzymes.

Figure 1   Structure of mitochondria 1948.

ology, originally published in 1941, was completely revised and reissued. The chapter devoted to cytoplasmic structures concerned itself only with mitochondria and the Golgi complex and was written by the editor, G. H. Bourne. Mitochondria or chondriosomes were defined as elements of definite form which lie freely in the cytoplasm, possess the power of independent movement, and may take the form of filaments, rods, or granules. It was believed that they could change their form and that in the pancreas they could eventually metamorphose into zymogen granules. Their motility in the cytoplasm was believed to be accomplished by a wriggling movement of the mitochondrion itself. This feat was said to be accomplished by periodic contractions of polypeptide chains of proteins which are present in them. A considerable discussion was devoted to staining properties and gross chemical composition of the structures, which consisted mainly of lipid and protein. Various cytologists reported seeing internal bodies contained within mitochondria. A generally accepted structure for the mitochondrion is indicated in Figure 1.

The following arguments were advanced about the function of mitochondria:

1. A role in cellular respiration based either on a redox system made up of Vitamin A and glutathione or on the presence of certain enzymes such as succinic dehydrogenase, cytochrome oxidase and Krebs cycle enzymes.

2. A cellular scavenger function based upon the association of hydrolytic enzymes with the structure and upon microscopic examination of mitochondria attaching to digestible substances in the cytoplasm.

3. A protein synthetic function based on microscopic examination of granules which appear to exit from the mitochondria.

4. A role in fat metabolism based on their content of fatty acid oxidases.

5. A role in maintaining a reserve store of metabolic material for the cell.

A relationship was believed to exist between mitochondria and the Golgi apparatus such that the former converts to the latter or is derived from the latter or that the mitochondria liberate material in the cytoplasm which passes to the Golgi apparatus.

The Golgi apparatus has been variously described as consisting of a fibrous reticulum, network, ring, or cylinder, a very irregular fenestrated plate, a more or less incomplete hollow sphere, vesicle or cup, a collection of small spheres, rodlets and platelets or discs, a series of anastomosing canals, a group of vacuoles and a differentiated region of homogeneous cytoplasm crossed by irregular interfaces. The network appearance has frequently been attributed to artifacts of fixation and staining. Thus silver impregnation, the favored means of revealing the network, is suspect because silver itself, as examined by high power microscopy, has the tendency to form a network structure. Many cytologists considered the structure to be composed of two parts, the outer being essentially lipidic in nature and the inner hydrophilic.

Hirsch (1939), a prominent cytologist, had very definite views on the structure. He declared that the solid granules of Golgi substance really constitute a "presubstance" of the apparatus, which could build up nets. The presubstance undergoes a transformation with the Golgi structure to gradually develop the double structure. The outer part, or cortex, was known as the externum and the inner part as the internum. The internum, or aqueous part, was believed to contain a reservoir of Vitamin C, a substance endowed with a wide variety of vital redox and cellular synthesizing properties. In the living cell the Golgi system was composed principally of spheroid bodies, the network appearance being prominent in fixed material. Various staining procedures and the effects of solubilizing and enzyme treatments on the cell strongly implicated lipid and protein as its main structural elements. The intimate association of Vitamin C with the whole structure and the assumed central metabolic importance of the vitamin provided the basis for a wide range of speculations concerning the staining properties and function of the organelle. The structure was believed to be involved in the synthesis of secretory substances, the bringing together of cytoplasmic metabolites for subsequent reaction, the concentration of products synthesized elsewhere in the cell and acquired by a process of diffusion through the cytoplasm, the active process of secretion, the synthesis of fats, and the production of lipoid droplets. Finally, Bourne conceded, "Although there are indications of the function which the Golgi complex may have, we are still in the position of having very little certain knowledge of its true role in cell physiology."

In 1950 the New York Academy of Sciences published the proceedings of a conference held two and one half years earlier on "Structure in Relation to Cellular Function." Only two papers on cytoplasmic structure were presented at this conference, one by Claude on the general cytoplasm (3) and one by J. Brachet on the localization of RNA in the cell (4).

Claude pointed out that other workers had previously isolated structural units from broken cells. Thus Miescher had obtained a crude preparation of nucleic acid in 1913. Warburg obtained cytoplasmic granules by centrifugation and was able to show that they were responsible for most of the oxygen uptake of cell-free extracts of guinea pig liver. In 1934, Bensley and Hoerr reported the isolation of a mitochondrial preparation.

Claude then summarized the principles of his refinement of the method of isolation of cell organelles by the process of differential centrifugation of a tissue homogenate. The cytoplasm of cells analyzed by these procedures yielded three main fractions:

1. A large granule fraction with particles ranging in size from approximately 0.5 to 2.0 μ in diameter.

2. A microsome fraction consisting of submicroscopic elements approximately 50 to 200 mμ in diameter.

3. A supernatant fraction containing dissolved material and particles less than 50 mμ in diameter.

The elements of the large granule fraction appeared to be bounded by semipermeable membranes as revealed by swelling and shrinking induced by changes in osmotic tonicity of the medium and by electron microscopy. The presence of cytochrome oxidase, succinoxidase, and cytochrome c in the large granule fraction had recently been shown by Claude, Schneider, and Hogeboom. Mitochondria were identified as major constituents of these fractions. The microsomal particles were much less characterized and the presence of a limiting membrane for these units was not definitely established.

Claude concluded by suggesting that the mitochondria can be thought of as power plants of the cells since respiratory enzymes seem to be localized in these granules. He also pointed out that a large fraction of the cytoplasmic RNA occurs in the microsome fraction and that Brachet and Caspersson linked the high concentration of RNA in tumor and embryonic tissues to their high rates of protein synthesis. Claude suggested that the correlation in these tissues might just as well have been linked to their active anaerobic respiration.

Brachet in the same conference discussed the inference that RNA in the cell appears to be localized in the nucleolus of the nucleus and in cytoplasmic ribonucleoprotein granules which could be isolated by highspeed centrifugation. Unfortunately, most of the observations discussed by Brachet were obtained by studying a fraction composed of both mitochondria and microsomes and so are not particularly revealing about either fraction *per se*. In an attempt to relate his and Caspersson's suggestions of the importance of RNA for protein synthesis, he reviewed the information available at the time concerning these two subjects. At that time it was thought that protein synthesis involved a reversal of the hydrolytic reactions as catalyzed by the same enzymes (see pages 10–12). Brachet reviewed that correlating data did exist to support his proposal. Thus RNA occurred in the small granules and dipeptidases and cathepsin, which can synthesize peptide bonds, also occurred in these granules. He also stated that the essential respiratory enzymes occurred in this fraction, but this was due to the fact that his preparations were mixed with mitochondria. Several other arguments were also stated by Brachet, but future developments have shown that these particular supporting observations were not valid.

The technique of isolation of cellular components by fractional differential centrifugation was perfected by Hogeboom, Schneider, and Palade (5). These investigators learned that the morphological, staining, and enzymatic characteristics of mitochondria could be preserved by homogenizing and fractionating rat liver in hypertonic sucrose solutions (0.88 M) rather than in the usual type of saline medium.

THE BEGINNINGS OF FINE STRUCTURAL STUDIES

The first clear fine structural studies of cytoplasmic elements were published in 1953 independently by Sjöstrand and Palade. These studies were the results of new and improved techniques in tissue embedding, thin sectioning and fixation, and the superb skill of these electron microscopists.

Sjöstrand showed that mitochondria from retinal rods of the guinea pig, the exocrine cells of mouse pancreas, and the proximal convoluted tubules of mouse kidney were all similar in structure and possessed a double membrane boundary (6). Furthermore, in the interior of the mitochondria there were a great number of transversely-oriented double membranes with different spacings in the different types of cells studied. He also demonstrated a rather highly developed system of double membranes in the cytoplasm of the exocrine cells of the mouse pancreas. This system was revealed by preparing the tissue for microscopy both by the technique of freeze-drying and by a process involving osmium fixation.

Palade presented a comprehensive fine structural study of the mito-chondrion (7) in which he confirmed the existence of the limiting double membrane and in which he further elaborated on his own recent report (8) of the internal transverse membrane system which he named "cristae mitochondriales." Electron micrographs were presented of mitochondria from liver parenchymatous cells, proximal convoluted tubules of the kidneys, spermatocytes in a seminal tubule, and striated muscle cells from tissues of the rat and from a mesophyll cell in the leaf tissue of the plant *Nicotiana tabaccum.*

Palade went to considerable lengths to determine whether the cristae represented internal lamellae (sheets) or filaments (tubules). He con-cluded that for the most part lamellae would be the accurate description.

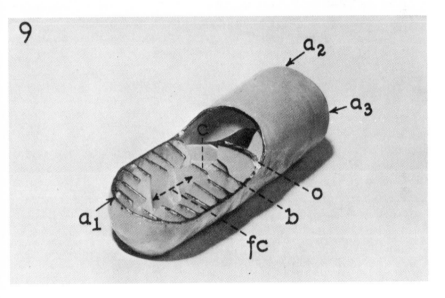

Figure 2 A photograph of the tri-dimensional model of a mitochondrion. The mitochondrial membrane and the cristae were made out of paraffin plates of equal thickness and the model was cut open to show the arrangement of the cristae as deduced from electron micrographs. Note how the apparent thickness of the cristae varies with the angle of the section: the crista marked *o* appears thicker than that marked *c* because the first was sectioned obliquely and the other normally. A branching crista can be seen at *b*, and the central free channel at *fc*. The crista marked *c*, which appears "free" in the middle of this mitochondrial profile, is actually anchored down in the membrane but its attachment is obscured by the other cristae. From G. E. Palade, "An Electron Microscope Study of the Mitochon-drial Structure," *J. Histochem. Cytochem.*, **1**, 188 (1953).

Next, Palade considered whether the lamellae went completely across the mitochondrion-like walls or septa dividing it into a series of compartments or whether they only extended part way from the inner mitochondrial membrane as a series of ridges leaving a free central channel. As a result of a careful consideration of many electron micrographs cut at different angles through mitochondria, he concluded that they were like ridges rather than septa (Figure 2). Although the same general features of mitochondria seem to apply for the wide variety of tissues studied, individual variations exist for the relative amounts of cristae and inner matrix (Figures 3–6).

The Discovery of the Endoplasmic Reticulum

The first hint of this intracytoplasmic reticular structure was obtained in 1945 when Porter, Claude, and Fullam demonstrated that single cells from a culture of chick embryo tissue, spread on a thin plastic film and examined directly in the electron microscope, showed a fine lacelike reticulum after fixing with osmium (9).

Although it was over three hundred years ago that Hooke saw the first cells in plants, our current knowledge of cytoplasmic structures was acquired only since the early 1950's. For example, in 1953 Porter reviewed the evidence for the existence of a basophilic component in the cytoplasm which was below the resolving ability of the light microscope (10). The basophilic substance was so named because of its striking affinity for basic dyes and it has been variously referred to as chromophilic material, ergostoplasm, and Nissl's substance. Claude had shown that this substance was concentrated in his high-speed small-particle or microsomal fraction. It was widely believed that microsomes represented discrete cellular organelles and the term for many years was loosely used to describe noncharacterized materials which were isolated after highspeed centrifugation of broken cell homogenates. Porter found that cells grown in tissue culture were suitably thin for the direct observation of fine structural organization by electron microscopy. He was able to reveal a reticular system in the inner regions of the cytoplasm of various animal cells which he called endoplasmic reticulum. Porter examined by electron microscopy a microsomal fraction isolated in a sucrose medium, in collaboration with Hogeboom, and reported upon the similarity in appearance of this preparation to elements of the endoplasmic reticulum. An identification of the microsomal fraction with the basophilic component of cytoplasm had also been made by Slautterback (11), Claude (12), Brenner (13) and others.

Palade and Porter examined a series of animal cells by two techniques—one by whole mounts of single cells grown in tissue culture,

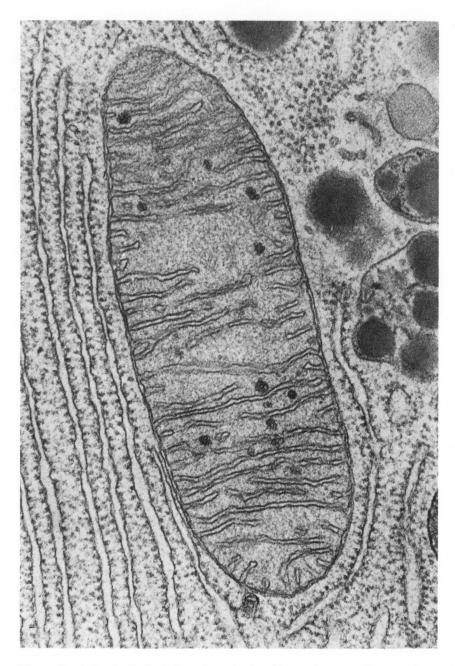

Figure 3 A longitudinal section of a mitochondrion and surrounding cytoplasm from the pancreas of a bat. (Micrograph courtesy of Dr. K. R. Porter.) Magnification 78,000 X. From D. W. Fawcett, *An Atlas of Fine Structure,* Saunders, Philadelphia, (1966).

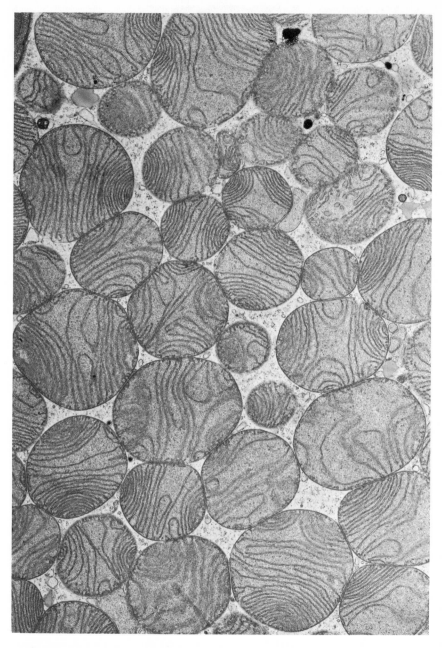

Figure 4 Mitochondria from interscapular brown adipose cell from the bat, *Eptesicus fuscus*. Collidine buffered osmium fixation. Lead citrate staining. Magnification, 11,500 X. From D. W. Fawcett, *An Atlas of Fine Structure*, Saunders, Philadelphia, (1966).

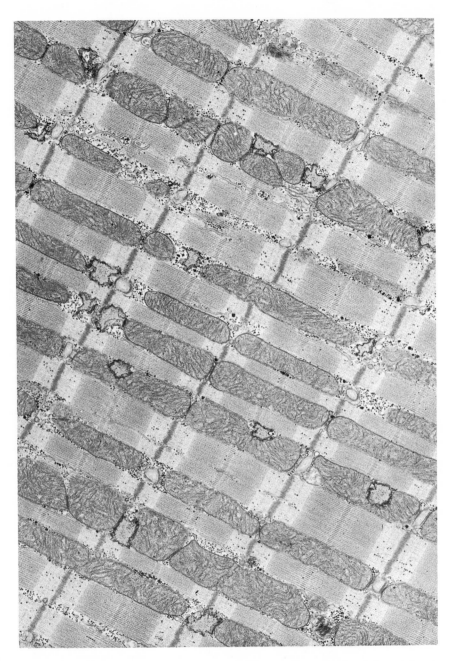

Figure 5 Cardiac muscle from right ventricular papillary muscle of the cat heart. Note arrangement of mitochondria, here and in figure 6, for efficient utilization of energy produced. Phosphate buffered osmium fixation. Lead citrate staining. Magnification 11,000 X. From D. W. Fawcett, *An Atlas of Fine Structure*, Saunders, Philadelphia, (1966).

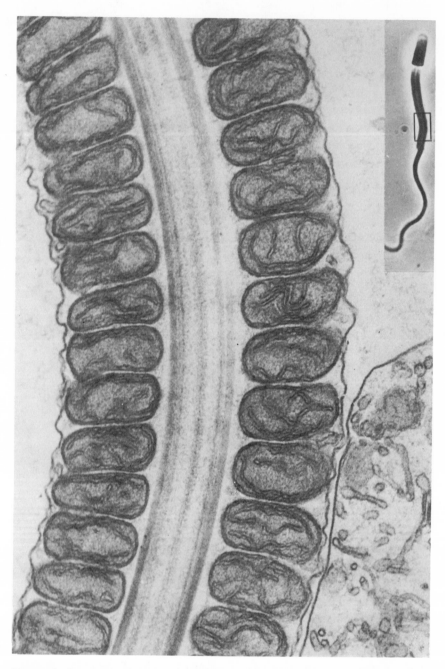

Figure 6 Spermatozoon from the epididymis of the bat, *Myotis lucifugus*. Collidine buffered osmium fixation. Lead citrate staining. Note apposition of mitochondria to the contractile longitudinal fibrils of the tail. Magnification, 82,000 X. From D. W. Fawcett and S. Ito, *Amer. J. Anat.* **116,** 567, 1965.

and one by ultra-thin sections of tissue embedded in n-butyl methacrylate (14). It was found that, although of somewhat varied appearance, the endoplasmic reticulum was present in cells prepared by either method. The observations showed that the reticulum is a network of cavities bounded by membranes which may enlarge into relatively vast, flattened vesicles which were described as cisternae. Most of the electronmicrographs lack the detail that began to appear in later publications but in one fairly sharp picture of an acinar cell of the parotid gland in a newborn rat attention was drawn to numerous small granules which covered the outside surface of the limiting membranes of the endoplasmic reticulum and which were arranged in an orderly pattern around the nucleus. These granules were not discussed at this time but they were soon to be brought to a position of prominence in the next few years as the ribosomes and polysomes of central importance to protein biosynthesis. Palade extended observations of the endoplasmic reticulum to a wide variety of cell types and demonstrated a continuity for the membranes of the endoplasmic reticulum with that of the outer nuclear membrane. Attention once more was drawn to the dense granules attached to the cytoplasmic side of part of the endoplasmic reticulum (15).

Palade Granules or Nucleoprotein Granules or Ribosomes

In 1955 Palade wrote a paper entitled, "A Small Particulate Component of the Cytoplasm (16)." These granules, which for a while after the appearance of this work, were known as Palade Granules, were dense bodies measuring 100 to 150 A in diameter after fixation. Although their dense appearance was enhanced by osmium fixation, they were also demonstrable after fixation with formaldehyde and were therefore not the result of the osmium treatment. They were found in forty different cell types and were mostly attached to the membranes in contact with the cytoplasmic matrix—the endoplasmic reticulum and outer nuclear membranes. They occurred singly, in random scatter, in linear series, loops, spirals, circles, and rosettes, or closely packed. Their appearance and occurrence in various tissues could be correlated with the small granules obtained by centrifugation which are very rich in RNA and seem to account for the basophilic character of the cytoplasm. Although the term ergastoplasm had been used to describe the basophilic substance of the cytoplasm, Palade suggested that this term was not appropriate for the endoplasmic reticulum since its basophilic properties seemed to depend on the amount of attached small granules. The terminology preferred by Palade, and which has since come into general use, is rough- or smooth-surfaced endoplasmic reticulum, depending on the presence or absence of attached granules.

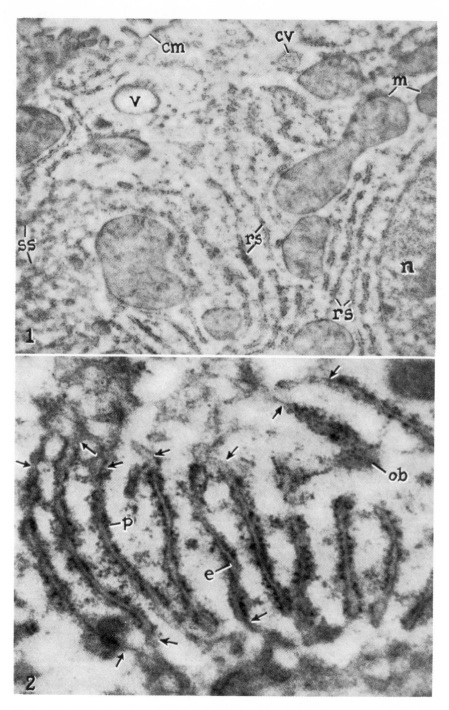

Figure 7a Endoplasmic reticulum *in situ* in rat liver. The top figure shows an electron micrograph of part of a parenchymatous liver cell at a relatively low

Although the smooth-surfaced endoplasmic reticulum is frequently found in continuity with the rough-surfaced system, the smooth variety rarely forms flattened cisternae, but usually occurs in close-meshed networks of branching tubules. In electronmicrographs its profile appears most often as isolated ellipses, circles, or short segments of branching tubules. In liver the system seems to be concerned with detoxification mechanisms and with lipid and cholesterol metabolism. The administration of certain lipid-soluble drugs causes a marked increase in the amount of smooth endoplasmic reticulum (Figure 11).

The Identity of Microsomes

A now classic study in which the power of electron microscopy was blended with biochemical analysis to characterize a cell fragment known to be important in protein synthesis was reported by Palade and Siekevitz (17). These authors examined the microsome and postmicrosome fractions obtained by differential centrifugation of rat liver homogenates. They further fractionated the microsomes by treatment with Versene, deoxycholate, and ribonuclease. Their studies showed that the microsomes were derived from the endoplasmic reticulum either by a pinching-off process or by fragmentation of the membranes followed by a healing process, and therefore the microsomes appeared as a collection of closed vesicles (Figure 7). Both the smooth- and rough-surfaced

---

magnification. The nucleus appears at *n* and the cell membrane, limiting a bile capillary, at *cm*. A number of mitochondrial profiles (*m*) can be seen in the cytoplasm that contains, in addition, numerous profiles of the endoplasmic reticulum. The latter vary in shape from circular to elongated and belong to both the smooth surfaced (*ss*) and the rough surfaced variety (*rs*). The vacuolated structure at *v* probably derives from a dense, peribiliary body. From G. E. Palade and P. Siekevitz, "Liver Microsomes," *J. Biophys. Biochem. Cytol.*, **2**, 171 (1956). The bottom electron micrograph shows a small field in the cytoplasm of a parenchymatous liver cell with an array of 10 elongated profiles, one of them marked *e*, which in three dimensions correspond to a pile of preferentially oriented cisternae. The profiles belong to the rough surfaced variety; that is, bear small dense particles (*p*) attached to the outer surface of their limiting membrane. The normally sectioned profiles show clearly their limiting membrane and their content. An obliquely sectioned cisterna (*ob*) offers a side view of its particle-dotted membrane. The upper left and the lower right corners of the figure are occupied by short, contorted profiles of smooth surfaced variety. Their distribution indicates that in three dimensions they correspond to a tightly meshed, randomly oriented reticulum. Rough-surfaced and smooth-surfaced profiles are found in continuity in many placed (arrows), a finding that indicates that the two profile varieties are not different unrelated structures but represent local differentiations within a common continuous system, the endoplasmic reticulum. Magnification 56,000 X. From G. E. Palade and P. Siekevitz, *J. Biophys. Biochem. Cytol.*, **2**, 171 (1956).

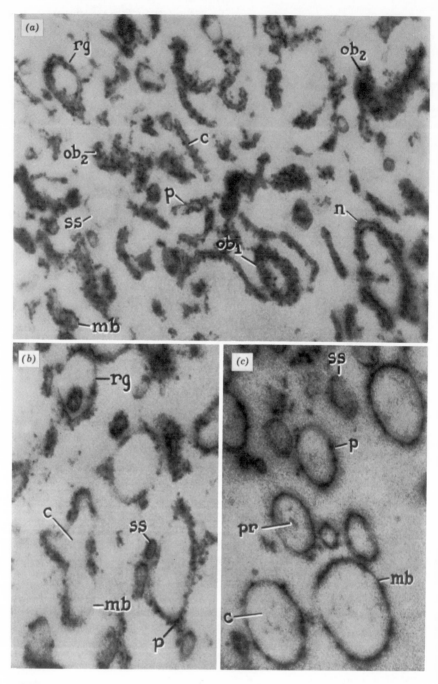

Figure 7b Elements of the isolated microsome fraction from rat liver. These figures illustrate the osmometer-like behavior of isolated microsomes. All corresponding

forms of the endoplasmic reticulum contributed, since smooth- and rough-surfaced profiles were present in the microsomal fraction.

In a less comprehensive study three years prior to the results reported in this work, Slautterback had seen the elements of the microsome fraction but had not realized their relation to each other or to the endoplasmic reticulum (18). The unit components of the microsome fraction are the membranes of the endoplasmic reticulum and the dense granules attached to some of the membranes. Versene could cause the release and destruction of the granules; metal ions therefore appeared to be involved in holding the granules together and binding them to the membranes. Deoxycholate, a surface active agent, causes the release of the particles by depolymerizing the membranes. Chemical analysis showed that the granules released by 0.5% deoxycholate contained somewhat less RNA than protein and about 10% by weight of phospholipid. Nearly all of the associated phospholipid, DPNH-cytochrome c-reductase and an alcohol-soluble hemochromogen were solubilized by deoxycholate (DOC) and these substances appeared to be contained by the membrane component. Since studies on the biochemistry of protein synthesis had implicated microsomes (see page 17) and these studies showed that microsomes existed in the cell in the form of the endoplasmic reticulum, concepts of the cell's involvement in protein synthesis should be linked to this structure.

An analogous study (19) using guinea pig pancreas was reported by Palade and Siekevitz soon after the report of this first investigation

microsome pellets were isolated from liver homogenates prepared in 0.88 M sucrose. All were fixed *in toto* in 2 per cent OsO_4 dissolved in 0.88 M sucrose for the pellet in (a) in 0.028 M Na acetate—0.028 M Na veronal for the one in (b) and, finally, in distilled H_2O for the pellet in (c). (a) The electron micrograph shows that microsomes fixed in the presence of 0.88 M sucrose retain the characteristic, flattened appearance of intracellular cisternae and appear to have a content slightly denser than the embedding plastic. (b) The microsomes appear as relatively large vesicles with a content similar in density to the embedding plastic. A comparison with (a) indicates that these microsomes have swollen in the hypotonic fixative without losing the small, dense particles attached to their membrane. (c) Fixation in the presence of distilled water has caused further increase in size and complete enspherulation of the microsome vesicles (their oval appearance is due to compression by the microtome knife). They retain their attached particles, but the latter appear less distinct than in (a). Some of the microsomal vesicles contain a faint, fine precipitate (pr). In all three figures, mb indicates the membrane of the microsomes; c, their content; p, attached particles; and ss, smooth surfaced profiles. For (a) and (b), rg indicates ring-shaped profiles. In (a), n is a normally sectioned fragment of the endoplasmic reticulum, ob_1 and ob_2, fragments cut at increasing degrees of obliquity. All three figures 52,000 X. From G. E. Palade and P. Siekevitz, *J. Biophys. Biochem. Cytol.*, **2**, 171 (1956).

with rat liver. Essentially similar results were obtained in that the microsome fraction derived from this tissue was composed of small closed vesicles of granule-studded membranes which had the same morphological appearance as the endoplasmic reticulum of the intact cell. Some smooth vesicles were also encountered. In pancreas, significantly more granules occurred free in the cytoplasm than were found for liver. These granules appeared very similar to those attached to membranes. A third component was recognized in the microsomal fraction from guinea pig pancreas and this was identified as intracisternal granules. Pancreatic microsomes were found to have less phospholipids and hemoprotein than liver microsomes and an apparent lack of DPNH-cytochrome c-reductase activity. Also, the stability of the pancreatic microsomes to various treatments differed from that of liver; for example, they were more disrupted by aging, their membranes were solubilized by lower DOC concentrations, and the granules were more easily removed by ribonuclease. A tendency of the granules to associate into chains and clusters was also observed.

At a symposium held in 1955 at the Henry Ford Hospital in Detroit, Palade presented a clear and comprehensive report on the status of cytoplasmic structures (20). There were several differences in interpretation between Palade and Sjöstrand on the inner structure of mitochondria; for example, they differed as to whether the cristae were complete septa (partitioning walls) or incomplete partitions as favored by Palade. Palade presented strong evidence in support of the concept that the cristae are discontinuous partitions which leave a centrally open mitochondrial channel or matrix.

In terms of light microscopy an ill-defined structure called the centrosphere was often mentioned. The structure was depicted as surrounding the centrosome. More recently electron microscopy has revealed piles of smooth-surfaced cisternae packed in a characteristically tight manner and surrounded by swarms of small and large vesicles in this area. Palade considered that these formations represented a modification of the endoplasmic reticulum, whereas other microscopists, principally Dalton (see page 168), considered this structure to be a distinct cytoplasmic entity and, in fact, to be the Golgi apparatus. Brief mention was made by Palade of lysosomes, whose existence had just been postulated by de Duve et al. (21).

Lysosomes

The discovery of these cytoplasmic organelles was the unexpected result of a biochemical investigation. The later demonstration of their cytological existence was a confirmation of the initial indications of their presence reported by Appelmans, Wattiaux, and de Duve (22). By using techniques of differential centrifugation these authors found that the enzyme acid phosphatase sedimented partly with the mitochondria, as characterized by their associated oxygen uptake ability or cytochrome oxidase activity, and partly with the microsomes, characterized by their associated glucose-6-phosphatase. Under the best conditions one third of the respiratory activity was isolated practically free of acid phosphatase, and more than 70% of the cytochrome oxidase was separated in association with less than 10% bound phosphatase. Other fractions containing 10 to 20% of the original bound phosphatase of the extract contained only traces of cytochrome oxidase. From the sedimentation characteristics of these enzyme activities the authors proposed the existence of granules somewhat smaller in size than the mitochondria but larger than microsomal units.

These studies were pursued in greater detail by de Duve, Pressman, Gianetto, Wattiaux, and Appelmans, who studied the sedimentation characteristics of 13 enzymes, some of which were known to be valid markers for mitochondria and microsomes (21). They separated a rat liver homogenate by differential centrifugation into nuclei, heavy mitochondria, light mitochondria, microsomes, and supernatant fractions. The recoveries in terms of enzyme activities and nitrogen were excellent. According to their distribution pattern, the enzymes fell into four groups. The first group comprising cytochrome oxidase, succinate-cytochrome c-reductase, rhodanese, 40% of the total fumarase, and the antimycin A-sensitive DPNH and TPNH-cytochrome c-reductases, represented the true mitochondria and was concentrated mostly in the heavy mitochondrial fraction. A second group comprising glucose-6-phosphatase, antimycin A-insensitive DPNH and TPNH-cytochrome c-reductases, cytochrome b_5, part of the fumarase activity and a special type of β-glucuronidase, was found in the microsomal fraction. The third group of enzymes including acid phosphatase, ribonuclease, deoxyribonuclease, cathepsin, and most of the β-glucuronidase, sedimented principally in the light mitochondrial fraction and added considerable support to the idea that the enzymes were contained in granules other than mitochondria. The name lysosome was proposed to emphasize their richness of content in hydrolytic enzymes. The enzyme uricase had some unique distribution characteristics and the possibility of a granule intermediate between lysosomes and microsomes was considered. A summation of characteristics of lysosomes was presented by de Duve (23). A collection of hydrolytic enzymes having generally an acid pH optimum enclosed by a lipoprotein membrane and occurring in a variety of cell types adequately describes the structures which, assuming a spherical shape, would measure about 0.4 μ in diameter. Procedures which destroy the lipoprotein membrane release the various enzymes in a parallel fashion. Using a sucrose density gradient centrifugation technique, the enzyme uricase was found in a different region from that of acid phosphatase. Once again the possible existence of another group of particles was considered (see page 177). The morphological identification of lysosomes was attempted by electron microscopy on fractions enriched with lysosomes (Figure 8). They appeared as dense bodies very similar to structures previously described as peribilary bodies by Rouiller (24).

An excellent summation by Palade of the characteristics of the endoplasmic reticulum appeared in 1956 (25). He pointed out that although spread preparations of single cells allowed one to see the continuity and reticular characteristics of the system, the newer techniques with ultra-thin sections of 20 to 40 mμ (which is smaller than the mesh size of the reticulum and even of the diameter of its vesicles and tubules) produce only profiles of the system. The reticulum has been found in all cells examined with the exception of the mature erythrocyte. Cells of more recent biochemical interest, such as rabbit reticulocytes and *E. coli*, also appear to lack such a system. The endoplasmic reticulum, however, has different characteristics in different cell types and even in the same kind of cell under different conditions. The variations include the extent of the system, its volume, the content of attached granules, and its arrangement in the cytoplasm. Basically, the reticulum appears to be an interconnected network of tubules, vesicles, and flattened sacs or cisternae, and has continuity with the outer nuclear membrane. A connection with the plasma membrane has been suggested but is not established. For a time other cytologists called it a double membrane system, but Palade explained that such images would result from folded membranes, membranes in apposition, and flattened vesicles. It is an interesting thought that the space within the two nuclear membranes and that enclosed within the endoplasmic reticulum may at times be in direct communication with the outside surface of the cell and separated by a membrane from the inside of the cytoplasm. Palade also emphasized that the Golgi elements which appear to be continuous with the endoplasmic reticulum may represent a local differentiation of the system. Figures 9–12 show recent electron micrographs of the endoplasmic reticulum.

Modern Concept of the Golgi Body

Studies initiated in 1953 by Dalton, Felix, Schneider, and Kuff lifted the Golgi apparatus out from a background of uncertainty and skepticism and firmly established it as an important cytoplasmic structural

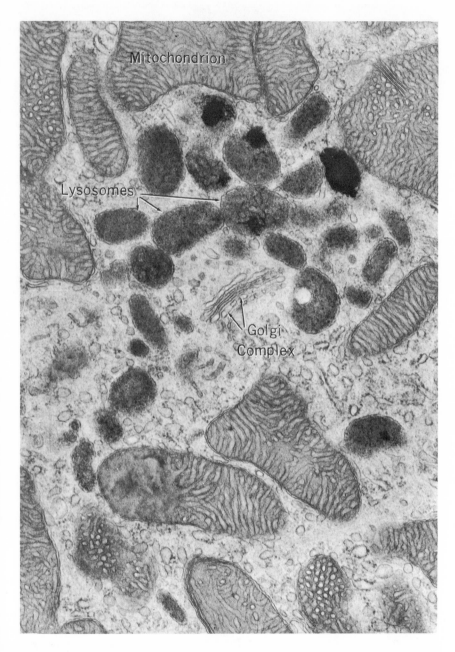

Figure 8 Hamster suprarenal cortex, collidine buffered glutaraldehyde and osmium fixation. Lead citrate staining. Magnification, 23,000 X. From D. W. Fawcett, *An Atlas of Fine Structure,* Saunders, Philadelphia, (1966).

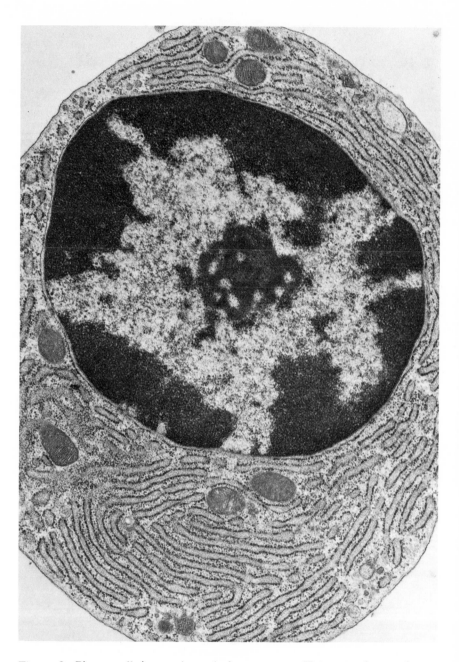

Figure 9 Plasma cell from guinea pig bone marrow. Note extensive rough surfaced endoplasmic reticulum throughout cytoplasm. Cacodylate buffered glutaraldehyde and osmium fixation. Uranyl acetate and lead citrate staining. Magnification, 19,000 X. From D. W. Fawcett, *An Atlas of Fine Structure*, Saunders, Philadelphia, (1966).

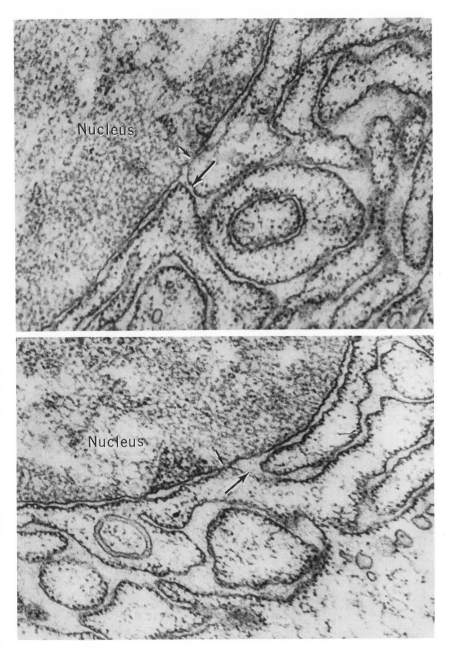

Figure 10 Serous cells of the mouse salivary gland. Veronal acetate buffered osmium fixation. Lead hydroxide staining. (Micrographs courtesy of Dr. Harold Parks.) Note continuity of endoplasmic reticulum with outer nuclear membrane (large arrows), nuclear pores (smaller arrows) and ribosomes on outer nuclear membrane. Magnification, 67,000 X. From D. W. Fawcett, *An Atlas of Fine Structure*, Saunders, Philadelphia, (1966).

171

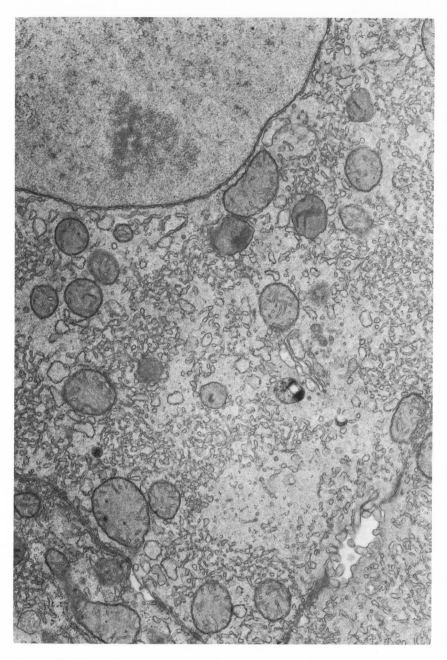

Figure 11 Liver of a hamster fasted three days and given daily injections of 80 mg/kg of phenobarbital. Tubular profiles of smooth (agranular) endoplasmic reticulum are clearly seen. Phosphate buffered osmium fixation. Lead citrate staining. Magnification, 16,000 X. From D. W. Fawcett, *An Atlas of Fine Structure,* Saunders, Philadelphia, (1966).

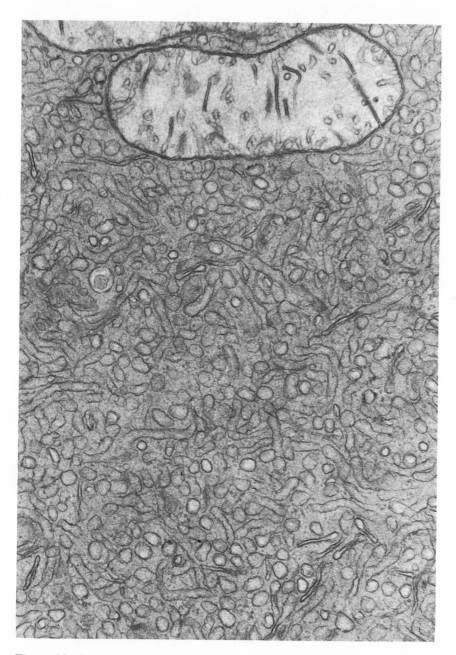

Figure 12 Agranular reticulum from opossum testicular interstitial cell. Collidine buffered glutaraldehyde and osmium fixation. Lead citrate staining. Magnification, 38,000 X. From D. W. Fawcett, *An Atlas of Fine Structure*, Saunders, Philadelphia, (1966).

entity with a respectable fine structural image. Dalton and Felix reported on studies of epithelial cells of the duodenum and head of the mouse epididymis (26). Observations were reported of the tissues in the fresh state with the phase contrast microscope and in supravitally stained preparations with both the phase contrast and light microscopes, and with fixed preparations using the phase contrast, light, and electron microscopes. In reference to arguments that the appearance of this structure was the result of the previous treatment of the tissue, the authors stated, "The comparison of the appearance of the Golgi substance in the fresh state with that in fixed impregnated specimens, both with the light and electron microscopes, indicates that the form of the Golgi substance has been affected little if at all by the classical methods used to visualize it." Contrary to the observations of other investigators with other cell types, the Golgi substance was not stainable with neutral red or methylene blue. In the same year Schneider, Dalton, Kuff, and Felix reported on the isolation of Golgi structures from homogenates of rat epididymus using a centrifugation technique in a tube of layered sucrose solutions of increasing densities (27). In the following year, Dalton and Felix

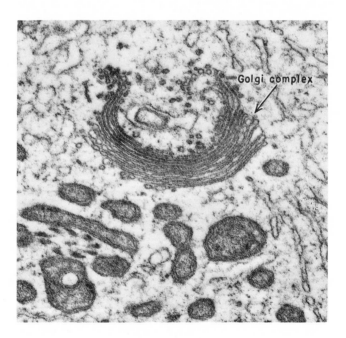

Figure 13 Golgi complex in spermatid of Helix pomatia. Mitochondria also are present. Fixed in chromosmium and stained with lead hydroxide. Magnification, 38,000 X. Courtesy of Dr. A. J. Dalton, National Cancer Institute.

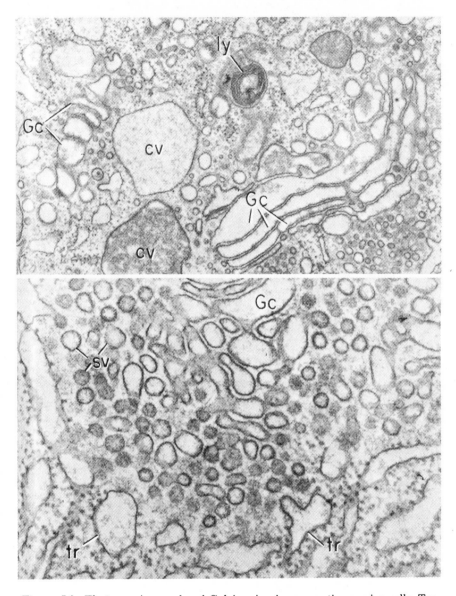

Figure 14 Electron micrographs of Golgi region in pancreatic exocrine cells. Top. All elements of Golgi complex at low magnification 23,000 X. Bottom. Peripheral elements of complex at higher magnification (55,000 X). Symbols: *tr*, transitional elements; *sv*, small peripheral vesicles; *Gc*, Golgi cisternae, *cv*, condensing vacuoles; *ly*, lysosomes. From G. E. Palade, *J. Amer. Med. Assoc.*, **198**, 815 (1966).

described experiments which cytologically followed the release of the Golgi substance during disruption of the epithelial cells by crushing or homogenization, and further described the cytological and staining properties of the organelle *in situ* and in isolation (28). Schneider and Kuff, in an accompanying paper (29), described in greater detail their experiments on the isolation and biochemical characterization of the Golgi substance (Figures 13 and 14). (For putative functions of the Golgi system, see page 272.)

The Nuclear Membrane

The ultrastructure of the nuclear membrane was first revealed in 1950 by Callan and Tomlin who studied stretched isolated nuclear membranes from amphibian oocytes (30). These studies revealed two layers of nuclear membranes, the inner of which was believed to be continuous, and the outer porous with its tightly packed pores bordered by rings called annuli. Many workers with other tissues have confirmed these gross features. Afzelius studied the ultrastructure of the nuclear membrane of the sea urchin oocyte by means of the newer ultra-thin sectioning technique (31). Although Callan and Tomlin thought of the two membranes as distinct and separable entities, Afzelius concluded that the system was a double membrane with discontinuities (pores) encircled by walls (annuli). The central portion or inner diameter of the pore is 300 to 400 A and the outer diameter measures about 1000 A.

The nature of the central portion has been the subject of much debate. Is it an open channel? Is it covered by a separate membrane or by the continuity of one of the two nuclear membranes or is it plugged in some manner? After considering various of the arguments and observations, R. W. Merriam tended to favor the idea of a diaphragm that is part of the pore structure itself but distinct from the membranous elements of the envelope (32). In another study Merriam pointed out that pores are not a common feature of membranes and are even absent from the endoplasmic reticulum which is continuous with the outer nuclear membranes (33). Therefore, pore formation is imposed on the membranes by some property of the nuclear surface. Merriam reviewed reports on the existence of a central globule within the pore and speculated on its possible association with the diaphragm deduced in his earlier work.

Watson described the pore as being formed by a joining of the outer and inner membranes around it (34). Because the outer membrane is continuous with the endoplasmic reticulum, he had suggested earlier that the nuclear envelope might be considered as a specialized element of the endoplasmic reticulum. This latter thought is interest-

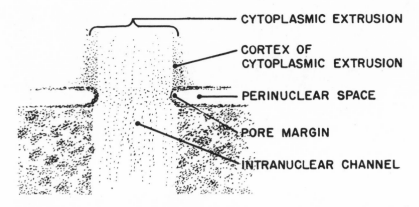

CYTOPLASMIC EXTRUSION

CORTEX OF
CYTOPLASMIC EXTRUSION

PERINUCLEAR SPACE

PORE MARGIN

INTRANUCLEAR CHANNEL

Figure 15 Schematic diagram of the pore complex as seen in sections perpendicular to the nuclear surface. From M. L. Watson, *J. Biophys. Biochem. Cytol.*, **6**, 147 (1959).

ing insofar as it would be impossible to isolate intact nuclei free of cytoplasmic contamination. Watson's studies with animal cells were entirely negative with respect to the existence of a diaphragm or membrane covering the pore. His investigations revealed a spacing of 200 to 300 A between the two nuclear membranes, and pore diameters of about 1000 A. He described a structure which he called the pore complex (Figure 15). This unit appears to be a cylindrical channel of about 1200 A in diameter which runs through the center of the pore and penetrates deep into the nucleus where it gradually loses its identity by anastomosing and blending with the nucleoplasm. On the cytoplasmic side it extrudes for several hundred angstroms. Crescents and spirals of ribonucleoprotein particles frequently appear on the cytoplasmic side of the nuclear envelope around the annuli and may possibly be associated with the pore complex, although no such association was actually demonstrated.

Figures 16 and 17 show two views of nuclear membranes and pores.

Microbodies

Microbodies seem to have been first noticed in 1954 by Rhodin who described these organelles in his dissertation on ultrastructural organization and function in the brush border kidney cells of the mouse. Rhodin considered the units as entirely distinct from mitochondria. Two years later these organelles in hepatic cells were the objects of a study of Rouiller and Bernhard (35). They were described as ovoid or, less frequently, as round dense granules between 0.1 and 0.5 mμ in diameter,

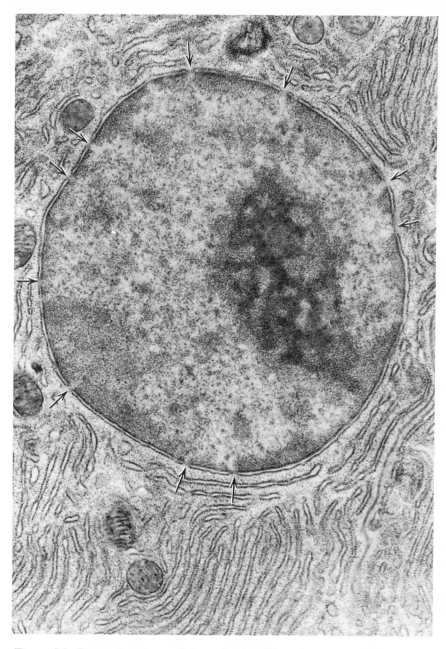

Figure 16 Pancreatic acinar cell from the bat. Note the two membranes of the nuclear envelope and the frequent occurence of pores (arrows). Collidine buffered osmium fixation. Lead hydroxide staining. Magnification, 18,000 X. From D. W. Fawcett, *An Atlas of Fine Structure*, Saunders, Philadelphia, (1966).

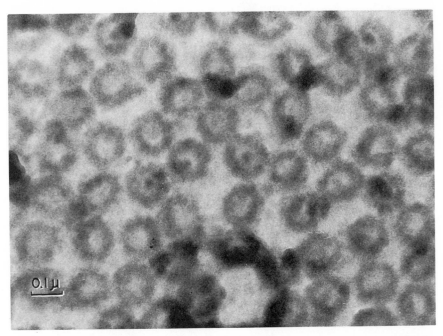

Figure 17 Nuclear envelope of immature frog oocyte after fixation in OsO_4 showing annuli. Magnification 78,000 X. (From R. W. Merriam, *J. Cell Biol.* **12**, 79 (1962).

limited by a single well-defined membrane and containing a dense, finely granulated substance. A densely opaque and homogeneous core occupies the center. Although they were easily distinguishable from mitochondria, analogies were drawn between the structure and cellular locations of these bodies and those of the mitochondria. The authors suggested that transition stages between microbodies and mitochondria exist and that, in fact, microbodies are the cytological precursors of mitochondria. Novikoff discarded this interpretation of Rouiller and Bernhard and pointed out the similarity in nature of microbodies and lysosomes (36). In a preliminary note (37) de Duve *et al.* described the results of improved fractionations of a rat liver homogenate in a glycogen gradient. Lysosomes were concentrated at the bottom of the tube but enzyme activities for uricase, catalase, and D-amino acid oxidase occurred together, suspended in the gradient.

More recently (38) Hruban and Swift identified the dense crystalloid inclusion of the hepatic microbody as crystalline uricase and thus provided evidence for a suggestion of Beaufey and de Duve that microbodies

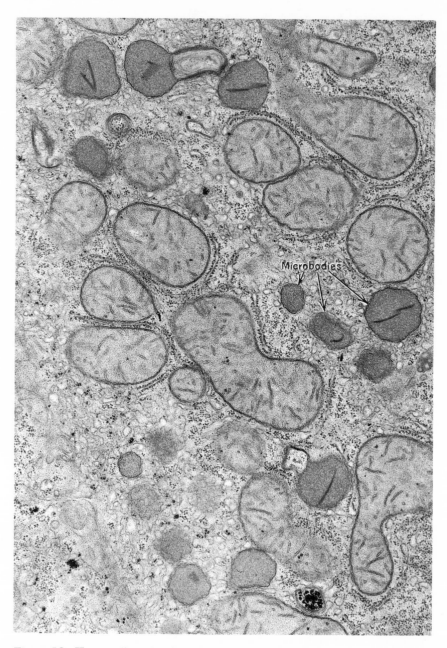

Figure 18 Hamster liver showing microbodies and thin dense crystals within them. Phosphate buffered osmium fixation. Lead citrate staining. Magnification 20,000 X. From D. W. Fawcett, *An Atlas of Fine Structure*, Saunders, Philadelphia, (1966).

contain uricase and, by inference, catalase and D-amino oxidase as well. The origin and function of these bodies, however, still remain a mystery (Figure 18).

Microtubules

From time to time in asymmetric or elongated cytological formations such as flagella, cilia, or sperm cells, dense filaments were observed and described (39, 40). Ledbetter and Porter pointed to the wide occurrence of such filaments and microtubules in asymmetric cells and in the so-called fibrils common to the mitotic spindle of dividing cells. These microtubules have an outer diameter of 230 to 270 A and they are formed from slender filamentous subunits which have center-to-center spacings of 45 A and which are themselves tubular in nature. They suggested that these units comprised a cytoplasmic structure of wide occurrence in the cells of both plants and animals (Figure 19). Ledbetter and Porter stressed that older techniques of tissue fixation in osmium tetraoxide did not preserve these structures, but that a preliminary treatment with gluteraldehyde followed by osmium tetraoxide resulted in clearer images of this cytological unit. Ledbetter and Porter emphasized that microtubules are primarily associated with portions of moving cytoplasm such as flagella and mitotic spindle-fibers, and they proposed that bending, shortening, or elongation of the fibers could set up an undulatory motion for the surrounding ground substance of the cytoplasm. Sandborn, Koen, McNabb, and Moore proposed that microtubules were common to all areas of the cytoplasm and were in all cells of the rat in particular and mammalian cells in general (41). They also stressed the importance of preliminary fixation in glutaraldehyde in order to preserve the structure for electron microscopy.

Fawcett and Witebsky described a series of about 25 microtubules which formed a bundle that encircled the nucleated erythrocyte of the toadfish just beneath the plasma membrane. They interpreted the findings in terms of the maintenance of the discoid shape of this type of cell (42). This study thus provided additional evidence for a suggestion made by Meves in 1904 after he observed such a rigid structure with the light microscope.

Mitochondrial Fine Structure

An important step in the elucidation of mitochondrial fine structure was taken by H. Fernández-Morán (43). Improved methods of specimen preparation and electron microscopy developed mainly by Fernández-Morán were applied in this study. The principal developments employed were improved low-temperature or cryofixation techniques, and use of

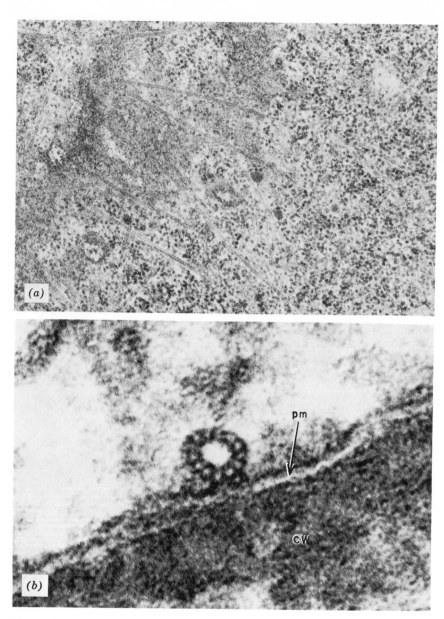

Figure 19 (a) Electron micrograph of section cut nearly parallel to and including a portion of the end wall (gray area at upper left) and adjacent cell cortex from root tip of Phleum pratense. The long slender microtubules, unlike those found in parallel array along the side walls, show an apparently random orientation within the cell cortex adjacent to the end wall. Magnification, 43,000 X. (b) Transverse section of microtubule from cortex of Juniperus chinensis root-tip cell, showing circular subunits which compose the wall. (pm, plasma membrane; cw, cell wall.) Electron micrograph. Magnification, 40,000 X. From M. C. Ledbetter and K. R. Porter, *Science*, **144**, 872 (1964). Reproduced with permission from the American Association for the Advancement of Science. Copyright 1964.

wet or partially hydrated preparations which were enclosed between impermeable films in vacuum tight microchambers and which were examined with microbeams of very low intensity in specially cooled chambers. Standard osmium-fixation which produces positive images (dark structures against a light background), and negative staining with phosphotungstate revealed that the inner mitochondrial membrane, particularly in cristae, possessed a repeating superficial structure (Figure 20). This structure was shaped like a balloon on a stalk. The diameter of the spherical part was 80 to 100 A. Fernández-Morán referred to the structure as "elementary particle (EP)" with the explicit assumption that it constituted both a basic building block of the membrane and that it contained the entire integrated assembly of elements of the respiratory chain. This declaration started in motion a controversy between Fernández-Morán, D. E. Green, and their collaborators at the Institute for Enzyme Research in Madison, Wisconsin, and other groups, the most notable being those of Britton Chance at the University of Pennsylvania and Donald Parsons at the Cancer Institute in Canada. At issue was the biochemical function of this newly discovered cytological subunit. (This argument is reviewed in some detail in Chapter 6.)

Shortly after the report of Fernández-Morán, Parsons published electron micrographs of negatively stained preparations of disrupted mitochondria from mouse and rat liver (44). These photographs showed in clear detail the close packing of subunits on the surface of the inner mitochondrial membranes (Figure 21). The dimensions reported were 75 to 80 A for the spherical heads, and 30 to 35 A for the width of the stems, which were 45 to 50 A long. The center-to-center spacing for adjacent units was 100 A. In all, eleven types of mitochondria were examined from normal rat liver, kidney, and hepatoma; normal mouse liver, kidney, and brain; leukemic mouse thymus and lymph nodes; two types of mouse plasma-cell tumors, and normal bovine pancreas. All showed the projecting subunits. Parsons also reported the existence of a new subunit which he found on tubular projections of the outer membrane surface. This structure was interpreted to be a hollow cylinder 60 A tall and 60 A wide with a central channel of 20 A diameter and the packing density was such that the center-to-center distance for adjacent units was 80 A. More complete details of the structure of the subunit or the cristal surface and of the arguments linking it to the respiratory assembly were provided in a paper by Fernández-Morán, Oda, Blair, and Green (45). Working with beef heart muscle mitochondria, these investigators reported that 2000 to 4000 particles covered a square micron of membrane surface and they estimated that 10,000 to 100,000 particles could be contained per mitochondrion, depending

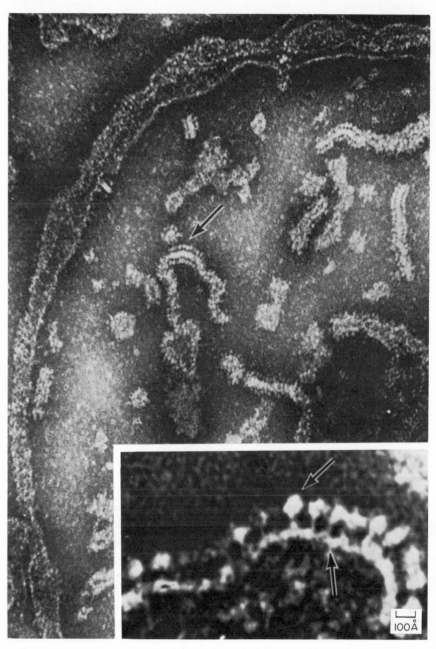

Figure 20 Isolated beef heart mitochondrion embedded in thin phosphotungstate (PTA) layer. Prepared by microdroplet cross-spraying procedure involving only

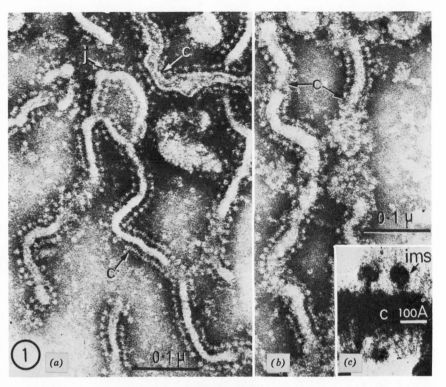

Figure 21 The subunit associated with the inner membranes or cristae. (*a*) Part of a mitochondrion from a negatively stained preparation of mouse liver. A few of the cristae (*c*) are shown. The cristae consist of long filaments which branch at some points (*j*). The surfaces of the cristae are covered with projecting subunits. Magnification 163,000 X. (*b*) Negatively stained cristae (*c*) prepared by spreading isolated lysed rat liver mitochondria. The subunits on the cristae appear similar to those of (*a*). Magnification, 163,000 X. (*c*) Higher magnification—a few subunits from the same preparation as (*b*). The spherical heads are 75 to 80 A diameter and the stems 30 to 35 A wide and 45 to 50 A long. The center-to-center spacing is 100 A. Reversed print. Magnification 650,000 X. From D. F. Parsons, *Science,* **140,** 985 (1963).

brief interaction of mitochondrial suspension in 0.5 M sucrose with 1 per cent PTA. Notice characteristic paired arrays of elementary particles in profiles of fragmented cristae which are readily distinguishable from the envelope of the mitochondrion. Magnification, 85,000 X. Inset: Enlarged segment of crista in a specimen similar to that shown in the main figure. Demonstration of three parts of the elementary particle: headpiece, stalk, basepiece. Invariant association of the arrays of headpieces *EP* with the underlying dense layer of crista. Segmentation and knoblike protruberance of dense layer at point of attachment of stalk. Magnification 420,000 X. From H. Fernández-Morán et al., *J. Cell Biol.,* **22,** 63 (1964).

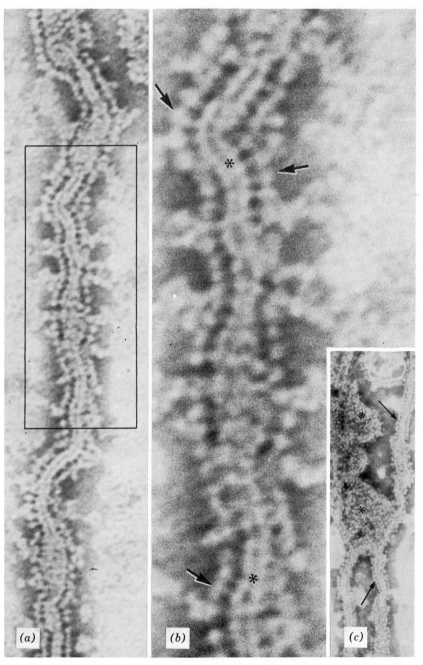

Figure 22 a   A portion of a ribbon from a disrupted sarcosome (muscle mito-chondrion) negatively stained with potassium phosphotungstate. The electron-opaque stain reveals a series of spherical particles, associated with the axis of the

on its size and type. In addition to the headpiece and stalk, this group felt that a basepiece in the membrane, with approximate dimensions of 40 × 112 × 112 A could be discerned. The existence of the projecting subunit on the inner mitochondrial membrane has been confirmed by many laboratories (Figure 22) (46–50).

The structure of muscle cells. Muscle cells contain an intricate and unique type of cytological structure, with a highly differentiated form of smooth-surfaced endoplasmic reticulum and a fine transverse tubular (T) system. It is beyond the scope of this brief cytological survey to cover this work; the interested reader may pursue the details in the classic papers describing these systems (51–53).

Chloroplasts

An excellent general description of the architecture of chloroplasts is provided by Ruth Sager (54). She summarized the conclusions drawn from light microscopy:

1. All the chlorophyll of the cell is contained in chloroplasts or chromatophores.

---

ribbon. The laterally placed particles are most clearly seen, but they are also present over the more expanded regions of the structure [cf. (*c*)]. As more clearly seen in *b* (representing the boxed area at higher magnification), the particles appear to be attached by stalks to the axial structure of the ribbon. Magnification, 250,000 X. From D. S. Smith, *J. Cell Biol.,* **19,** 115 (1963). **b** A portion of the field shown in (*a*) at higher magnification. The particles associated with the axis of the ribbon are spherical or slightly ovoid, 80 to 95 A in diameter, and the negative stain clearly indicates that (as in the regions indicated by arrows) each particle is produced into a stalk, 40 to 50 A in length and 30 to 40 A in diameter, apparently linking the "head" of the particle to the axial structure. There is some indication (*) that the axis of the ribbon is differentiated at the region of attachment of each "stalk" into a "dumbbell" configuration. Similar stalked particles also are seen in association with more or less intact cristae (from which the ribbons are thought to be derived), and on the fragments believed to represent the outer membranes of the sarcosome. It seems likely that all the particles visualized in these negatively stained preparations are associated with a substrate membrane in this manner. Magnification, 600,000 X. From D. S. Smith, *J. Cell Biol.,* **19,** 115 (1963). **c** A low-power field of a membranous ribbon from a negatively stained sarcosome preparation, isolated in deionised water, from flight muscle of the butterfly Nymphalisio. In addition to the sarcosomes of Calliphora and the species illustrated here, sarcosomes from the beetle Tenebrio were also examined and in each instance the appearance of the sarcosome fragments, after suspension in potassium phosphotungstate solution, was found to be similar. In this figure particles flanking the ribbon axis are seen (arrows) and particles are also distributed over the more expanded portions of the ribbon axis (*). Magnification, 90,000 X. D. S. Smith, *J. Cell. Biol.,* **19,** 115 (1963).

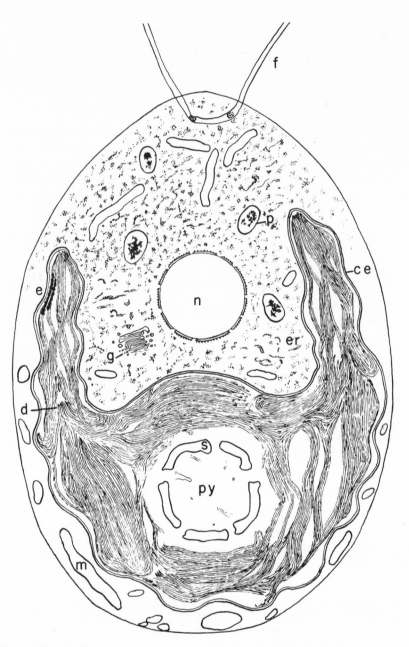

Figure 23a Diagrammatic sketch of normal green *Chlamydomonas* as seen at low magnification in the electron microscope. The chloroplast is shown surrounded by the double chloroplast envelope (*ce*) within which the eye-spot (*e*), pyrenoid (*py*), and starch plates (*s*) are located, as well as the paired lamellar membranes

2. Some chloroplasts contain round bodies visible under favorable conditions in the light microscope; these have been called grana.

3. Chloroplasts exhibit negative uniaxial form birefringence, suggestive of a submicroscopic laminated structure.

4. Chloroplasts have osmotic properties, swelling or shrinking in solutions of appropriate tonicity, suggesting the presence of a surrounding membrane.

Sager and Palade studied the fine structure of chloroplasts in the flagellated unicellular organism *Chlamydomonas* (55). This cell contains a single large chloroplast which occupies about half the total volume of the cell (Figure 23). The organelle is bounded by a double membrane envelope much the same as for the mitochondrion minus the cristae. The grana of light microscopy were revealed to be stacks of flattened membrane vesicles which the authors described as discs. The space surrounding the grana and permeating between the discs themselves was referred to as the matrix and represented the area called stroma by the light microscopists. Grana are connected to each other by other lamellae. The space within each disc in a granum was called inner space and that between the two membranes of the envelope was named outer space, an intended pun in deference to its unexplored nature. In addition, the structure may contain granules, vesicles, lipid droplets, and starch granules within the matrix. These features, with minor variations, are common to all species reported in the literature. Other features that show more variation from species to species are unpaired lamellae, eyespots, and pyrenoids. The pyrenoid contains a network of tubules which connect with lamellar discs between starch plates at the periphery of the pyrenoid. The chlorophyl seems to be contained in the membrane structure of the discs of the grana.

The ultramicro structure of starch-free chloroplasts of fully expanded leaves of *Nicotiana Rustica* was described in detail by T. Elliot Weier (56). The chloroplast is bounded by a double membrane which may have pores in it. The grana within a chloroplast are connected to each other by a fine series of flexuous channels called a fretwork (Figure 24). More recent electron microscopic studies have revealed re-

---

arranged as discs. At low magnification the disc arrangement is clearly seen only in occasional well-oriented regions. The cytoplasm also contains other systems of organelles, including mitochondria (*m*), Golgi material (*g*), endoplasmic reticulum (*er*) consisting of membranes and RNA-containing granules, and vacuoles containing metaphosphate (*p*). The nucleus (*n*) is surrounded by a double membrane with pores and a dense coating of RNA-containing granules on its outer surface. From R. Sager, *Brookhaven Symposia in Biology*, No. 11, 1958, p. 101.

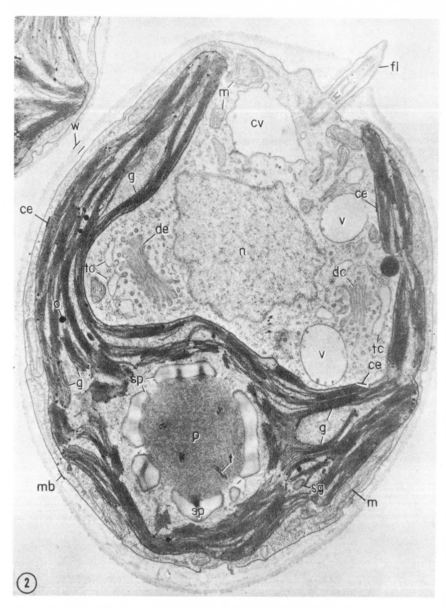

Figure 23 b General morphology of a mutant cell of *Chlamydomonas reinhardi* grown in the light. Cell fixed in 2 per cent OsO₄ in phosphate buffer and stained with UO₂ acetate and lead citrate. The figure shows the profile of a medially sectioned cell whose cell wall is marked *w* and cell membrane *mb*. The cap shape of the chloroplast and the location of the pyrenoid (*p*) within the thickened posterior part of the plastid are clearly shown. The chloroplast is bounded by a chloroplast

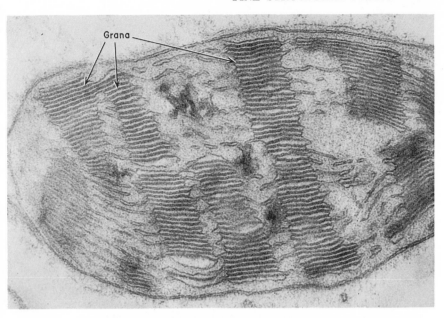

Figure 24 Oval plastid from a leaf of *Nicotiana Rustica* showing double membranous envelope, grana (membrane stacked discs), fretwork (membranous connections between grana) and stroma (matrix bathing the grana). 18,000 X. From T. E. Weier, *Amer. J. Botany,* **48**, 615 (1961).

peating units on the surface of chloroplast membranes analogous to the repeating units of the inner mitochondrial surface referred to as elementary particles (see page 183). Speculation is growing about the possibility that the integrated enzyme systems concerned with photosynthesis and photophosphorylation may be contained in these structural subunits (57, 58).

envelope (ce) and contains many grana (g), that is, stacks of fused discs connected by one or more free or fused discs. Osmiophilic globules (o) and a few starch granules (sg) are scattered among the grana. The pyrenoid (p) appears as a large finely granular mass of polygonal profile surrounded by discontinuous shell of starch plates (sp) and penetrated by a system of tubules (t). Within the chloroplast cup are located the nucleus (n), dictyosomes (dc) and their associated vacuoles (v), and endoplasmic reticulum cisternae of transitional type (tc). Mitochondria (m) are concentrated at the anterior pole of the cell and between the chloroplast and the cell membrane. A contractile vacuole is marked cv and a flagellum fl. Magnification 16,000 X. From I. Ohad, P. Siekevitz, and C. E. Palade, *J. Cell Biol.,* **35**, 521 (1967).

Bacterial Fine Structure

The bacterial cytoplast is also bounded by a plasma membrane. In gram-positive organisms this membrane is covered with a thick, rigid wall composed largely of mucopeptides and teichoic acids (frequently polyglycerophosphate or polysugar phosphate with esterified alanyl esters), polysaccharides, and sometimes protein. Although much thinner than from gram-positive organisms, the surface covering of gram-negative organisms is cytologically more complex. Kellenberger and Ryter proposed that the surface is composed of a multilayered system with the cytoplasmic membrane fitted to the inner layer (59). A detailed summary of information concerning this outer coat for *E. coli* is presented by Bayer and Anderson (60) (Figure 25). The outermost layer is made of lipoprotein and presents surface structural features of protrusions, humps, and channels. Just under this coat is a layer or double layer of lipopolysaccharide. The next layer in is described as a rigid

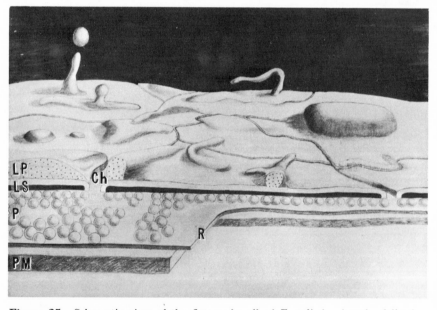

Figure 25   Schematic view of the flattened wall of *E. coli* showing the following features: *LP* = lipoprotein layer with protrusions and humps—it is dotted when seen in cross section; *LS* = lipopolysaccharide, containing the channels Ch; *P* = protein elements, covering the rigid glucosaminopeptide layer R; *PM* = the protoplasmic membrane. From M. E. Bayer and T. F. Anderson, *Proc. Nat. Acad. Sci.*, **54**, 1592 (1965).

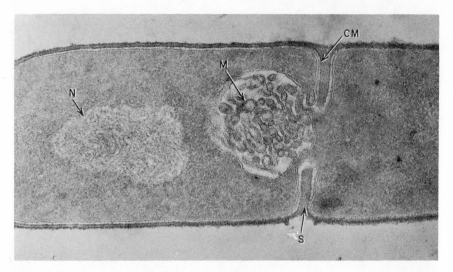

Figure 26   Part of a section of a rapidly dividing cell of *B. subtilis* strain SB 19. Osmium-fixed, uranyl acetate-treated. The nuclear area (N), mesosome (M), transverse septum (S), and cytoplasmic membrane (CM) are visible. 120,000 X. From F. A. Eiserling and W. R. Romig, *J. Ultrastruct. Res.*, **6,** 540 (1962).

glucosaminopeptide layer whose surface is covered with a material that can be digested by proteolytic enzymes. The rigid layer is quite sensitive to lysozyme and is responsible for maintaining the rodlike shape of the bacterium. When the cells are treated with lysozyme this layer is attacked and the cells assume a spherical and labile form (protoplasts). Just under the rigid layer is the plasma membrane itself. Bacteria do not seem to possess many of the cytoplasmic organelles common to higher organisms. Thus the nuclear membrane, mitochondria, endoplasmic reticulum, and Golgi apparatus are missing as far as discrete structures are concerned. Nuclear contents, however, are present and mitochondrial-like respiratory systems are located in the plasma membrane. Ribosomes appear to associate with the plasma membrane and with an undefined fibrillar network (see page 302). A fascinating type of cytoplasmic extension of the plasma membrane occurs principally in gram-positive organisms. This unit appears as a system of closely packed membranous whorls or as clusters of vesicles and tubules and has been called mesosomes by Fitz-James (61) and chondrioids by Van Iterson (62). Many functions have been attributed to them, such as in the formation of spore septa (61), mitochondrial functions (62), control points for DNA replication (63) and involvement in transverse septum forma-

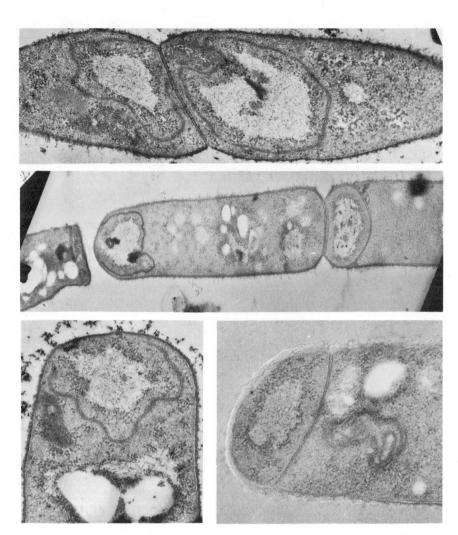

Figure 27 Electron micrographs of advanced stages of spore formation in members of the genus *Bacillus*. Note membranous structures (mesosomes) in the vicinities of the spores. In the center of the developing spore the nuclear material presents a characteristic reticular appearance. The clear areas represent vacuoles that contained lipid. Magnifications are from 25,000 X to 43,000 X. Electron micrographs kindly furnished by Dr. P. C. Fitz-James.

Figure 28 Bacillus treated with potassium tellurite. Series of four sections through a chondrioid showing the latter's connection to the cell envelope. (*a*) On the lower side of the picture, a little underneath the electron-opaque deposit of reduced tellurite on the left, is the lower part of the tube connecting the chondrioid with the cell envelope. (*b*) The vesiculo-tubular structures in the chondrioid are connected to the cell periphery by a tube. (*c*) In the original print it can be observed that the chondrioid is enveloped by a continuation of the plasma membrane and furthermore

194

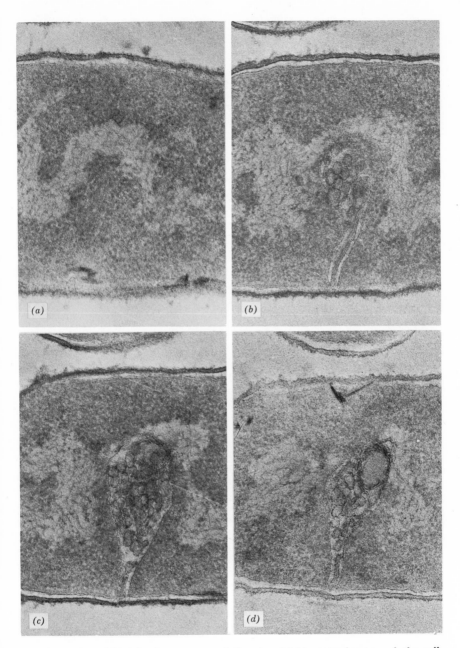

that the tube within the lower space of the chondrioid protrudes toward the cell wall through the empty space between the cell wall and the plasma membrane caused by the latter's retraction. The empty-looking space inside the chondrioid and the space between the plasma membrane and the cell wall are thus interconnected. (d) In the chondrioid are concentric membranes which accumulated reduced tellurite 95,000 X. From W. Van Iterson, Bacteriol. Revs., **29**, 299 (1965).

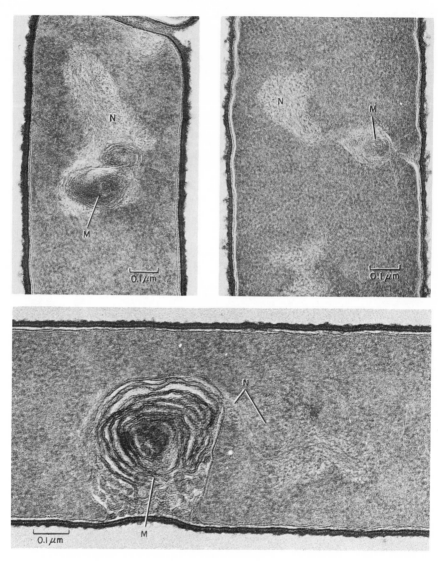

Figure 29   Sections of growing *B. subtilis,* showing mesosomal structures (*M*) and nuclear bodies (*N*). The nuclear bodies are connected to the membrane by the intermediacy of mesosomes. From F. Jacob, A. Ryter, and F. Cuzin, *Proc. Roy. Soc.* (London), **B-164,** 267 (1966).

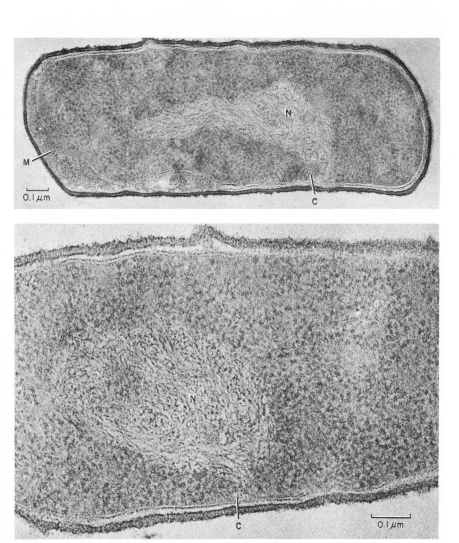

Figure 30 Sections of growing *B. subtilis* after 30 min. in 0.5 M sucrose. Mesosomes have been expelled from cytoplasm [one (*M*) can be seen as a vesicle at the left pole of the top picture]. Nuclear bodies then appear directly connected with the membrane in region *C*. From F. Jacob, A. Ryter, and F. Cuzin, *Proc. Roy. Soc. (London)*, **B-164**, 267 (1966).

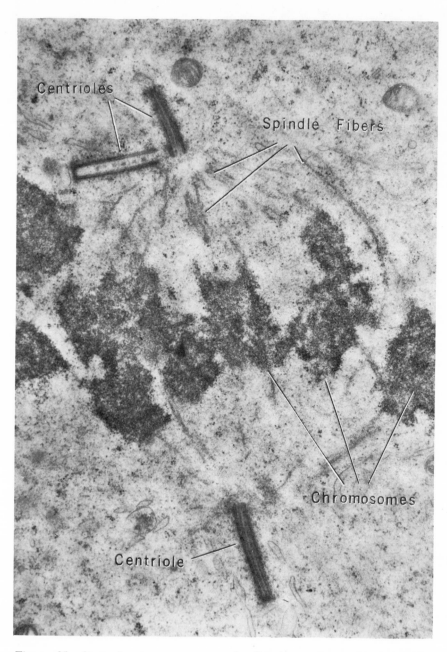

Figure 31 Spermatocyte from cock testis. Collidine buffered osmium fixation. Lead hydroxide staining. Micrograph courtesy of Dr. T. Nagano. Magnification, 20,000 X. From D. W. Fawcett, *An Atlas of Fine Structure,* Saunders, Philadelphia, (1966).

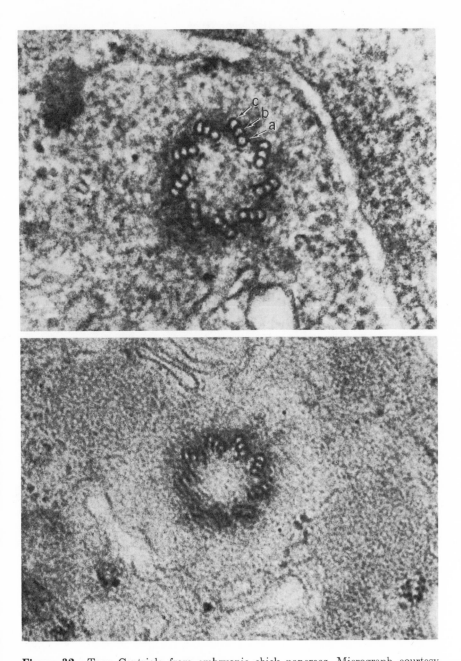

Figure 32 Top: Centriole from embryonic chick pancreas. Micrograph courtesy of Dr. Jean André. Magnification 100,000 X.
Bottom. Centriole from fish pancreatic islet cell. Micrograph courtesy of Dr. Arthur Like. Magnification 100,000 X. From D. W. Fawcett, *An Atlas of Fine Structure,* Saunders, Philadelphia, (1966).

tion for cell division (64). Figures 26–30 show mesosomal structures and Figures 31 and 32 show centrioles.

REFERENCES

1. A. Claude, *Science,* **97,** 451 (1943).
2. A. Claude and E. F. Fullam, *J. Exp. Med.,* **81,** 51 (1945).
3. A. Claude, *Annals N.Y. Acad. Sci.,* **50,** 854 (1950).
4. J. Brachet, *Annals. N.Y. Acad. Sci.,* **50,** 861 (1950).
5. G. H. Hogeboom, W. G. Schneider, and G. E. Palade, *J. Biol. Chem.,* **172,** 619 (1948).
6. F. S. Sjöstrand, *Nature,* **171,** 31 (1953).
7. G. E. Palade, *J. Hist. Cytochem.,* **1,** 188 (1953).
8. G. E. Palade, *Anat. Rec.,* **114,** 427 (1952).
9. K. R. Porter, A. Claude, and E. F. Fullam, *J. Exp. Med.,* **81,** 233 (1945).
10. K. R. Porter, *J. Exp. Med.,* **97,** 727 (1953).
11. D. B. Slautterback, *J. Appl. Physics,* **23,** 163 (1952).
12. A. Claude, *Biol. Symp.,* **10,** 111 (1943).
13. S. Brenner, *So. African J. Med. Sci.,* **12,** 53 (1947).
14. G. E. Palade and K. R. Porter, *J. Exp. Med.,* **100,** 641 (1954).
15. G. E. Palade, *J. Biophys. Biochem. Cytol.,* **1,** 567 (1955).
16. G. E. Palade, *J. Biophys. Biochem. Cytol.,* **1,** 59 (1955).
17. G. E. Palade and P. Siekevitz, *J. Biophys. Biochem. Cytol.,* **2,** 171 (1956).
18. D. B. Slautterback, *Exp. Cell Res.,* **5,** 173 (1953).
19. G. E. Palade and P. Siekevitz, *J. Biophys. Biochem. Cytol.,* **2,** 671 (1956).
20. G. E. Palade, in *Enzymes: Units of Biological Structure and Function,* Ed., O. H. Gaebler, Academic Press, New York, 1956, p. 185.
21. C. de Duve, B. C. Pressman, R. Gianetto, R. Wattiaux, and F. Appelmans, *Biochem. J.,* **60,** 604 (1955).
22. F. Appelmans, R. Wattiaux, and C. de Duve, *Biochem. J.,* **59,** 438 (1955).
23. C. de Duve, in *Subcellular Particles,* Ed., T. Hayashi, Ronald Press, New York, 1954, p. 128.
24. C. Rouiller, *Compt. Rend. Soc. Biol.,* **148,** 2008 (1954).
25. G. E. Palade, *J. Biophys. Biochem. Cytol.,* Suppl. **2,** 85 (1956).
26. A. J. Dalton and M. D. Felix, *Amer. J. Anat.,* **92,** 277 (1953).
27. W. C. Schneider, A. J. Dalton, E. L. Kuff, and M. Felix, *Nature,* **172,** 161 (1953).
28. A. J. Dalton and M. D. Felix, *Amer. J. Anat.,* **94,** 171 (1954).
29. W. C. Schneider and E. L. Kuff, *Amer. J. Anat.,* **94,** 209 (1954).
30. H. G., Callan and S. G., Tomlin, *Proc. Roy. Soc. London,* **137B,** 364 (1950).
31. B. A. Afzelius, *Exp. Cell. Res.,* **8,** 147 (1955).
32. R. W. Merriam, *J. Biophys. Biochem. Cytol.,* **11,** 559 (1961).
33. R. W. Merriam, *J. Cell Biol.,* **12,** 79 (1962).
34. M. L. Watson, *J. Biophys. Biochem. Cytol.,* **6,** 14 (1959).
35. C. Rouiller and W. Bernhard, *J. Biophys. Biochem. Cytol.,* Suppl. **2,** 355 (1956).
36. A. O. Novikoff, in *The Cell,* Eds., J. Brachet and A. E. Mirsky, p. 299. Academic Press, New York, 1961.
37. C. de Duve, H. Beaufay, P. Jacques, T. Rahman-Li, O. Z. Sellinger, R. Wattiaux, and S. DeConinck, *Biochim. Biophys. Acta.,* **40,** 186 (1960).

38. Z. Hruban and H. Swift, *Science,* **146,** 1316 (1964).
39. M. H. Burgos and D. W. Fawcett, *J. Biophys. Biochem. Cytol.,* **1,** 287 (1955).
40. M. C. Ledbetter and K. R. Porter, *Science,* **144,** 372 (1964).
41. E. Sandborn, P. F. Koen, J. D. McNabb, and G. Moore, *J. Ultrastruct. Res.,* **11,** 123 (1964).
42. D. W. Fawcett and F. Witebsky, *Z. Zellforsch.,* **62,** 785 (1964).
43. H. Fernández-Morán, *Circulation,* **26,** 1039 (1962).
44. D. F. Parsons, *Science,* **140,** 985 (1963).
45. H. Fernández-Morán, T. Oda, P. V. Blair, and D. E. Green, *J. Cell Biol.,* **22,** 63 (1964).
46. D. S. Smith, *J. Cell Biol.,* **19,** 115 (1963).
47. W. Stoeckenius, *J. Cell Biol.,* **17,** 443 (1963).
48. F. S. Sjöstrand, *Nature,* **199,** 1262 (1963).
49. J. T. Stasny and F. L. Crane, *J. Cell Biol.,* **22,** 49 (1964).
50. A. L. Lehninger, in *The Mitochondrion,* W. H. Benjamin, New York, 1964.
51. K. R. Porter and G. E. Palade, *J. Biophys. Biochem. Cytol.,* **3,** 269 (1956).
52. K. R. Porter, *J. Biophys. Biochem. Cytol.,* **10,** 219 (1961).
53. C. Franzini-Armstrong and K. R. Porter, *J. Cell. Biol.,* **22,** 675 (1964).
54. R. Sager, *Brookhaven Symp. in Biol.,* **11,** 101 (1958).
55. R. Sager and G. E. Palade, *J. Biophys. Biochem. Cytol.,* **3,** 463 (1957).
56. T. E. Weier, *Amer. J. Bot.,* **48,** 615 (1961).
57. T. Oda and H. Huzisige, *Exp. Cell. Res.,* **37,** 481 (1965).
58. R. B. Park and J. Biggins, *Science,* **144,** 1009 (1964).
59. E. Kellenberger and A. Ryter, *J. Biophys. Biochem. Cytol.,* **4,** 323 (1958).
60. M. E. Bayer and T. F. Anderson, *Proc. Nat. Acad. Sci.,* **54,** 1592 (1965).
61. P. C. Fitz-James, *J. Biophys. Biochem. Cytol.,* **8,** 507 (1960).
62. W. Van Iterson, *Bact. Rev.,* **29,** 299 (1965).
63. A. Ryter and F. Jacob, *Compt. Rend.,* **257,** 3060 (1963).
64. P. C. Fitz-James, *Bact. Rev.,* **29,** 293 (1965).

Chapter 5

CELL MEMBRANE STRUCTURE

In this chapter I discuss some features and historical background of concepts of membranes and membrane structure. On various occasions in earlier chapters the potential significance of the membranes of cells was discussed in connection with protein synthesis. From this point on we shall consider in greater detail how membranes may be vital elements in many of the cells' complex functions, including protein synthesis.

If this account had been written a few years ago it would have been possible to give a clearer picture of cell membranes and the fine details of their molecular structure. At that time we were much more confident of the picture that had been slowly assembled by electron microscopy, X-ray diffraction, crystallography, physical chemistry, and biochemistry.

In the light of a current critical reevaluation of concepts about membranes, it seems quite possible that the picture of the membrane painstakingly constructed during the last 40 years is no more than an ingenious assembly of available information into one of several possible pictures.

It must be admitted at the outset of a chapter devoted to a consideration of cell membranes that we do not know the details of their molecular structure and therefore we do not know how they accomplish the many functions attributed to them. This change in evaluation of our knowledge about such an important subject once again emphasizes the importance of encouraging constant questioning of accepted concepts by those who do not find the existing picture satisfactory, and this is especially true the more acceptable the hypothesis in question. In this chapter we trace the development of our recent and current concepts of membrane structure.

The recognition of the importance of lipids in the structure of the limiting membranes of cells is attributed to E. Overton who in 1895 arrived at this conclusion because of the relative ease with which certain lipoid substances penetrated cells (1). In 1917 Langmuir and Harkins et al. studied the arrangements of various molecules in monomolecular films spread on water surfaces (2, 3). Langmuir advanced the idea that lipids at an air-water interface arranged themselves in a monolayer with the polar ends of the molecules in register at the water interface and the nonpolar ends in register at the air interface. Studies with soap bubbles by Perrin (4) and with fatty acid crystals by Bragg and Bragg (5) led to the concept of the packing arrangement for such molecules being in layers with the hydrophobic portions in apposition.

The next major conceptual development occurred in 1925 when Gorter and Grendel tried to correlate the quantity of membrane lipid in erythrocytes with the surface area of these cells (6). This was a bold step because several nonevaluable assumptions had to be made: (a) that all of the extractable lipids were employed in the limiting membrane structure; (b) that the method of extraction (acetone) extracts all of the lipids involved in this structure; (c) that the method of measuring the surface area of these extracted lipids (the use of a Langmuir trough) produces a valid measure of the area being covered in a living cell; (d) that they could measure the actual surface of the cells (average diameter was measured with the aid of a microscope and assuming a disc shape for the cells the area per cell was calculated from a formula). The authors examined erythrocytes from a number of mammalian species including man and found the ratio of area of extracted lipids to cell area to be 2:1. They concluded that the erythrocyte is covered by a membrane of lipid two molecules thick with the polar groups facing outward toward polar environments on both sides.

These experiments serve as a strong support for subsequent developments of membrane concept and so it is important to evaluate their soundness at this time. We now know that the first assumption is probably correct and that essentially all of the lipid of the erythrocyte is contained in the plasma membrane. Assumption (b) was incorrect and only 70 to 80% of the total lipid was extracted (7). The Langmuir tray appears to provide an adequate means to measure surface area of lipids but a decision must be made as to how much compression of the surface film should be applied before making the measurement. Finally, more recent data show that the surface area for human erythrocytes is 145 ± 8 μ^2, not 99 μ^2 as calculated by Gorter and Grendel (8).

Since the original investigators erred in two ways (incomplete lipid extraction and too small a value for the area of erythrocytes), to a similar

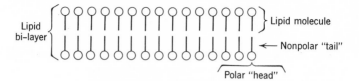

Figure 1   Lipid bimolecular leaflet.

quantitative extent the errors compensated for each other and the same result of 2:1 for area of membrane lipids to area of the cells can be obtained today. Others have calculated the area of lipids on the basis of measurements of phospholipid-water mixtures by X-ray diffraction and measurements of lipids spread in Langmuir trays and have arrived at values as low as 1.4:1 for the areas involved. Bar et al. suggest that these measurements are obtained with the lipids at some arbitrary state of compression in the Langmuir tray or in a form not representative of the condition in a spread bi-layer (7). We therefore have no basis for rejecting (or accepting) the ratio of 2:1 for the human erythrocyte membrane.

In the late 1920's thoughts about cell membranes were influenced by the concept of a lipid bi-layer (see Figure 1).

In the early 1930's Harvey and Shapiro measured the surface tension of oil droplets from marine eggs (9) and Cole measured the surface tension of starfish eggs (10). Pure lipids were then expected to exhibit a surface tension of five or more dynes per cm.[1] Harvey and Shapiro, however, found values of only 0.2 dyne per cm and Cole's values were even less. Soon afterward Danielli and Harvey concluded that the low values were due to the presence of some surface active agent (12). They decided that this was protein and that lipid polar surfaces were covered by at least one protein monolayer.

The various observations and ideas just discussed were compounded by Danielli in the form of a detailed model for membrane structure which has survived in essence during the past 31 years (through the time of writing this account) and has been the major influence on chemical and physical concepts of membranes. The model was proposed in a paper by Danielli and Davson, but a footnote in the paper made it clear that the merits or deficiencies in the postulation for membrane structure should be attributed to Danielli (13). This classic and artistic representation is reproduced in Figure 2.

The layer of protein was considered to be present as an unavoidable

[1] Recent studies with pure lipid bi-layers by T. E. Thompson yield surface tension values of one dyne per cm (11).

EXTERIOR

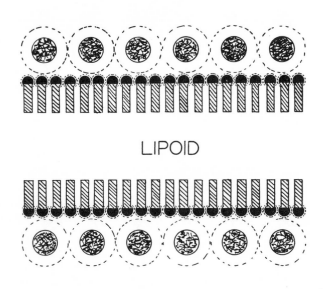

LIPOID

INTERIOR

Figure 2   The structure of a cellular membrane according to Danielli and Davson. From J. F. Danielli and H. Davson, *J. Cell. Comp. Physiol.*, **5**, 495 (1935).

consequence of adsorption dependent solely on the surface activities of the lipid and protein. The presence of a protein layer on the external surface of the membrane was casually included in the model as follows: "We may note in passing that, unless cells are subjected to rigorous washing, there will also be a similar adsorbed layer of protein on the outside of the membrane." The proteins were represented as globular units with dense cores surrounded by less dense heavily hydrated layers. This was based on recent (at the time) X-ray diffraction studies of Bernal and Crowfoot using crystalline pepsin (14). The spacing of protein could give the membrane selectivity on the basis of size of molecules that could penetrate the structure at particular points. The electric charges of the proteins could influence selectivity for the passage of charged substances. Various other properties of this hypothetical structure were considered in relation to the suspected properties of cellular membranes and it was concluded that such a model could explain many of the phenomena exhibited by intact cells.

X-ray diffraction techniques can reveal the presence of repeating units in crystalline structures. Schmitt, Bear, and Clark studied the diffraction patterns of a variety of nerves and discovered repeating frequencies in the lipid myelin sheath[2] with fundamental spacings of 171 A in a radial direction outwards from the nerve (15). A spacing of 4.7 A in the tangential direction was attributed to interchain distances of the lipid molecules. The longer radial repeat distance was attributed to lipid molecules arranged in cylindrical smectic[3] fluid-crystalline layers wrapped concentrically about the axon. In the meantime newer information had indicated the presence of thin sheets of neurokeratinogenic protein intercalated between the lipid layers. Schmitt, Bear, and Palmer re-examined the question on the basis of the newer information (16). They concluded that the myelin sheath is composed of concentrically wrapped layers of mixed lipids, alternating with thin, possibly unimolecular, layers of neurokeratinogenic protein material. Within the layers the lipid molecules are oriented with paraffin chains extending radially and with polar groups in aqueous interfaces, loosely bonded to polar areas of the protein. Four possible representations of the membrane structure were given (see Figure 3). The protein was considered to be present as networks or fabrics of polypeptide chains with polar side chains loosely bonded to the polar groups of the lipids. The possibility of the protein being present as globular molecules was discounted, especially because of the resistance of the structure to alteration induced by heat and detergents. When fresh and dry specimens of amphibian and mammalian spinal roots were examined, it was found that the major long spacing shrunk about 26 A on drying. In dry mammalian preparations, the long-spacing identity period was 159 A. Allowing $2 \times 67 = 134$ A for the dimension contributed by the bimolecular lipid layer, 25 A were left for the protein. Water was believed to occupy space between protein-lipid or lipid-lipid polar groups.

Further support for the general concept of cell membranes, as developed so far, was provided by measurements of electrical capacitance and impedance of cell surfaces to high frequency alternating current (17). The values determined for these parameters were consistent with a layered structure of 50 to 100 A thickness, composed of lipid and protein.

---

[2] Studies of myelin structure form the basis for developing a general concept of membrane structure. It is therefore necessary to review the experimental support for the model proposed for myelin membranes. The reader who is not familiar with the formation of the myelin sheath should refer to Figures 5 and 6 at this point.

[3] "Smectic" is a term used to describe a fluid state in which the molecules appear to be associated in planes or layers.

MYELIN SHEATH STRUCTURES
(RADIAL DIRECTION)

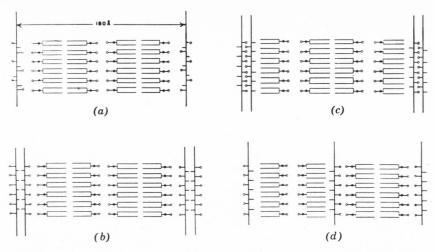

Figure 3 Schematic representation of several possible modes of packing of lipid and protein in the nerve myelin sheath (radial direction). From F. O. Schmitt et al., *J. Cell. Comp. Physiol.* **18**, 31 (1941).

The postulated layers and their exceptionally regular concentric arrangement in the myelin sheath were subsequently observed with the aid of the electron microscope by Fernández-Morán, Sjöstrand, and Geren and Raskind (18–20). Sjöstrand described the pattern in osmium-fixed mouse sciatic nerve as consisting of 25 A thick principal lines spaced apart at 119 ± 1.5 A. Halfway between these thicker lines he observed an intermediate line that was much finer. The thick line was interpreted as representing protein and the clear zone as lipid. The fine intermediate line which cleaved the clear zone was believed to represent the ends of the lipid molecules in mutual contact. Sjöstrand noted that this interpretation placed the heavy lines slightly farther apart than would be expected from recent X-ray data obtained by Finean (21) and electron microscope data obtained by Fernández-Morán (22).

In 1957 Robertson showed that when frog peripheral nerve fibers were fixed with $KMnO_4$, the single dense line that represented the membranes of axons and Schwann cells could be resolved into structures consisting of two single dense lines separated by a single light space (23). The three parts of this membrane sandwich each measured about 25 A and therefore each membrane measured about 75 A across. When an axon was engulfed by a Schwann cell, the membrane of each cell was separated by a clear gap of 150 A. The nature of the gap substance was completely unknown. In the formation of myelin the two apposing membranes of the Schwann cell come together, completely obliterating the gap area (see also Geren 24). The major dense line in myelin results when two cytoplasmic sides of the Schwann cell membrane fuse and the lighter intraperiod line results from a blending of the external surfaces (Figure 4).

Two interesting questions were posed by this work. First, although the Schwann cell membrane by itself appeared symmetrical, why did either a dark or light line result, depending upon whether two internal or external surfaces blended? Second, if the myelin sheath were formed in the manner indicated, why did the blending of two ∼75 A units produce a repeating unit of 120 A between two dense bands instead of 150 A?

The individually important techniques of electron microscopy and

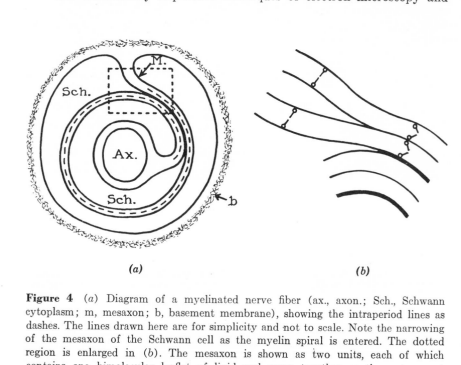

(a)   *(b)*

Figure 4 (*a*) Diagram of a myelinated nerve fiber (ax., axon.; Sch., Schwann cytoplasm; m, mesaxon; b, basement membrane), showing the intraperiod lines as dashes. The lines drawn here are for simplicity and not to scale. Note the narrowing of the mesaxon of the Schwann cell as the myelin spiral is entered. The dotted region is enlarged in (*b*). The mesaxon is shown as two units, each of which contains one bimolecular leaflet of lipid and come together as the outer most myelin lamella, which in contact with the next in the spiral produces the major dense line. The polar ends of phospholipid molecules are indicated by the circles and the nonpolar carbon chains by the lines attached to the circles. From J. D. Robertson, *J. Biophys. Biochem. Cytol.*, **3**, 1043 (1957).

low-angle X-ray diffraction can be blended to form a powerful experimental approach for the study of the repeating membrane subunits of myelin. This is especially true when the collaboration is between two experts of these respective fields. Such a study was performed by Fernández-Morán and Finean (25). While for electron microscopy it is necessary to desiccate the specimen and subject it to fixation and staining procedures, X-ray diffraction can be performed on the fresh specimen under near physiological conditions. It is also possible to follow the changes in X-ray diffraction properties of the specimen as it is taken through the stages in preparation for electron microscopy. A further advantage of X-ray diffraction is that it gives an average for repeating units in a fairly large sample, whereas in electron microscopy individual minute areas are examined in each single exposure. The advantage of electron microscopy, however, is in the visualization of structural details of macromolecular dimensions.

It was found that in the normal course of osmium-fixation for electron microscopy an alteration in the structure of nerve myelin occurred which caused a shrinkage of 20 A in the radial repeat unit that normally measures about 170 A for amphibians and about 180 A for mammalian specimens. There was also a drastic modification in the distribution of X-ray scattering power within the treated sample. A further shrinking of 20 A occurred when the osmium-treated tissue was freeze-dried. When dehydration was accomplished through a graded series of alcohol concentrations, shrinkage was also noted but when the specimen was then embedded in methacrylate for sectioning it re-expanded to yield an X-ray diffraction pattern resembling the original moist osmium-fixed preparation. The electron microscopic image showed a spacing for the repeating distance 10 to 20 A lower than was revealed by X-ray diffraction on the same methacrylate-embedded specimen. In contrast to osmium-fixation the shrinkage produced by $KMnO_4$ treatment was relatively small; $KMnO_4$-fixed specimens were better preserved when embedded in gelatin for sectioning than when embedded in methacrylate.

The outer and inner surfaces of the Schwann cell membrane show differences when encountered in the myelin sheath. This difference is revealed by the alternation of heavy dense lines and less dense intermediate lines after treatment with osmium. This difference is further revealed by the greater intensity of the first order reflection in the low-angle X-ray diffraction pattern. The difference was attributed to unknown "difference factors."

The interesting observation was made that partial extraction of the lipids with acetone did not much affect the ability of the dense lines to be stained with osmium. Therefore osmium images do not necessarily reveal the location of lipids since the dense lines may represent the location of proteins. Some indication was obtained that the lipoprotein layers may have further structural divisions of 60 to 80 A dimensions in the lateral direction. Another significant observation was that freezing and thawing of fresh whole nerve could induce the dissociation of the layer components in the myelin sheath, causing a halving of the radial repeat unit as determined by X-ray diffraction. The significance of this observation is further discussed on page 213.

THE CONCEPT OF A UNIT MEMBRANE

The term "unit membrane" implies a basic architectural design for membranes which is more or less constant throughout nature. The champion of this idea has been J. D. Robertson. In 1959 Robertson painstakingly developed the arguments which he felt supported the case for the unit membrane concept (26).

In the days of light microscopy, when resolution was limited to about 1800 A, the cell membrane was thought of as the thinnest visible external line surrounding the cell. When improved techniques of electron microscopy were available, it was seen that this external region could be resolved into discrete structures which sometimes included bundles of collagen fibrils and other less well-defined external structures. The resolving power of the electron microscope is so great that in expertly prepared specimens a dark line 25 A wide can be seen at the periphery of cells. This extends the confusion about the cell border over two magnitudes.

Robertson emphasized the importance of arriving at criteria for exactly defining which visualized structure should be considered the cell membrane. Physiologists speak of a cell membrane as a definite aggregate of molecular species which will serve as a barrier having definite physical and biological properties.

Frequently it happens that two membranes occur side-by-side giving the appearance of a double membrane. Such configurations are seen surrounding nuclei and mitochondria and in certain configurations of the endoplasmic reticulum and the Schwann cell. The dimensions most often encountered are two approximately 100 A thick membranes separated by a clear space of 100 to 150 A.

Much of the basic experimental work which helped develop the unit membrane model was obtained in studies of the myelin sheath and cell membrane of Schwann cells. Nerves are frequently surrounded by or engulfed in Schwann cells, as indicated in Figure 5.

In unmyelinated nerve fibers a simple embedding process occurs. When the axon is completely surrounded, the two areas of the Schwann

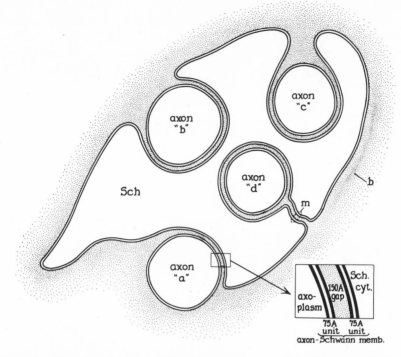

Figure 5 Diagram of an unmyelinated nerve fiber. A single Schwann cell (Sch.) is shown with four associated axons. Axon (a) is merely in apposition with the Schwann cell and axon (b) is partly embedded in it. Axon (c) is more deeply embedded and axon (d) is completely surrounded. The two overlapping lips of the Schwann cell extending around axon (d) come together to make a double membrane leading to the outside referred to as a "mesaxon" or by the more general term "meso." The stratification of the membrane of the axons and the Schwann cell are indicated in the figure. A portion of one axon-Schwann membrane is enlarged to the lower right. The entire Schwann cell is surrounded by a matrix material indicated by the stippling. This material extends into the gaps of the double membrane and is sometimes aggregated in a layer near the external surface of the Schwann cell as shown to the right. When this material is aggregated in this fashion, it is referred to as "basement membrane." From J. D. Robertson, Biochemical Society Symposia No. 16, *Subcellular Components*, 1959, pp. 3–43.

cell plasma membrane come close together to form a double membrane structure which Robertson called a mesaxon. The mesaxon consists of two 75 A membranes separated by a gap of 150 A. Each 75 A membrane could be further resolved into two dense lines just under 25 A each in thickness and separated by a light interzone.

Myelinated nerve fibers were first presented in high resolution micrographs by Fernández-Morán and Sjöstrand, and Geren and Raskind. After OsO₄ fixation a dense line of 25 A is seen repeating approximately every 120 A. In the center of the light zone a thin broken intraperiod line is frequently seen; KMnO₄-treatment in place of OsO₄ consistently shows the intraperiod line and after prolonged fixation it can appear as prominently as the major dense line.

The formation of myelin is illustrated in Figure 6 which is based on the studies of Geren (24). The large gap of the mesaxon seems to disappear or be greatly reduced in the compact myelin. The light ∼30 A zone of the Schwann cell surface membrane is seen to be continuous with the myelin. The two surfaces of the Schwann surface membrane react differently to OsO₄-fixation, the layer next to the cytoplasm staining much more intensely, whereas KMnO₄-fixation stains both layers to a more comparable degree.

When lipid molecules are dispersed in water they tend to form layers

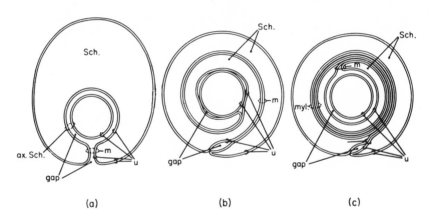

(a)   (b)   (c)

Figure 6   Three stages in the development of nerve myelin are indicated. The earliest recognized stage is shown in (a). Here a single axon is embedded in a Schwann cell (Sch.). The enveloping lips of the Schwann cell come together around the axon to form a double membrane known as the mesaxon (m). There is usually a small gap between the Schwann cell membranes (u) forming the mesaxon, which is continuous with the gap between the axon and the Schwann-cell membranes making up the axon-Schwann membrane (ax-Sch.). At a later stage of myelination represented by (b) the mesaxon is elongated in a spiral around the axon and its central gap is largely obliterated. The gap of the axon-Schwann membrane is also partially obliterated. At a later stage (c) the mesaxon is elongated further and the cytoplasmic surfaces of the spiral mesaxon loops come together to form the major dense lines of compact myelin (myl). From J. D. Robertson, Biochemical Society Symposia No. 16 *Subcellular Components,* 1959, pp. 3–43.

of bimolecular leaflets as revealed by X-ray diffraction patterns. Peculiar cylindrical wormlike forms 1 to 2 μ in diameter, called myelin forms, are also found in such lipid water systems. Lipids extracted from peripheral nerve form systems of bimolecular leaflets with an average spacing of 60 to 70 A. This natural tendency of lipids to line up as bimolecular leaflets is one of the basic arguments in favor of a similar orientation in myelin membranes. The most likely place for protein in such a structure was considered by Robertson and many others to be along the hydrophilic surfaces of the leaflets. The X-ray diffraction studies of Schmitt et al. (16) discussed on page 206 were cited in support of these interpretations. Four possible structures for myelin sheath membranes were proposed by Schmitt, Bear, and Palmer (Figure 3).

As a result of the studies of Finean et al. (27, 28), which showed that under certain conditions the prominent repeating unit of myelin can be halved, an altered model (Figure 7) was suggested. In this model each bimolecular leaflet is bounded by its own extended protein layer and halving is readily visualized. Using all of the facts discussed so far, Robertson explained myelin structure in terms of the continued apposition of Schwann cell membrane as it wrapped itself around the axon. The difference factor between the cytoplasmic and exterior surfaces showed up as an alternation of heavy dense lines and lighter intraperiod lines in myelin. The heavy lines represented the apposition of two cytoplasmic surfaces and the intraperiod line results from the apposition of the external surfaces (Figure 8). The fact that such a melding of two ~75 A Schwann cell membranes should yield a 150 A repeating unit in myelin and that electron micrographs revealed a spacing of only 120 A was attributed to shrinkage effects caused by the electron beam.

When nerve fibers are soaked in hypotonic solutions the myelin period of X-ray diffraction increases from between ~170 and 190 A to ~270 A (29). Under the electron microscope it is seen that the myelin is broken up into layers ~150 A across which have the appearance of two Schwann cell membranes stuck together at their cytoplasmic surfaces and separated at their external surfaces. Furthermore, the gap of the mesaxon is greatly expanded, adding further weight to the concept that the external side of the Schwann cell membrane is much more hydrophilic than the cytoplasmic surface. It was proposed that the external surface may contain a monolayer of polysaccharide, which would be the chemical basis of the difference factor discussed by Finean.

These considerations from electron microscopic and correlated physical studies show that the compound membranes of the myelin sheath can be explained in terms of the basic ~75 A tri-layered membrane structure of the Schwann cell plasma membrane.

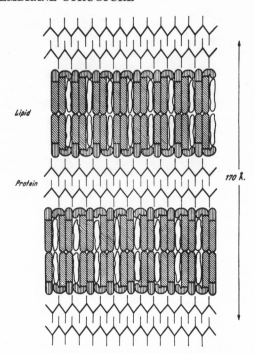

Lipid

Protein

170 Å.

Figure 7   Diagram reproduced with permission from *Biochemical Problems of Lipids,* edited by G. Popjak and E. LeBreton and published by Butterworth and Co. This is a representation by Finean of his conception of the arrangements of the lipid bimolecular leaflets and the protein monolayers in the radially repeating unit of myelin. The nonpolar portions of the lipid molecules are indicated by diagonal lines and the polar portions by vertical lines. Cholesterol molecules are inserted between every other lipid molecule (without cross-hatching). The backbone chains of the protein monolayers are indicated by the zig-zag lines and the side-chains by the related vertical lines. According to this model, each bimolecular leaflet of lipid is bounded on its polar surface by a monolayer of protein.

The next big step taken by Robertson was to extend these observations and conclusions reached from the Schwann cell case to membranes in general. The primary basis of this move was the similar tri-layered appearance for other membranes observed with the aid of the electron microscope and the great similarity ($\sim$75 A) for membrane thickness. In cases in which discrepant dimensions were reported ($\sim$100 A), Robertson cautioned that methods of calibration, measurement, and fixation might be responsible. The membranes of the endoplasmic reticulum, Golgi complex, nucleus, cell vacuoles, microbodies, and mitochondria were included in the generalization.

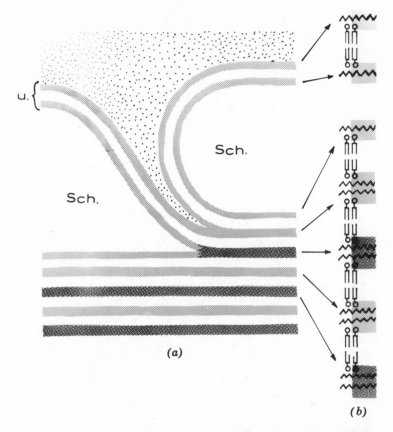

Figure 8 This diagram illustrates the relationships of the enfolded Schwann cell membrane in a mesaxon (or meso) of an adult myelinated fibre. The two cell membranes come together along their outside surfaces to form the outermost myelin lamella. The inside surfaces of the mesaxon loops are in apposition at the major dense lines in compact myelin. Three complete repeating units of myelin are indicated in (*a*). The molecular structure of the outer two repeating lamellae is indicated in (*b*). At the top the molecular structure of the Schwann cell membrane is indicated. The molecular diagrams are based on Finean's model shown in Figure 7. The stippling superimposed on the molecular model indicates the densities that are observed after permanganate fixation. From J. D. Robertson, Biochemical Society Symposia No. 16, *Subcellular Components*, pp. 3–43, 1959.

Newer Information on Liquid-Crystalline
Structures in Lipid-Water Systems

The concept that lipid dispersed in aqueous systems is most stable in a lamellar structure was revised as a result of detailed studies involving improved X-ray diffraction techniques published by Luzzati and Husson (30). They demonstrated that many liquid-crystalline structures do in fact exist, the lamellar organization being only one among the several possibilities. For a potassium palmitate-water system, a phase diagram was presented which showed that various stable structures ex-

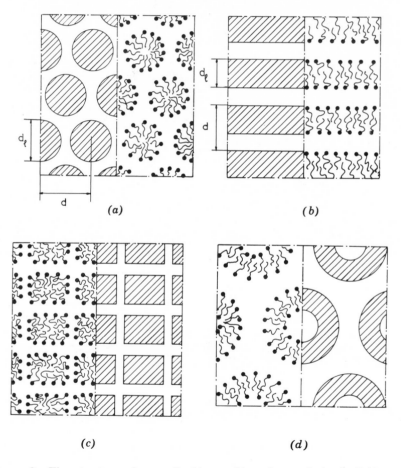

(a) (b)

(c) (d)

Figure 9   The structure of some liquid-crystalline phases of simple lipid-water system. (a) Middle. (b) Neat. (c) Rectangular. (d) Complex hexagonal. From V. Luzzati and F. Husson, *J. Cell Biol.*, **12**, 207 (1962).

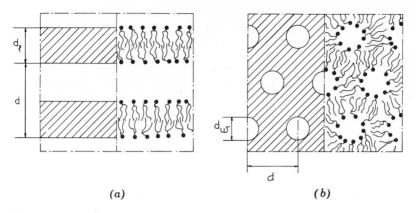

(a)   (b)

Figure 10 The structure of the liquid-crystalline phases of the phospholipid-water system. (a) Lamellar. (b) Hexagonal. From V. Luzzati and F. Husson, *J. Cell Biol.*, **12**, 207 (1962).

isted for particular combinations of temperature and concentration of the soap. There was a very viscous gel-like (coagel) phase at the lowest temperatures and an isotropic micellar phase at the highest temperatures. At intermediate temperatures, several different phases were encountered, as illustrated in Figure 9.

Many different lipid systems representative of anionic, cationic, and nonionic varieties were studied. Although individual systems presented their own characteristics, the general phenomenon of phase transitions was apparent. One system examined was an ether extract of human brain which was a mixture rich in phospholipids. Over a limited concentration range three main phases were encountered: the coagel, lamellar, and hexagonal. The hexagonal phase was continuous for the nonpolar regions and discontinuous for the polar regions (Figure 10).

For all systems, small changes in temperature and concentration could induce drastic changes in the structures. The authors cautioned that liquid-crystalline structures existing *in situ* before fixation, drying, and embedding for electron microscopy would not be likely to withstand all of these operations. Therefore, the finished micrographs might represent artifacts in certain cases. They also pointed out the potential biological significance of a hexagonal phase such as was encountered in the brain phospholipid system. The long, narrow water channels could introduce remarkable permeability properties. Further physiological connotations of these findings were discussed, such as the possibility that controlled phase transitions in membranes could alter their functional capacity. Stoeckenius was able to fix preparations of phospholipid with

OsO_4 and thereby obtain electron micrographs of both the lamellar and hexagonal forms (31). Stoeckenius explained that for a phospholipid-water mixture, the hexagonal phase is stable only at low-water concentrations and elevated temperature. If tissue from a warm blooded animal is exposed to a cold solution of OsO_4 for fixation, as is commonly done, one may wonder if some phase transition to the lamellar form might be induced. If water penetrates the specimen faster than OsO_4, the tendency toward phase transition may be even greater.

Since the formulation of the unit membrane concept by Robertson (see pages 210–214), much interest and discussion of the proposed generality of membrane structure has been generated. Robertson has been asked to continually summarize the current status of the unit membrane, sometimes several times a year. The picture as of June, 1963 was summarized for the Twenty-second Symposium of the Society for the Study of Development and Growth (32). Much of the material summarized by Robertson is a review of the earlier arguments from which the unit membrane hypothesis was drawn. Newer information was also available, however, to supplement the picture.

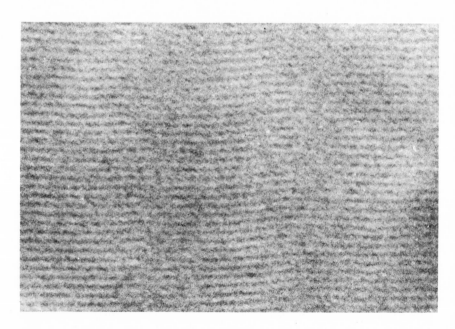

Figure 11   Section of egg cephalin fixed with OsO_4, embedded in Araldite, and sectioned. Magnification 720,000 X. In Michael Locke, Ed., *Cellular Membranes in Development*, Academic, New York, (1964).

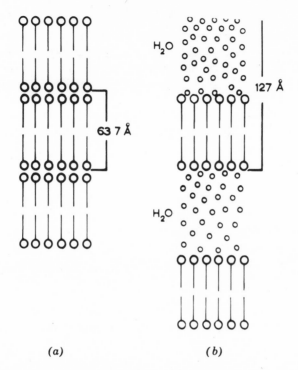

(a)   (b)

Figure 12   This diagram [from F. O. Schmitt, et al., *J. Cell. Comp. Physiol.* **18,** 31 (1941)] shows a lipid in the smectic state in (a) with the repeating period of 63.7 A detected by X-ray diffraction indicated. In (b) the system has imbibed water along the polar surfaces of the bimolecular leaflets of the lipid, increasing the period to 127 A. In Michael Locke, Ed., *Cellular Membranes in Development,* Academic, New York, (1964).

Purified egg cephalin was fixed with OsO_4, embedded in Araldite and examined by electron microscopy. A lamellar structure of alternating 20 A thick light and dark lines was observed (Figure 11). The question existed as to whether the dark lines represented the aligned polar ends of the lipid molecules or the nonpolar carbon chains. An answer for this question was reached when it was found that the addition of water to the system altered the appearance of subsequent electron micrographs. Water should enter between the polar heads of the molecules, as illustrated in Figure 12, and cause a separation into bimolecular leaflets.

If the dark lines represent the deposition of osmium in the polar regions, we should see two dense strata for each leaflet. If the nonpolar region is stained, a single dense line should be seen for each separate leaflet. The results obtained were consistent with the interpretation that the dark lines represented polar regions of the lipid molecule. When isolated crystalline proteins were studied by the same methods, electron micrographs with homogeneous densities were obtained. Hence, Robertson concluded, that it seems reasonable to regard the dense strata in myelin as combined lipid polar surfaces and protein (or other nonlipid) monolayers.

I would like to emphasize at this point that the weakest link in the generalized unit membrane concept is the step from the myelin sheath to other cellular membranes. The justification for extending the concept of membrane structures from myelin to other membranes rests on the similarity of the tri-lamellar structure seen in electron micrographs and in the similarity of the dimensions of this structure from membrane to membrane. Robertson disposed of those cases where the unit membrane was not seen after fixation with OsO_4 as problems of technique. Although $KMnO_4$ could be depended on to reveal the unit membrane appearance of membranes from a variety of sources, the use of OsO_4 was much more troublesome in this respect. Not only does Robertson argue for the complete generality of the unit membrane structure for all cellular membranes, but he also feels that the various membranous systems in a cell are continuous and interconnected. This concept has not drawn the support or interest that the generalized unit membrane hypothesis has drawn and it is not my purpose to consider the arguments brought forward by Robertson to support this idea. The interested reader should investigate reference 26 for the details of this proposal.

Other reported differences in dimensions of various membranes were also attributed by Robertson to differences in technique and to individual differences of the components making up particular bimolecular leaflet structures.

The concept of a simple coated bimolecular leaflet as universal for membrane structures was somewhat shaken when reports of subunit patterns in the plane of the membrane appeared from several laboratories. Robertson himself found indications of a subunit division in synaptic discs of the Mauthner cell club endings of goldfish medulla (33). When two unit membranes were united in apposition, a five-layered structure resulted (external-compound-membrane, ECM) which showed dense granules repeating at a period of about 90 A along the line of union of the outside surfaces of the two membranes (see a and b, in Figure 13). Vague transverse densities seem to run across the light central zones of each membrane, resulting in a scalloped appearance. Upon tilting the membrane with respect to the electron beam in the microscope, the transverse densities became lines, and dots appeared between the lines (c through e). In a completely tilted position (90°), the top surface

Diagrams of synaptic discs

Observed   Hypothetical

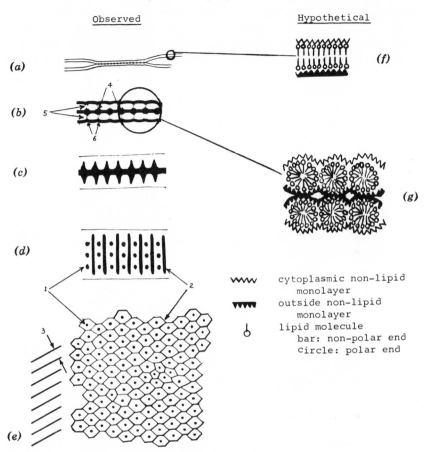

www   cytoplasmic non-lipid
           monolayer
vvvv   outside non-lipid
           monolayer
      lipid molecule
       bar: non-polar end
       circle: polar end

Figure 13 Schematic diagrams illustrate the appearances of synaptic discs in the goldfish Mauthner cell club endings in a through e; f shows the general molecular pattern of one of the unit membranes making up the discs; g is a possible molecular interpretation of observed pattern in b. Reasons are given in the text for believing that this arrangement of lipid molecules is not generally present in the living state in unit membranes and probably not responsible for the particular image in b in this specific case (according to J. D. Robertson). From J. D. Robertson, *Ann. N.Y. Acad. Sci.*, **137**, 421 (1966).

of the membrane seemed to exhibit a granulo-fibrillar substructure (e). Although the appearance of membrane subunits could be the basis of rejection for the universal tri-layered unit membrane hypothesis, Robertson suggested two alternative possibilities. The scalloped subunit pattern could be an optical artifact resulting from the tilting of the membrane specimen to present semi-edge-on views of the detailed outside surface structure of the membrane. Another possibility is the occurrence of a phase transition of the type described by Luzatti and Husson, and Stoeckenius, but going from a native-layered smectic state to a cylindrical state.

In a paper presented two years later Robertson strengthened the basis of his rejection of the reality of a subunit structure in these and other membranes (34). Reports of an intrinsic globular fine structure in unit membranes of frog retinal rod outer segments (ROS) have been published (35, 36). Robertson emphasized that this type of tissue was particularly susceptible to damage induced by normal preparative procedures. Furthermore, X-ray diffraction patterns obtained from both fixed and unfixed fresh retinal rods failed to give evidence for a clear-cut subunit structure in the membranes. These latter objections, however, may have been overcome by more recent work of Blaisie et al. (see page 229 and reference 61).

Robertson described another optical phenomenon with retinal rod outer segments that turned out to be a straightforward image artifact. In longitudinal sections there was an appearance of "cross-over" regions between adjacent membranes, giving the impression of continuity from one lamella to another (Figure 14). Robertson published this work with the impression that the images were real. However, further study with models showed that discontinuous membranes, when tilted, would produce an image-overlap phenomenon (37). This experience led Robertson to consider that transverse densities seen in unit membranes may also be overlap artifacts of a similar nature, though on a reduced scale. To test this idea he made electron micrographs of a few lamellae of ROS in transection and then of the same lamellae after tilting the plane of the section through 5°. He interpreted the micrographs as showing no transverse densities before tilting, but showing clear-cut transverse densities after tilting (Figure 15). This satisfied Robertson that evidence of globular substructures in membranes may be attributed to image-overlap phenomena induced by tilting or natural curvature of the membranes. It is unfortunate that the micrographs published by Robertson in support of this argument (not reproduced here) do not show this phenomenon convincingly enough for the skeptics.

The evidence indicating subunit globules in membrane structure was

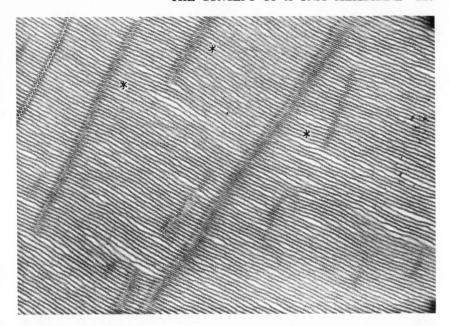

Figure 14 Longitudinal section of a portion of a retinal rod outer segment showing an incisure to the upper left and several sets of "cross-overs" at the asterisk symbols. Magnification 38,000 X. From J. D. Robertson, *Ann. N.Y. Acad. Sci.*, **137**, 421 (1966).

taken much more seriously by Sjöstrand, who earlier had described 50 A globular components in mitochondrial membranes of potassium permanganate-fixed kidney tissue. Similar globules were seen in α-cytomembranes (rough-surfaced endoplasmic reticulum) and mitochondrial membranes of mouse pancreas exocrine cells fixed by freeze-drying (38). Sjöstrand believes that the similar appearance of these subunits by either $KMnO_4$ or freeze-drying fixation adds substance to their suspected reality. He has suggested further that the globules may be separated from each other by layers of protein.

The view of Sjöstrand on discrete globular subunits in mitochondrial and smooth-surfaced cytomembranes was stated even more emphatically in a paper in the *Journal of Ultrastructure Research* in 1963 (39). The total thickness of the mitochondrial membranes was measured as 50 to 60 A. Actual septa separating the globular elements were as thin as 10 A. The globules are pictured as rounded globular micelles separated by protein molecules which, in the case of the mitochondria, could constitute enzymes of the electron transport chain and structural protein. The

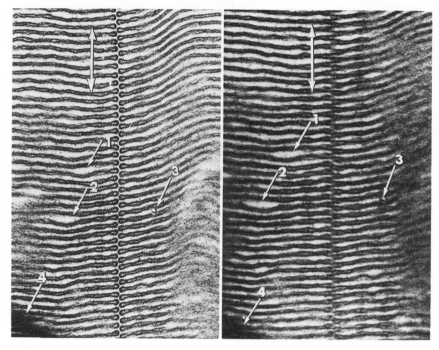

Figure 15   Stereoscopic micrograph of an area in a longitudinal section of a retinal rod outer segment showing (in left-hand figure) a clear-cut incisive. In the right-hand figure the cross-over image has been produced by tilting the section through a large angle on an axis parallel to the fissure shown in the left-hand figure. The same reference points (1, 2, 3, 4) are indicated for the two views. Magnification 65,000 X. From J. D. Robertson, *Ann. N.Y. Acad. Sci.*, **137**, 421 (1966).

plasma membrane was distinguished by the absence of observable globular units within its structure.

Sjöstrand's attack on the unit membrane hypothesis of Robertson resumed some 200 pages later in the same volume of the *Journal of Ultrastructure Research* which carried the description of globular sub-units already discussed. Careful measurements of the membranes in proximal convoluted tubule cells of mouse kidney after either potassium permanganate or osmium fixation were made (40). In marked contrast to the ~75 A considered to be characteristic of the uniformity of unit membranes, Sjöstrand reported that membranes come in categories of sizes. The thinnest membranes measured 50 A after osmium fixation or 60 A in potassium permanganate-fixed tissue. Mitochondrial membranes and α-cytomembranes (rough-surfaced endoplasmic reticulum) belong to this category. An intermediate group containing smooth-surfaced cytomembranes and Golgi membranes measured 60 A after osmium-fixation and 70 to 80 A when potassium permanganate was the fixative. The third and thickest class of membranes are represented by the plasma membrane and the membrane surrounding zymogen granules, which measure 90 to 100 A in thickness. These variations in size, as well as the appearance of globular subunits in Sjöstrand's micrographs, satisfy him that the unit membrane hypothesis is untenable.

Robertson, however, argues (as of December 1966) that reported variations in size of membranes, as well as the appearance of subunit images, may be due to artifacts of preparation of the specimens and image-overlap phenomena (41) (pages 220–222). Insofar as membrane thickness variation is concerned, the strength of this objection is weakened by the fact that Sjöstrand's observations of different sized membranes were obtained in single electron micrographs containing the different membranes.

Lucy and Glauert added a new dimension to the possibilities for conceiving of lipid arrangements in cellular structure. Working with simple binary or ternary lipid mixtures composed of lecithin, cholesterol, and/or saponin, they found that a basic globular micellar structure 35 to 40 A could be formed (42). This basic structure was quite versatile in that, depending upon its exact composition (i.e., lecithin-cholesterol or saponin-cholesterol) and the conditions of mixing and time of standing, the micelles could combine to form lamellae, tubules, hexagonal arrays, stacked discs, and double helices. The subunit structure of the tubules which contained lecithin and cholesterol was found to be quite similar in appearance and dimensions to that found for bacterial flagella which are built up from protein molecules. Helices which were formed from globular subunits in mixtures of ovolecithin, cholesterol, and saponin resembled the helical structures comprising tobacco mosaic virus. All of these observed configurations were revealed with the aid of negative staining techniques employing potassium phosphotungstate. The authors properly considered how much of the observed phenomena could be influenced by the use of this technique. Various controls were performed. The lipid suspension was allowed to stand without phosphotungstate, then at the last minute either potassium or calcium phosphotungstate was added. Normal helices were still formed. Uranyl acetate, however, caused less perfect helices to be formed. Osmium tetroxide was also detrimental towards helix preservation and was able to cause disruption of existing helices.

The fact that short helices are normally observed after the aqueous-lipid dispersions are first prepared and that longer helices develop upon standing indicates that the more complex structures are present in the wet state and do not form simply as a result of the drying process in the presence of the phosphotungstate. It must also be emphasized that the simple lipid mixtures used in this work are not representative of the lipid composition of any known membrane and, furthermore, real membranes contain nonlipid components as well. The hitherto little appreciated ability of lipid mixtures to form complex structures from simple globular micellar subunits, however, adds an important new dimension to consideration of biological structure.

Nilsson found evidence of globular substructure in outer segment disc membranes from both rods and cones of tadpole retinae (36). The globules appeared as spherical light regions with diameters approximately 20 to 30 A. They were found between the two opaque layers of the membrane and separated by opaque septa 15 to 25 A thick. A similar globular substructure was also found in the plasma membrane of the outer segment and in the mitochondrial membranes, α-cytomembranes, and Golgi membranes of the inner segment. The globular appearance was more prominent after section staining with osmium tetroxide than when other methods using uranyl acetate and lead hydroxide were employed. Nilsson pointed out that when such small subunits are involved, even very thin sections can cause serious superimposition artifacts. This complication does not compromise the possibility that the membrane is built from these globular subunits but makes observation of detail more difficult. Nilsson considered that the globules may well represent important structural units in the membranes where they have been observed and he drew support from the work of Sjöstrand and Elfvin (43) where similar structures have been seen in freeze-dried specimens.

Benedetti and Emmelot contributed more information bearing on the potential complexity and diversity of membrane structure (44). They examined negatively stained plasma membranes isolated from rat liver. Two types of local membrane differentiation were observed. A hexagonal array of subunit facets forming the membrane layer was occasionally seen. The extent of this phenomenon could be correlated with the temperature to which the membranes were exposed during the preparation for microscopy. The micellar arrangement of hexagonal facets was absent in untreated membranes stained at 2°, occasionally present in membranes pretreated at 37° and stained at 2°, and very abundantly present in untreated or pretreated membranes stained at 37°. This suggests that the hexagonal pattern is a temperature-dependent phase transition of the type studied by Luzzati and Husson and Stoeckenius (see pages 216–218). Another structure seen was a globular unit 50 to 60 A in diameter attached to the membrane surface by a short stalk 20 A long. These units were distributed in arrays over the membrane surface but were never found in the same area which exhibited the hexagonal pattern. The differences in size between these attached globules and that of the elementary particle of Fernández-Morán (see pages 181–187) were emphasized. Even if both types of attached globules were artifactually induced by the preparation treatment, the authors argued that the difference in appearance between the plasma membrane units and the mitochondrial membrane subunits would still indicate intrinsic structural differences between the two types of membranes.

The concept of unit membrane has been running into firm resistance in the case of chloroplasts. These organelles are surrounded by a double membrane. Inside the structure are stacks of packed membranes called grana. The grana contain a series of flattened membrane vesicles or discs which are also referred to as thylakoids. In thin sections the membranes of the thylakoid give a tri-layered appearance with two outer dense osmiophilic bands and an internal, less dense region. The over-all thickness of this structure is about 160 A. On the basis of studies employing polarization optics and X-ray diffraction (45–50) a thylakoid membrane model was proposed in which the outside layer was made of protein particles and the inside was a layer of lipid.

Shadowed preparations of isolated chloroplast membranes have revealed a repeating structure first described by Steinman (51) and by Frey-Wyssling and Steinman (52). Since then, other investigators have presented evidence of structural subunits in chloroplast membranes. Park and Pon have proposed a structure which contains functional units of small flattened spheres approximately 200 A in diameter attached to the inner surface of the osmiophilic layer that surrounds the thylakoid (53, 54). These subunits then pack against each other in such a way that the combined thickness of the two layers, comprising a single thylakoid, is less than the sum of the thicknesses of two such particles. Park and Pon believe that the units have a functional significance in photosynthesis similar to the role proposed for the electron transport particle or elementary particle of mitochondria (see pages 267–270). The functional unit of photosynthesis has been named a quantasome (55).

Weier and Benson propose a structure for chloroplast membranes which consists of sheets made up of globular lipoprotein subunits approximately 60 A in diameter (56). Their electron micrographs clearly reveal a subunit component in the plane of the membrane. This unit has a light core of about 37 A diameter and a dark outer rim. Specific details, such as the location of chlorophyl and associated membrane lipids, are discussed by Weier and Benson in their review. In their view of this membrane structure, they break away from the tri-lamellar unit membrane of Robertson.

The new and powerful electron microscopic technique of freeze etching has been brought to bear on this problem in an attempt to establish the fine structure of chloroplast membranes. In this technique the object is not chemically fixed and dehydrated as in conventional electron microscopy. Instead, the isolated chloroplast is quick-frozen, in a medium containing 20% glycerol, by means of liquid freon (−150°). The frozen block of tissue is then cleaved with a cutting device. The plane of cleavage cuts through the specimen. The specimen can be partially dehydrated by sublimation to produce an etching effect around structures that cannot be dehydrated. The etched subcellular surface is then covered with a thin platinum-carbon layer which forms a stable fine structural replica. Finally, the replica is removed from the tissue by dissolving away the organic material. The technique has been applied by both Mühlethaler and by Branton and Park (57, 58).

For the nonmicroscopist, it is reassuring that these two laboratories have both obtained such clear and essentially identical results. They both show subunits embedded in lamellar matrices. The subunits described by Mühlethaler have a diameter of about 60 A. Those described by Branton and Park were of two types and measured approximately 110 A and 175 A in diameter. More recently, Mühlethaler has seen two sizes of subunits; one approximately 40 A in diameter he feels may be a basic structural protein, and a large type could represent enzyme complexes. The size of these subunits is not the critical point. Mühlethaler interprets his freeze etching micrographs as revealing the external surface of the chloroplast membranes with the globular units making up the outer surface. Branton and Park believe that the freeze etching technique has actually cleaved the membrane itself so that the particles revealed represent subunits comprising the inner membrane structure. Thus by the same technique, Mühlethaler supports the basic concept of a unit membrane in chloroplasts and Branton and Park completely reject the possibility of a simple tri-lamellar unit membrane for chloroplasts. With so much attention focused now on this problem, it is hoped that the structure of the chloroplast membrane may soon be resolved in terms that can help to settle the current controversy on the concept of universality of membrane structure.

In 1966 several authors made a complete break from the restrictions and concept of the unit membrane. Parsons explained his views in a lengthy discussion at the Seventh Canadian Cancer Research Conference held in Ontario in June of 1966 (59). Parsons noted the observations of Sjöstrand that classified biological membranes in terms of thickness.

Mitochondrial membranes and the endoplasmic reticulum measure 50 to 60 A in thickness, whereas the plasma membrane of most cells measures 90 to 100 A. Parsons, Williams, and Chance described methods for isolating outer mitochondrial membranes separate from inner membranes (60). The two membranes were markedly different in phospholipid content, buoyant density, functional activity (phosphorylation-linked electron transport was associated with the inner membranes only), and the presence of attached protein subunits on the inner membrane which were absent from the outer membrane.

Parsons emphasized the diversity of various kinds of neutral lipids and phospholipids that are found in membranes and the fact that different classes of membranes are strikingly varied in their particular composition of lipids. There are also marked differences in total lipid content of membranes, varying from the inner membrane of mitochondria which contain 22% phospholipid to myelin membranes which contain 70 to 75% lipid. Attention was called to the work of Blaisie, Dewey, Blaurock, and Worthington (61) who examined oriented pellets of outer segment membranes of frog retinae. After mild drying, diffraction patterns gave evidence of 40 A subunits packed in square array of side length 70 A. Electron microscopy of both negatively and positively stained preparations also indicated 40 A subunits packed in the same way. Two reservations were stated for the relevance of these findings to membrane fine structure. The subunits were not observed in nondried material and it was not clear whether these subunits were contained in the lipoprotein layer of the membrane or in a surface layer.

The current debate over the applicability of a unit membrane concept to all membranes is reminiscent of the debate which raged for years over the assumed existence of peptide intermediates in protein synthesis. At first the idea was taken as extremely reliable and over a period of many years several laboratories reported results consistent with the idea of their necessity for protein synthesis (see pages 35–39). When newer ideas became prominent and the concept of peptide intermediates no longer seemed useful, their existence was denied. In Chapter 3, I went to some length to show that out of the vast accumulation of experimental data published in support of their existence or non-existence there was no really substantial proof to establish the case in either direction. Further developments in protein synthesis have progressed the field very satisfactorily without having to rely on the concept of free peptide intermediates.

In the present context we are now moving from a position where the unit membrane concept seemed fairly secure to one where many would have the whole idea discarded with other scientific fashions that have passed. The analogy to the case already cited for peptide intermediates is that the unit membrane concept was never firmly established and the arguments cited by others to discard it are equally inconclusive.

Parsons argues against the unit membrane because of the wide variations in composition, enzymatic activity, permeability characteristics, and thickness. The simple lipid bimolecular leaflet, coated on both sides with nonlipid material, is exceedingly adaptable. The aforementioned differences could be built into the structure the same way that 20 different amino acids can be built into proteins with widely different functions, shapes, and sizes. Thickness could be influenced by the extent of interdigitation or fluidity of the hydrocarbon portions, or the amount of interaction of the various electron microscopic stains with the components of the membrane. Preparative artifacts could also be a factor in distorting some membranes more or less than others. Overlap and other optical artifacts as well as possible phase transitions of the components during the preparation for observations also could complicate the picture.

Parsons cited the work of Fleischer, Fleischer, and Stoeckenius who were able to extract at least 95% of the lipid of the inner mitochondrial membrane and still observe the tri-layered unit membrane appearance in the electron microscope (62). This obviously weakens the argument that the tri-layered appearance tells anything about the arrangement of the lipids in this membrane, but it is also possible that because of the lipids which subsequently were extracted the protein layers were forced into the railroad track arrangement which produces a tri-layered image.

An ingenious alternative model for membrane structure was proposed by Parsons; namely, that of a flat (two-dimensional) micellar network packed in hexagonal array. The thickness of such a sheet of phospholipid would be only 9 A.

Korn entered the growing controversy in 1966 with a critical analysis of the basis upon which the unit membrane hypothesis was constructed (63). He further emphasized incompatibilities that he saw between current biochemical concepts and the idea of a unit membrane. The unit membrane hypothesis is based on two assumptions: (a) there is one basic structure for all biological membranes; (b) the basic structure is the unit membrane described by Robertson.

Korn reviewed the X-ray diffraction and electron microscopic studies of myelin, both of which revealed the spacings of the heavy dense lines and intraperiod lines separated by clear zones (see pages 207–216). He emphasized, however, that these studies say nothing of the orientation of lipid layers. The unit membrane hypothesis flows from the studies with myelin in the following manner. The continuity of the myelin sheath with the Schwann cell plasma membrane and the similarly layered appearance of fixed preparations in the electron microscope are interpreted to mean that deductions concerning the structure of myelin are also applicable to the plasma membrane. Since membranes in other cells, under certain conditions of fixing and staining, also show the tri-layered structure, they too would have the same structure as myelin. Since myelin is chemically, metabolically, and functionally different from all other membranes for which data are available, Korn questioned the important step which extends the studies on myelin to all other membranes. Differences among membranes reported in the literature were also recounted. Variations in thickness (see pages 223–225) and the appearance of globular subunits (see pages 222–229) were important among these. The recent evidence of Blaisie, Dewey, Blaurock, and Worthington was considered particularly significant (61). These investigators used electron microscopic and X-ray diffraction techniques to examine isolated and pelleted outer segment membranes of frog retina. Their work indicated the presence of globular subunits 40 A in diameter in the membrane within unit cells 70 A in width. Mindful of the susceptibility to structural alteration of their specimen, they monitored its birefringence before and after the X-ray treatment. No change in birefringence was observed. The natural tendency of lipids to aggregate in ordered micellar forms as well as bimolecular leaflets (see pages 216–218) was discussed.

The uncertainty of what is actually stained by $KMnO_4$ or OsO_4 and what kinds of structural alterations may be induced, particularly by the latter compound, precludes an unambiguous interpretation of fixed and stained electron microscopic images. The interpretation of the dark lines as polar portions of the lipids rests on assumptions that these are the areas of deposition of the heavy metal. Hydration experiments, which cause a separation of myelin forms along the dark lines, however, add further support to these interpretations (see pages 219–220). The chemistry of OsO_4 reactions with unsaturated fatty acids has not been properly studied in relation to possible influences of this chemical on subsequent structural images of treated membranes. Korn's laboratory has reported that OsO_4 tends to join together unsaturated hydrocarbon side chains of fatty acids through osmic ester bridges. On the basis of these uncertainties, Korn concludes that the dense lines in membranes fixed with OsO_4 reveal nothing about the molecular orientation of the original membrane. This conclusion was reinforced by the findings that the usual tri-layered structure can be seen after OsO_4 fixation of mitochondria from which nearly all of the lipid was previously extracted.

Next Korn considered data relating to the quantitative aspects of amounts of lipid and protein in membranes in terms of the surface areas of each and the surface area to be covered by a particular membrane. The data, assumptions, and errors of Gorter and Grendel (pages 203 and 204) relating to the membranes of erythrocytes were reviewed. Korn felt that newer information might lower the original reported value from the 2 molecules of lipid per unit of cell surface reported by Gorter and Grendel to 1.4. This conclusion, it should be noted, is partially contradicted by recent calculations of Bar et al. (page 204), who claim that compensatory errors made by Gorter and Grendel would tend to leave the value of two intact. Korn constructed a table showing the contents of protein and lipid for seven species of membrane from microbial and higher ordered biological sources. There was a wide variation in relative amounts of protein, phospholipid, and cholesterol. If we assume that the component molecules always contribute a constant partial area, the ratios of calculated areas of protein to lipid for these membranes vary over a range of one order of magnitude from 0.43 to 5.4. He emphasized, however, that these interpretations were uncertain because of possible contamination of the preparations with extraneous protein. Furthermore, the excess protein may be in a globular rather than an extended form. What was apparent was that myelin was distinguished from the other membranes by its extremely low value of 0.43. He also presented a table showing the wide variation in the nature of the lipids comprising individual membranes from different sources. This fact, however, is not necessarily incompatible with the unit membrane hypothesis if different lipids could fit into the basic pattern, much as different amino acids can be used to form all the different proteins.

In the final section of his critique Korn raised additional arguments against the unit membrane hypothesis, but in pursuit of the analogy I drew between this question and the long unresolved question of peptide intermediates, I maintain that none of these considerations is a strong indictment against the existence of the unit membrane.

Since membranes are believed to be involved in the most varied kinds of biological processes, it is unlikely that a basic unit structure should exist. Furthermore, he stated that it is difficult to envision how a structure whose fundamental properties are largely dictated by the physical-chemical characteristics of its lipid components could carry out these many activities. I think that the basic unit membrane could find variety in the nature of its different lipids and proteins and that the varied biological roles could be handled by varied coatings of different enzyme assemblies.

As to the manner of biosynthesis of unit membranes, Korn argued that the first step would undoubtedly be the formation of the lipid bimolecular leaflet. Since contemporary molecular biology does not make provision for specifying lipid arrangements, he doubts that such unit membranes could have a unique means of biosynthesis. I think that this problem can also be reconciled with the unit membrane. Let genetic information specify the membrane proteins which assemble themselves in a unique fashion and the lipids can then be specified by configurations and electrostatic charge influences of the protein surface to the extent that the lipid arrangement must be specified. This proposal is not very different from Korn's suggestion that lipoproteins are first synthesized and specifically aggregate under direction of the protein components. Although Korn is entirely correct in maintaining that the unit membrane hypothesis is based upon the barest minimum of experimental support and questionable assumptions, I do not think any or all of the amassed contradictory considerations provide a sufficient basis for entirely discarding the concept at this stage of development. It is a distinct service to point out the apparent incompatibilities of this model and to consider other alternatives, but it would be wrong to take a firm position with the data available at this time.

Green and Perdue discounted the unit membrane hypothesis as an outmoded historical development which should be replaced by the new concept of basic membrane structure that they proposed (64). In addition to all of the incompatibilities that I have been discussing in these last six pages, they claimed that the binding of protein to phospholipids in various membranes is hydrophobic rather than electrostatic (as assumed in the unit membrane hypothesis) and is much more resistant to dissociation than would be expected for the thesis that lipids and proteins form separate phases joined by electrostatic interactions.

They proposed that all membranes are composed of macromolecular lipoprotein repeating units, and felt that the inner membrane subunits of mitochondria are attached to individual base plates and that the whole "tripartite" unit consisting of a headpiece, stalk, and postulated basepiece, comprises the repeating structural unit from which inner mitochondrial membranes are built. Evidence for the existence of base plates was not given. Experiments were cited where membranes have been dissociated by detergents and then, upon dialysis of the bile salts, membrane configurations were again encountered. The interpretation suggested was that these experiments are explained by the dissociation of membrane structural building blocks and their reassociation after removal of the bile salts. Since projecting subunits similar to the mitochondrial inner membrane subunits have been reported for other kinds of membranes, the authors believed that the subunits also are attached to detachable baseplates and that this mode of membrane formation and basic structure is universal. The idea is interesting, but at the moment of writing it must be considered no more than an exciting speculation.

Some support for this possibility can be found in the work of Cunningham and Crane (65). They published electron micrographs of negatively stained preparations of various kinds of membranes from rat liver, rabbit liver, red blood cell, beef heart, onion stems, and spinach chloroplasts. In practically all cases, structural detail was observed in and on the membranes. These were described as sheaths, fringes, globules, granules, pebbles, and bumps. To those untrained in electron microscopy it is difficult to decide how much credence these images should command. They may very well be indicators of structural subunits from which individual membranes are built. This is certainly a matter that will provide interesting historical reading some years from now.

A discussion of membrane structure written in the middle 1960's should contain some mention of structural protein. First described in 1961 by Green, Tisdale, Criddle, Chen, and Bock (66), it was thought to be the major structural assembly protein of the mitochondrial membrane, and indeed seemed to account for approximately 50% of its total protein. It is recognized largely on the basis of its peculiar solubility characteristics (insoluble at near neutrality and soluble at elevated pH and in the presence of detergent), its lack of enzymatic activity, and its ability to form soluble complexes with cytochromes a, b, c_1, c and myoglobin, and to bind P_i, ATP, DPN, and phospholipid.

In the most recent publications available from Green and coworkers it is reported that structural protein is not needed for electron transport activity and is, in fact, a superfluous component. They further report that it is not needed in order to form or reconstitute membranes (pages 269–270). The molecular weight of the minimum combining subunit is $\sim$23,000 and it is believed that all subunits are the same. This would seem to eliminate the possibility that specificity is contributed to membranes by this potential building unit.

Woodward and Munkres have reported the isolation of similar structural protein from mitochondria, nuclei, microsomes, soluble fractions, and mycelia from *Neurospora* (67).

The eventual elucidation of the biological role of this widespread species of protein will no doubt be an important achievement of future research. In the meantime the suspected role of structural proteins in membranes remains entirely unsupported by the accumulated experimental data at this time.

In conclusion it must be noted that although we have come to recognize the potential importance of membranes in relation to the biochemistry of the living cell, we are also beginning to recognize that our previous simplified concept of their structure is inadequately supported by experimental observations. In view of the growing interest and activity in this field, the situation will hopefully be vastly improved in the coming years.

REFERENCES

1. E. Overton, *Vjschr. Naturforsch. Ges. Zurich,* **40,** 159 (1895).
2. I. Langmuir, *J. Amer. Chem. Soc.,* **39,** 1848 (1917).
3. W. D. Harkins, F. E. Brown, and E. S. H. Davies, *J. Amer. Chem. Soc.,* **39,** 354 (1917).
4. J. Perrin, *Annals Phys.,* Series 9, **X,** 160 (1918).
5. W. H. Bragg, and W. L. Bragg, in *X-Rays and Crystal Structure,* 4th ed., rev., Harcourt, Brace, New York, 1924.
6. E. Gorter and F. Grendel, *J. Exp. Med.,* **41,** 439 (1925).
7. R. S. Bar, D. W. Diamer, and D. G. Cornwell, *Science,* **153,** 1010 (1966).
8. M. P. Westerman, L. E. Pierce, W. N. Jensen, *J. Lab. Clin. Med.,* **57,** 819 (1961).
9. E. N. Harvey and H. Shapiro, *J. Cell. and Comp. Physiol.,* **5,** 255 (1934).
10. K. S. Cole, *J. Cell. and Comp. Physiol.,* **1,** 1 (1932).
11. T. E. Thompson, in *Cellular Membranes in Development,* Ed., M. Locke, Academic Press, New York, 1964, p. 83.
12. J. F. Danielli, and E. N. Harvey, *J. Cell and Comp. Physiol.,* **5,** 483 (1932).
13. J. F. Danielli and H. Davson, *J. Cell. and Comp. Physiol.,* **5,** 495 (1935).
14. J. D. Bernal and D. Crowfoot, *Nature,* **133,** 795 (1934).
15. F. O. Schmitt, R. S. Bear, and G. L. Clark, *Radiology,* **25,** 131 (1935).
16. F. O. Schmitt, R. S. Bear, and K. L. Palmer, *J. Cell and Comp. Physiol.,* **18,** 31 (1941).
17. R. Hober, in *Physical Chemistry of Cells and Tissues,* Blakiston, Philadelphia, (1945).
18. H. Fernández-Morán, *Exp. Cell. Res.,* **1,** 309 (1950).
19. F. S. Sjöstrand, *Experientia,* **9,** 68 (1953).
20. B. B. Geren and J. Raskind, *Proc. Nat. Acad. Sci.,* **39,** 880 (1953).
21. J. B. Finean, *Exp. Cell Res.,* **5,** 202 (1953).
22. H. Fernández-Morán, *Exp. Cell. Res.,* **3,** 282 (1952).
23. J. D. Robertson, *J. Biophys. Biochem. Cytol.,* **3,** 1043 (1957).
24. B. B. Geren, *Exp. Cell Res.,* **7,** 558 (1954).
25. H. Fernández-Morán and J. B. Finean, *J. Biophys. Biochem. Cytol.,* **3,** 725 (1957).
26. J. D. Robertson, *Biochem. Soc. Symp.,* **16,** 3 (1959).
27. J. Elkes and J. B. Finean, *Exp. Cell Res.,* **4,** 69 (1953).
28. J. B. Finean, *Exp. Cell Res.,* **6,** 283 (1954).
29. J. B. Finean and P. F. Millington, *J. Biophys. Biochem. Cytol.,* **3,** 89 (1957).
30. V. Luzzati and F. Husson, *J. Cell. Biol.,* **12,** 207 (1962).
31. W. Stoeckenius, *J. Cell. Biol.,* **12,** 221 (1962).
32. J. D. Robertson, in *Cellular Membranes in Development,* Ed., M. Locke, Academic Press, New York, 1964.
33. J. D. Robertson, *J. Cell. Biol.,* **19,** 201 (1963).
34. J. D. Robertson, *Ann N.Y. Acad. Sci.,* **137,** 421 (1966).

35. H. Fernández-Morán, in *Ultrastructure and Metabolism of the Nervous Systems*, Research Publications, Assoc. Res. Nerv. Mental Dis. **40**, 235 (1962).
36. S. E. G. Nilsson, *Nature*, **202**, 509 (1964).
37. J. D. Robertson, *Proc. Nat. Acad. Sci.*, **53**, 860 (1965).
38. F. S. Sjöstrand, *Nature*, **199**, 1262 (1963).
39. F. S. Sjöstrand, *J. Ultrastructure Res.*, **9**, 340 (1963).
40. F. S. Sjöstrand, *J. Ultrastructure Res.*, **9**, 561 (1963).
41. J. D. Robertson, Paper read at the 133rd A. A. A. S. Meeting, Dec., 1966, Washington, D.C.
42. J. A. Lucy and A. M. Glauert, *J. Mol. Biol.*, **8**, 727 (1964).
43. F. S. Sjöstrand and L. G. Elfvin, *J. Ultrastructure Res.* **10**, 263 (1964).
44. E. L. Benedetti and P. Emmelot, *J. Cell. Biol.*, **29**, 299 (1965).
45. W. Menke, in *Biochem. of Chloroplasts*, Vol. I, Ed., T. W. Goodwin, Academic Press, New York, 1966, p. 3.
46. W. Menke, *Biol. Zbl.*, **63**, 326 (1943).
47. W. Menke and G. Menke, *Z. Naturf.*, **10b**, 416 (1955).
48. J. C. Goedheer, doctoral thesis, University of Utrecht (1958).
49. W. Kreutz, *Z. Naturf.*, **18b**, 1098 (1963).
50. W. Kreutz, *Z. Naturf.*, **19b**, 441 (1964).
51. E. Steinmann, *Exp. Cell. Res.*, **3**, 367 (1952).
52. A. Frey-Wyssling and E. Steinmann, *Vierteljahresschr. Naturforsch. Ges. Zuerich*, **98**, 20 (1953).
53. R. B. Park and N. G. Pon, *J. Mol. Biol.*, **3**, 1 (1961).
54. R. B. Park and N. G. Pon, *J. Mol. Biol.*, **6**, 105 (1963).
55. R. B. Park and J. Biggins, *Science*, **144**, 1009 (1964).
56. T. E. Weier, and A. A. Benson, in *Biochem. of Chloroplasts*, Vol. I, Ed., T. W. Goodwin, Academic Press, N.Y., 1966, p. 91.
57. K. Mühlethaler, in *Biochem. of Chloroplasts*, Vol. I, Ed., T. W. Goodwin, Academic Press, New York, 1966, p. 49.
58. D. Branton, and R. B. Park, *J. Ultrastruct. Res.*, (1967).
59. D. F. Parsons, in *Proc. 7th Canadian Cancer Res. Conf.*, Pergamon Press, 1967, p. 193.
60. D. F. Parsons, G. R. Williams, and B. Chance, *Ann. N.Y. Acad. Sci.*, **137**, 643 (1966).
61. J. K. Blaisie, M. M. Dewey, A. E. Blaurock, and C. R. Worthington, *J. Mol. Biol.*, **14**, 143 (1965).
62. S. Fleischer, B. Fleischer, and W. Stoeckenius, *Fed. Proc.*, **24**, 296 (1965).
63. E. D. Korn, *Science*, **153**, 1491 (1966).
64. D. E. Green and J. F. Perdue, *Proc. Nat. Acad. Sci.*, **55**, 1295 (1966).
65. W. P. Cunningham and F. L. Crane, *Exp. Cell. Res.*, **44**, 31 (1966).
66. D. E. Green, H. D. Tisdale, R. S. Criddle, P. Y. Chen, and R. M. Bock, *Biochem. Biophys. Res. Comm.*, **5**, 109 (1961).
67. D. Woodward and K. Munkres, in *"Organizational Biosynthesis"* p. 489–502. Eds. H. J. Vogel, J. O. Lampen and V. Bryson, Academic Press, New York, 1967.

Chapter 6

MEMBRANE BIOCHEMISTRY

Biochemistry is a later stage in the evolution of a science that can trace its antecedents back through biology, organic chemistry, inorganic chemistry, and alchemy. In these earlier phases of development a keen appreciation for pure crystalline precipitates and clear aqueous solutions was developed. The beginning phases of biochemistry applied the techniques of organic chemistry to obtain crystalline forms of important biological substances. The attainment of crystalline enzymes were heralded with justifiable enthusiasm. When Stanley applied these same techniques to obtain crystals of tobacco mosaic virus, his work was awarded the Nobel Prize.

Cells were crushed in order to liberate their water-soluble components and enzyme systems, such as for the fermentation of sugars. The characteristics of enzyme-catalyzed conversions of aqueous metabolites to aqueous products occupied the mainstream of biochemical attention, with enzymology and intermediary metabolism as the most active forms of the science.

If we try to designate the major remaining problems of biochemistry, where knowledge has lagged far behind and .where the application of standard techniques and approaches has met with firm resistance, we must include the mechanisms of selective permeability and active transport, the nature of the assembly of respiratory enzymes, and the manner of linking respiration to phosphorylation, the mechanism of hormone action, the manner by which cells efficiently synthesize many different types of individual proteins and enzymes, and the manner by which cells reproduce their organelles and themselves; both under normal conditions and in the cancerous state.

The answers to these unsolved questions will probably be found in detailed studies of membrane systems, which may include a type of nonaqueous biochemistry, or at least the biochemistry of reaction systems fixed on lipoprotein matrices.

In this chapter I have attempted to sketch the states of knowledge of some of these systems which, in the sense just discussed, are at the threshhold of a new approach in biochemistry. Our studies of these systems so far have served to stress a lack of fundamental understanding and we seem to be poised for an assault on the detailed structures of oriented enzymes and their capabilities.

SELECTIVE PERMEABILITY AND ACTIVE TRANSPORT

Cells are separated from their environments by membranes. The concentrations of identical solutes on either side of this molecular barrier may vary several thousandfold. In seeming defiance of the laws of thermodynamics, particular solutes (related in structure to other solutes which are denied entrance) are selectively transported across the membrane from a solution of low concentration to one of high concentration. The cell membrane within certain limits may hold out a hostile environment and, at the same time, extract from it only such substances which will best suit the requirements of the cell.

This amazing facility of a membrane is simultaneously exerted on all forms of substances needed by the cell. Many laboratories and countless numbers of papers and books have been written on various expressions of the transport problem as directed toward inorganic ions, sugars, amino acids, fats, etc.

Although it would be impossible to cover this vast literature, I can state in a concise manner that the way in which cells accomplish this phenomenon is a mystery whose secret has so far been safely maintained. Nonetheless, certain things are known about the phenomenon. The following list represents some pertinent observations.

1. Solutes can be accumulated within a cell from a medium of much lower concentration.

2. Some solutes, although freely soluble in water, cannot be extracted from cells unless the cell membranes are broken.

3. Mutated cells are found which have lost the ability to concentrate a particular species out of a group of related compounds (e.g., the ability to concentrate a particular amino acid could be lost and growth would occur only in the presence of high concentrations of this one amino acid).

4. Certain types of mutations are known as cryptic; in such cases

it appears that an enzyme necessary to metabolize a particular substrate is missing. Nonetheless, when the cells are broken, the enzyme activity is shown to be present. Subsequent work has established that the missing ability of the cell was the one which caused the substrate in question to enter and be accumulated.

5. Accumulation can be highly specific where one kind of molecule is passed through the membrane while another structurally related molecule is excluded.

6. In many cases the uptake of a particular solute can be inhibited on a competitive basis by a structurally related molecule. This kind of phenomenon is especially evident in the case of auxotrophic bacteria which require the addition of a particular substrate for growth. Structurally related but different substrates can seriously inhibit the growth of these organisms by interfering with the uptake of the required metabolite.

7. The rate at which a solute enters a cell increases rapidly in proportion to its external concentration up to a certain point. After this concentration is reached, the increase in entry rate as a function of external concentration falls to a much lower relationship. This situation implies the participation of a mediating or carrier substance which is present in limiting amounts. This type of saturation kinetics adheres to the Michaelis-Menten equations for enzyme-catalyzed reactions. Similarly, the equilibrium concentration within a cell as a function of external concentration frequently obeys the laws of an adsorption isotherm.

8. Diffusion mediated by a carrier substance is reversible and, after a radioactive form of a particular metabolite is concentrated, the addition of the unlabeled form of the metabolite causes the radioactive transported substrate to be removed.

9. After the external and internal concentrations for a particular substrate have reached equilibrium in the sense that efflux from the cell equals influx, the addition of a structurally related molecule to one side of the membrane will frequently cause the substrate on the other side to move across the membrane against its own concentration gradient. This phenomenon, which was first discovered by Rosenberg and Wilbrandt (1), is called counterflow; it is accepted as virtual proof of the existence in the membrane of a mobile carrier which is used for both substrates.

10. The preloading of a cell with an unlabeled form of a particular substrate may enhance the subsequent uptake of a labeled form of the same substrate. This phenomenon, which was discovered by Heinz and Walsh (2), conforms to a situation which has been defined by Ussing as exchange diffusion.

11. The ability to form and maintain a concentration difference across a membrane is dependent upon the integrity of energy-forming processes in the cell and can be inhibited by standard poisons of energy metabolism.

12. Various hormones have been shown to stimulate processes leading to the transport and concentration of particular substrates.

These listed characteristics point to specific and mediated processes for transporting substances across biological membranes. A brief discussion of some of the approaches used and ideas developed in regard to the transport of sugars and amino acids may serve to describe the trend in this field of research.

One of the most frequently studied parameters is the relation of the initial velocity of uptake to external concentration. For a process of free diffusion, a straight line should be obtained for a plot of concentration versus velocity (v). For a mediated process involving a component in the membrane, a saturation curve is obtained which indicates that the amount of mediating substance becomes limiting (Figure 1). Either a plateau may be reached for cases where the only means of penetration is via the mediated process, or a steady linear slope may be established to indicate the continuing uptake by free diffusion. A plot of $1/v$ versus $1/$(external concentration) may yield a straight line. This technique, which is familiar to biochemists as a Lineweaver Burk Plot can be used to characterize enzyme catalyzed reactions.

In fact, the finding of such a relationship in the case of transport is frequently taken as a demonstration of the existence of a transporting enzyme or "permease" according to the original suggestion of Cohen and Monod. The derivation of the equations and an evaluation of the

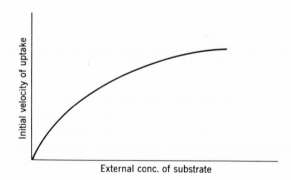

External conc. of substrate

Figure 1

assumptions would be helpful in an appreciation of the significance of the relationship.

Let S = substrate to be transported
M = mediator

Consider

membrane

$$\text{Side 1} \quad S + \overset{\overset{\displaystyle k_2}{\frown}}{\underset{\underset{\displaystyle k_1}{\smile}}{M}} SM \overset{k_3}{\to} S \quad \text{Side 2}$$

Assumption No. 1

$$S + M \rightleftharpoons SM \text{ is freely reversible}$$

Assumption No. 2

M is limiting

$$\frac{(S)(M)}{(SM)} = \frac{k_2}{k_1} = K_m$$

$$(SM) = \frac{(S)(M)}{K_m}$$

If the total amount of M is M_0, then the amount of free $M = M_0 - (SM)$.

$$(SM) = \frac{(S)[M_0 - (SM)]}{K_m}$$

$$K_m(SM) = (S)M_0 - (S)(SM)$$

$$(K_m + S)(SM) = (S)M_0$$

$$(SM) = M_0 \frac{(S)}{K_m + (S)}$$

The rate of penetration of S will be v where

$$v = k_3(SM) = k_3 M_0 \frac{(S)}{K_m + (S)}$$

It is evident that when M is fully saturated with S then $(SM) = M_0$ and the $v = k_3 M_0$ will be at its maximum or v_{max}

$$\therefore v = v_{max} \frac{(S)}{K_m + (S)}$$

which is the Michaelis-Menten equation and the reciprocal of both sides yields

$$\frac{1}{v} = \frac{K_m}{v_{max}} \frac{1}{S} + \frac{1}{v_{max}}$$

so that a plot of $1/v$ versus $1/S$ will yield a straight line; M may be an enzyme, but it does not have to be. It could represent a mobile carrier molecule of protein or nonprotein nature or a site on a membrane surface. The finding of an adherence to the Michaelis-Menten equation is compatible with the existence of a protein negotiating the transport process but in itself does not provide a strong case.

Similarly, if we check the derivation and assumptions for the expression which relates internal equilibrium concentration of a substrate within a cell to external concentration (Langmuir adsorption isotherm) it can be seen that no extrapolation as to the nature of the limiting mediating substance can be made.

$$S_{equil}^{int} = Y \frac{S_{ext}}{K_m + S_{ext}}$$

S_{equil}^{int} = internal concentration of transported substrate at equilibrium

Y = total number of binding sites for S

A priori it is reasonable to anticipate that a protein would be the mediator but this will require more direct proof.

There are many papers in the literature which describe the phenomenon of "uphill transport driven by counterflow." The theoretical basis in mathematical terms is discussed by Rosenberg and Wilbrandt (1). If a solute, for example a sugar such as galactose, is accumulated within a cell so that it achieves equilibrium with a given external concentration, a stable thermodynamic situation exists. The net flow of sugar in either direction would then require the expenditure of energy. This situation, however, is dynamic and if the mechanism of permeation involves a carrier equal numbers of sugar molecules traverse the membrane in both directions. If the same carrier can function for molecules other than galactose, for example glucose, an interesting phenomenon develops. When glucose is added to the medium, the carrier that has arrived at the external surface of the membrane discharges all of its galactose and then loads up with glucose and galactose on a statistical basis which is governed by the relative concentration of the two sugars and their affinities for the carrier. Therefore the carriers bring into the cell less than their full capacity of galactose. When they unload to the interior of the cell, they are again swamped with galactose. The process thus causes a net outward flow of galactose away from its equilibrium situation. The over-all situation, however, is thermodynamically sound since the glucose movement is "downhill." In this nonmathematical description it is apparent that a mobile carrier with definite structural requirements is involved. Any sugar for which the carrier has no affinity will not cause the counterflow. The carrier has to load on one side, discharge on the other and then reload and travel back to the other side to start another cycle. The demonstration of this phenomenon is taken as the strongest evidence in support of the concept of a mobile carrier.

Many papers demonstrate that the accumulation of a particular metabolite, such as an amino acid like valine, will be competitively inhibited by a structurally related amino acid like leucine or isoleucine. Once again, specific binding sites are inferred.

There are numerous papers postulating rather precise mechanisms for the accumulation phenomenon. One of the most quoted brought a new and controversial term to the field—"permease" (3). Cohen and Monod, in their use of this term, implied the existence of a protein with enzyme-like properties which directly interacts with an external solute and, although not transforming it chemically, transfers it physically from the outside of the cell to the inside. The specific permease mechanism was proposed as follows:

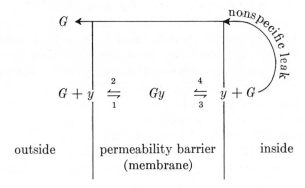

G represents the solute being transported and y the permease. For permeases that lead to the establishment of a concentration gradient, it is proposed that reaction No. 4 can be inhibited in the presence of a suitable energy donor. A nonspecific leak places a small drain on the internal concentration, but as the internal concentration rises this factor increases until at equilibrium it equals the influx rate. No permease has ever been found, but support for its existence was taken from Michaelis-Menten kinetics, genetic control of transport capabilities, inducibility of transport systems by appropriate substrates, and the sensitivity of the induction process to inhibitors of protein synthesis. All of these criteria indicate the importance of a protein in the over-all process but the protein could be involved in forming a nonprotein carrier or in maintaining some structural characteristic of the membrane.

Britten and McClure highlighted other deficiences in the formulation (4). They pointed out that the maintenance of a concentration gradient required the energy-dependent permease system balanced against the nonspecific leak. Therefore, in the absence of an energy supply to maintain the filling reaction, the pool should diminish. The pool should diminish also in the absence of an external supply of amino acid. But in the absence of glucose and/or amino acid in *E. coli*, pools are maintained. Britten and McClure in a similar manner pointed out other inconsistencies in the behavior of *E. coli* pools and predictions of the permease model. Britten proposed his own model which gave the cell a permeable cell membrane and a series of specific and mobile carriers in limited amounts which could combine with the substrate, circulate in the cytoplasm, and by an energy-dependent process transfer the substrate to fixed sites. This formulation, of course, requires that the concentrated substrates exist to a large extent in combined forms in the cytoplasm. No such combined forms have thus far been identified for the huge amounts of various components of cellular pools.

Other kinds of carrier models could be visualized which employ a protein or nonprotein agent that combines with substrate either with or without the aid of an enzyme. The energy of accumulation could influence the loading or unloading or mobility of the carrier or the state of the membrane or the state of the substrate once inside of the cell.

Through the course of the past 35 years various discrete mechanisms were proposed and considered to explain this vital ability of membranes to selectively regulate the traffic passing in and out of cells. In 1933, Lundsgaard proposed that glucose is concentrated as a consequence of the fact that phosphorylation of the glucose inside of the cell converts it from a freely diffusable substance (glucose) to one that no longer may pass the membrane (glucose phosphate) (5). This theory, although under attack from time to time, was widely accepted for about 20 years. Gradually, the number of incompatible observations gathered sufficient weight to collapse all support (6).

Another theory based on the postulated enzymatic conversion of glucose from a pyranose ring form to another more permeable form, catalyzed by the enzyme mutarotase, has been proposed (7) but has not received sufficient experimental support to remove the objections which have been raised against it (6).

Other theories which have been considered but which are not strongly supported by experimental observations include: (a) a specific type of pinocytosis in which the membrane, after being stimulated by the desired metabolite, forms a cup around the solution, engulfs a portion and then releases the captive substrate inside the cell; (b) a series of uniquely shaped pores in the membrane which will pass substrates that fit the pore and reject others; (c) a series of enzymes so oriented on the membrane that they will accept substrates from one side and discharge them from the other (permeases?); (d) specific contractile proteins which after attaching to a substrate will, under the influence of ATP, contract and by this motion transport the substrate across the membrane; (e) specific but poorly characterized pumps whose existence is inferred by the presence in the membranes of ATPases which are stimulated by ions that are normally transported and inhibited by substances which inhibit transport.

More recently it has been observed that the presence of Na^+ stimulates the transport of sugars and amino acids, and a proposal has been advanced that a carrier has a dual affinity for Na^+ and the other substrate. In the absence of Na^+ the carrier is immobile but in the presence of Na^+ it can transport the substrate and Na^+ to the interior of the cell. If such a model is made to operate in conjunction with a Na^+ pump that continually pumps Na^+ out of the cell at the expense of energy from ATP, then the Na^+ gradient will cause the influx of the substrate which shares the carrier with Na^+ (8).

Since the end of 1965 some entirely new and exciting experimental approaches to the problem of transport have been opened. Fox and Kennedy made use of previously known properties of the galactoside permease system of *E. coli* (9). Since galactoside transport is inhibited by sulfhydryl agents such as N-ethyl maleimide (NEM) and could be protected against this inhibition by transportable galactosides, these authors attempted to devise a specific method of labeling the permease protein with radioactive NEM and then to proceed with the purification of the labeled protein. The method developed was as follows: A culture of *E. coli* was induced to synthesize the proteins controlled by the lac operon, one of which controls the uptake of galactosides. Another portion of the parent culture was grown in the absence of galactosides and therefore did not develop the galactoside concentration system. Thiodigalactoside was then added to both the induced and uninduced cultures with the expectation that the galactoside would bind to and protect the sulfhydryl site of the postulated membrane protein or M protein according to the terminology of Fox and Kennedy. Unlabeled NEM was then added to react with all SH groups not protected by the galactoside and thus considered to be unconnected with galactoside concentration capabilities of the cell. The excess NEM was destroyed with mercaptoethanol and the two cultures were then washed. The washing presumably would remove the thiodigalactoside from the M protein SH sites but would not remove the NEM which was covalently bound to the nonspecific SH groups. The induced cells were then treated with C^{14}-NEM, which would be expected to label the M protein. The uninduced cells were treated with H^3-NEM. These cells should have little or no M protein for galactoside transport. Excess NEM was again destroyed by β-mercaptoethanol and the cells were mixed. By a process of detergent extraction of the membranes followed by chromatography on an ECTEOLA column a fraction enriched in C^{14}/H^3 was isolated. It is too early to know if the component possessing the high C^{14}/H^3 ratio is really a part of the transport system, but the approach is new and refreshing and the results so far seem promising.

Pardee, Prestidge, Whipple, and Dreyfuss found a low molecular weight sulfate-binding protein in cells of *Salmonella typhimurium* which were capable of transporting and concentrating sulfate (10).

A mutant of the organism was used which could not utilize sulfate but which contained a repressible system for transporting sulfate. When the transport system was repressed by the presence of cysteine during growth, very little sulfate could be bound by the organism. When the transport system was induced (or derepressed) by growing the organism in the presence of djenkolic acid, the sulfate-binding activity was enhanced. Transport-negative mutants also contained a repressible binding substance, although with a capacity of about one third of that of the transport-positive strain. These results, however, may be explained by some intracellular sulfate-binding protein rather than a permease in the membrane. An experimental attempt to cope with this difficulty was to apply techniques recently described by Neu and Heppel (11). Neu and Heppel showed that by either making spheroplasts of *E. coli* with EDTA and lysozyme, or by shaking intact cells of *E. coli* in cold water after a preliminary treatment with tris, sucrose, and EDTA, several enzymes are removed from the cells, whereas viability is not affected and many enzymes can be shown to be completely resistant to extraction. It has been widely assumed that only proteins and enzymes located near the surface of the cells are removable. Pardee et al. found that both of these procedures removed the sulfate-binding protein from a transport-negative mutant of *S. typhimurium*. When transport-positive cells were subjected to the osmotic shock procedure, both transport ability and sulfate-binding ability were reduced by 80%. The binding protein was found in the supernatant fluid. From these and other supporting observations, Pardee and his colleagues believe that the sulfate-binding protein is a component of the sulfate transport apparatus of these cells.

Kundig, Kundig, Anderson, and Roseman described a three protein system in *E. coli* that is capable of phosphorylating carbohydrates (12):

$$\text{phosphoenolpyruvate} + \text{heat stable protein} \overset{\text{Enz I}}{\rightleftharpoons}$$
$$\text{(HPr)}$$
$$\text{pyruvate} + \text{phospho-HPr}$$

$$\underline{\text{phospho-HPr} + \text{sugar} \overset{\text{Enz II}}{\longrightarrow} \text{HPr} + \text{sugar-P}}$$
$$\text{phosphoenolpyruvate} + \text{sugar} \rightarrow \text{pyruvate} + \text{sugar-P}$$

On lysis of the cells, Enzyme I and HPr are located in the soluble fraction and Enzyme II in the membrane fraction. Specificity of the system towards different sugars resides with Enzyme II. When galacto-side-induced cells are subjected to the osmotic shock procedure of Neu and Heppel (11), they lose the ability to accumulate galactosides and also the ability to concentrate methyl α-glucoside which activity is a con-stitutive system of the cells. The shock procedure also removes HPr which can be purified about ten thousandfold from the supernatant fluid. When HPr is added back to shocked cells, the ability to transport sugars is restored. HPr is inactivated by photo oxidation, trypsin digestion, and alkali treatment.

A somewhat similar situation was described by Piperno and Oxender who subjected cells of *E. coli* K-12 to the osmotic shock procedure of Neu and Heppel and found that the ability to take up leucine, isoleucine, and valine was lost (13). Other amino acid concentration systems were not affected. (There is some conflict here with the studies of Kundig et al. who reported that shocking *E. coli* W-2244 did not impair their ability to transport valine.) The supernatant fluid from the shocked cells contained a protein capable of binding leucine, isoleucine, and valine with the same affinity as was demonstrated by the transport system of the intact cells. It was not possible, however, to restore the transport ability of the shocked cells with respect to leucine, isoleucine, and valine by adding back purified binding protein.

As can be seen, many of these newer studies have made use of the osmotic shock treatment of Neu and Heppel. It is widely assumed that this procedure releases only surface components of cells and that the surface location combined with the specific binding properties present a strong case for believing the component to be part of the active trans-porting machinery of the cell. A note of caution was recently sounded by Neu, Ashman, and Price (14) who showed that the osmotic shock treatment of Neu and Heppel was capable of causing severe intracellular metabolic alteration which could cause the breakdown and digestion of 23s RNA. The concept that only surface components of the cells are affected, therefore, must be carefully re-examined by independent means.

The foregoing description of transport problems was not intended to present a comprehensive survey but rather to describe the main features of the problem and some of the experimental approaches that have been employed. The aggregate picture emerging from these newer studies is extremely exciting and full of promise. Transport capabilities of cells are undoubtedly mediated by surface components having just such properties as exhibited by the many contenders for this role described in the studies just cited. It does seem that we are on the verge of being able to define the complete system for the active uptake of some substances and it is hoped that newer developments in this field will soon realize this expectation.

OXIDATIVE PHOSPHORYLATION

An important process by which cells extract energy from ingested nutrients and channel this energy into soluble energy-storage metabolites includes a complex integrated network of oxidative and phosphorylating reactions collectively referred to as oxidative phosphorylation. The elucidation of the detailed manner by which the cell performs this activity has thus far not been accomplished despite all attempts at unraveling the process. The search has taken us to the membranes of cells and, as we conceptually encircle these structures involved in oxidative phosphorylation, we sense that the eventual solution to this problem is intertwined with some fundamental principles of membrane structure and function. Thus it would seem that oxidative phosphorylation is properly included in the area of membrane biochemistry.

The process of chemically reducing organic metabolites in the oxidative environment of the earth's atmosphere requires energy. Green plants, by another membrane-associated process that also is not completely understood, directly use the sun's energy to force the reduction of CO_2 by water, to sugars and fats. These substances, with the trapped energy of the sun, are eaten by animals and are systematically degraded in a stepwise manner to liberate their energy while being converted back to carbon dioxide and water. One mole of glucose, if burned in oxygen to CO_2 and H_2O, liberates 680 kcal of energy.

Usable energy in a cell is stored principally in the compound adenosine triphosphate (ATP). Under conditions existing in the cell, hydrolysis of the terminal phosphate of ATP liberates about 9 kcal of energy. The energy stored in ATP is used to power energy-requiring processes in the cell. One obvious measure of efficiency for energy-transformation mechanisms is related to the number of ATP molecules that could be synthesized from adenosine diphosphate (ADP) and inorganic phosphate (P_i) as a result of the oxidation of one mole of glucose. A 100% efficient process would allow the formation of approximately $680/9 = 75.5$ gram moles of ATP. It is also possible that some of the liberated energy might be directly used for work and not be indicated by ATP production.

By the classical and well understood glycolytic pathway which proceeds in the aqueous fraction extracted from broken cells, glucose

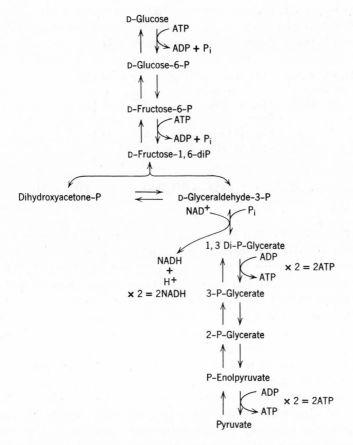

Figure 2 Glycolytic pathway from glucose to pyruvic acid. For glycolysis to be complete, pyruvic acid must be reduced to lactic acid.

can be converted to pyruvate (Figure 2). This process requires the priming energy of two molecules of ATP. One converts glucose to glucose-6-phosphate which is next converted to fructose-6-phosphate by the enzyme phosphohexose isomerase. The second required molecule of ATP phosphorylates fructose-6-phosphate to fructose-1,6-diphosphate. The further enzymatic conversion of fructose diphosphate via phosphoglycerate and phosphopyruvate yields a total of four molecules of ATP. The balance sheet up to this point shows a net production of two moles of ATP or an energy conservation of 18 kcal. The chemical conversion of glucose plus oxygen to two molecules of pyruvic acid plus water liberates 120 kcal of energy. Therefore, in one sense, the process is only $(18 \times 100)/120$ or 15% efficient. However the oxidant used in living cells is NAD (Coenzyme I) instead of oxygen and the end product accompanying pyruvate is reduced NAD instead of water. Therefore, instead of being lost as heat, an appreciable part of the energy originally in glucose becomes stored in NADH as shown below:

1) $C_6H_{12}O_6 + O_2 \rightarrow 2C_3H_4O_3 + 2H_2O$ $\Delta G = -120$
   (glucose)   (pyruvate)
2) $\phantom{C_6H_{12}O_6 + } 2H_2O \rightarrow O_2 + 2H_2$ $\Delta G = +112$

1 + 2) $C_6H_{12}O_6 \rightarrow 2C_3H_4O_3 + 2H_2$ $\Delta G = -8$
3) $2NAD + 2H_2 \rightarrow 2NADH + 2H^+$ $\Delta G = -8.7$
   (at pH 7.0)

1 + 2 + 3) $C_6H_{12}O_6 + 2NAD \rightarrow 2C_3H_4O_3 + 2NADH + 2H^+$
$$\Delta G \approx -17$$

With the realization that the actual ΔG values will be influenced by the particular cellular conditions it would appear that a process of 100% efficiency may function in the formation of two molecules of ATP from ADP, phosphate and the energy liberated in reaction $(1 + 2 + 3)$.

A by-product of the glycolytic sequence is reduced coenzyme I. This was involved in the oxidation of phosphoglyceraldehyde as shown in Figure 2. It is also an important component of the mitochondrial electron-transport system, but extramitochondrial NADH is not directly accessible to the electron-transport chain of mitochondria. One way that electrons of extramitochondrial NADH can enter the mitochondrial respiratory chain employs dihydroxyacetone phosphate. The cytoplasmic NADH reduces dihydroxyacetone phosphate to α-glycerophosphate. External α-glycerophosphate readily penetrates the mitochondria and is oxidized by a mitochondrial α-glycerophosphate dehydrogenase. This enzyme then inserts the electrons at the level of CoQ or cytochrome b. Therefore each pair of electrons handled in this manner yields two molecules of ATP from phosphorylation sites located between cytochrome b and oxygen (pages 255–257). During the formation of pyruvic acid from one mole of glucose, two moles of NADH are produced and if they enter the mitochondria as just described, four moles of ATP will be formed. The formation of this ATP, however, is a result of the functioning of the oxidative phosphorylating enzymes.

Pyruvic acid can be further metabolized via the series of oxidative reactions known as the Krebs cycle (Figure 3), named after Hans Krebs who established the details of its functional nature. This reaction sequence is also called the citric or tricarboxylic acid cycle because the tricarboxylic acids, citric, isocitric, cis-aconitic, and oxalsuccinic, are important early members of the cycle. Electrons removed in the oxidative steps of the Krebs cycle are funneled through the membrane-associated electron-transport chain, eventually reaching gaseous oxygen which, in the process, is reduced to oxygen in water. This process of oxygen consumption is known as respiration. The energy liberated in this manner from the electrons removed from two moles of pyruvate amounts to 526 kcal. Since the total process of

$$2 \text{ pyruvate} + 5O_2 \rightarrow 6CO_2 + 4H_2O$$

liberates 560 kcal, it is seen that the major energy release attends the oxidative steps *per se*. During the oxidation steps 30 moles of ATP are formed and so $30 \times 9 = 270$ kcal of energy are conserved. The efficiency of the process is therefore $270/560 \times 100 = 48\%$. The respiratory chain also produces four molecules of ATP from the oxidation of the NADH produced during glycolysis so that a total of 34 molecules of ATP are produced as a consequence of the function of the electron-transport system. Of the 2 plus 34 or 36 molecules of ATP synthesized during the metabolism of glucose to carbon dioxide and water, 34 were made via oxidative phosphorylation.

What can be said about this vital cellular activity that supplies the body with a steady flow of energy for carrying out the work of a living cell? Lavoisier, in 1780, showed that living cells require oxygen. Although the process of respiration was studied during the nineteenth century by many well-known chemists, it was only in the beginning of this century that the presence was revealed of enzymes in tissue that could catalyze the chemical reduction of colored dyes, most notably of which was methylene blue. Otto Warburg maintained that most if not all utilization of oxygen by cells was mediated by an iron-containing enzyme which he called Atmungsferment or respiration enzyme. Concurrently, in the early 1900's, Keilen studied iron-containing respiratory

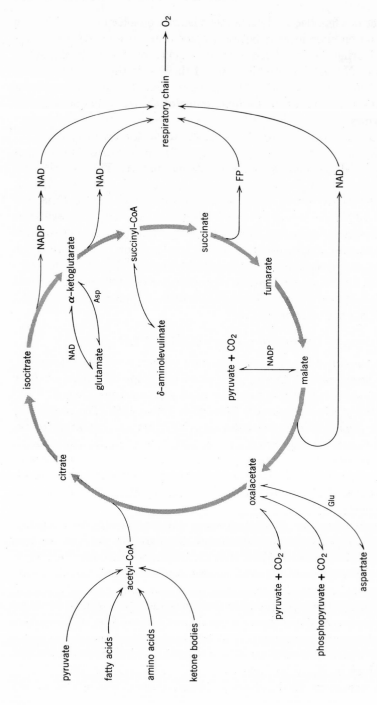

Figure 3 The Krebs citric acid cycle and associated reactions. Glu = glutamate; Asp = aspartate. Redrawn from A. L. Lehninger, *The Mitochondrion*, Benjamin, New York, (1964).

enzymes known as cytochromes. He showed in intact insect muscles that the cytochromes underwent reversible states of oxidation and reduction. It was in the 1930's that the respiratory coenzymes diphosphopyridine nucleotide (DPN) or nicotinamide adenine dinucleotide (NAD) and triphosphopyridine nucleotide (TPN) or nicotinamide adenine dinucleotide phosphate (NADP) were isolated and characterized largely through the efforts of Euler and Warburg. Also at this time flavin nucleotides and flavoproteins were discovered and studied by Warburg and Theorell. In the 1930's the general sequence of enzyme-catalyzed oxidation-reduction (redox) reactions was correctly formulated as:

substrate → dehydrogenase → NAD → flavoprotein →

cytochromes → oxygen

with the arrows indicating the direction of electron flow. But there was no appreciation of the concept of energy conservation through the integrated reactions leading to phosphorylation of ADP.

In 1931 Lohman discovered ATP in muscle tissue. The general role of ATP in cellular metabolism was not appreciated until between 1937 and 1941. During this time Warburg showed the formation of ATP coupled to the enzymatic oxidation of phosphoglyceraldehyde and Myerhoff showed the formation of ATP from phosphopyruvate. The concept of oxidative phosphorylation as we now think of it, coupled to respiration, is due to Kalckar. Belitser specifically linked phosphorylation to the process *per se* of electron transport from substrate to oxygen.

In the 1940's work on the isolation of cellular organelles by differential centrifugation of acellular homogenates led to the realization that mitochondria contained the enzymes for oxidation of cellular metabolites and the concomitant phosphorylation of ADP. The efforts of Claude, Porter, Palade, Hogeboom, Hotchkiss, and Schneider at the Rockefeller Institute in New York brought us to an appreciation of the central role of mitochondria in these processes. The full significance of the mitochondrial participation was revealed by the work of Leloir, Muñoz, Lehninger, and Kennedy in the 1940's. The last two investigators showed that mitochondria possessed sufficient capability to account for the total capacity of liver to oxidize fatty acids and Krebs cycle intermediates and, moreover, these reactions were accompanied by the phosphorylation of ADP.

The electron-transport chain receives electrons either directly from a substrate through catalysis by a specific dehydrogenase or electrons are passed to mitochondrial-bound NAD which directs them through the electron-transport chain.

The electron-transport chain is known to contain NAD, flavopro-

Table 1   Oxidation-reduction potentials of some mitochondrial systems; data* are for $pH \simeq 7.0$ and for temperature zone 25 to 37°

System	E'_0
Isocitrate ($\rightarrow$ α-ketoglutarate $+$ CO_2)	-0.48
Hydrogen	-0.42
NAD	-0.32
NADP	-0.32
D-β-hydroxylbutyrate ($\rightarrow$ acetoacetate)	-0.29
α-lipoate	-0.29
Peroxidase	-0.25
L-β-hydroxybutyryl-CoA ($\rightarrow$ acetoacetyl-CoA)	-0.24
Lactate ($\rightarrow$ pyruvate)	-0.18
Glutamate ($\rightarrow$ α-ketoglutarate $+$ NH_3)	-0.15
Old yellow enzyme	-0.123
Malate ($\rightarrow$ oxalacetate)	-0.10
Vitamin K_1	-0.06
NADH-cytochrome c reductase	0.0
Succinate	$+0.03$
Cytochrome b	$+0.04$
Ascorbate	$+0.06$
Butyryl-CoA dehydrogenase	$+0.187$
Cytochrome c	$+0.26$
Cytochrome a	$+0.29$
Coenzyme Q (in ethanol)	$+0.542$
Oxygen	$+0.82$

* The oxidized form of the couple is specified only when more than one product is known. Reprinted with permission from A. L. Lehninger, *The Mitochondrion*, W. A. Benjamin, New York, (1964.)

teins, coenzyme Q,[1] cytochrome b, cytochrome c, cytochrome c_1, cytochrome a, cytochrome a_3 and finally oxygen. Chance and Williams sequenced the order of reduction of these components by several criteria. First of all, the relative order is indicated by the redox potentials (Table 1), which provide a measure of their affinities for holding electrons. If an equal amount of reduced and oxidized NAD (redox potential -0.32 v) is mixed with an equilibrium mixture of reduced and oxidized cytochrome b (redox potential $+0.04$), in the presence of NADH dehydrogenase, electrons will be transferred from NADH to cytochrome b. In functioning mitochondria, they determined that a gradient existed for

[1] The role of coenzyme Q is not clearly settled and some non-heme iron components appear to be involved in electron transport. The composition and sequence of the electron-transport chain may vary for different organelles and cell types. The discussion here refers to mammalian mitochondria.

the percentage of each electron carrier in the reduced form. Thus it was found that 53% of NAD and NADP was reduced, 20% of the flavoprotein dehydrogenase, 16% cytochrome b, 6% cytochrome c, and less than 4% cytochrome a were in the reduced state. Furthermore, if the carriers were first fully reduced and oxygen was admitted, cytochrome a was found to be oxidized first, followed by cytochrome c, cytochrome b and then flavoprotein. The electron-transport chain at the present time with the various routes of entry for electrons is shown in Figure 4. A very helpful tool in studying this chain is the use of specific poisons which have been found to block the chain at various discrete points as indicated in the figure.

When two electrons removed from a substrate enter the electron-transport chain at the level of NAD or NADP and eventually are passed the whole length of the chain to oxygen, the free energy released by the process is 52.6 kcal. This energy release is the source of supply for forming most of the ATP that powers cellular chemistry. It is possible to measure the number of electrons traversing the whole chain by measuring the uptake of oxygen which is transformed to water as it accepts the exiting electrons. At the same time, ATP formation can be measured by studying the disappearance of inorganic phosphate and the concomitant appearance of ATP. Thus many investigators have confirmed the observation that when electrons enter the chain at the level of NAD, three moles of phosphate are incorporated into ATP per gram atom of oxygen consumed. This is referred to as a P/O ratio of 3. When succinate is oxidized the P/O ratio is 2. Therefore somewhere between NAD and oxygen there must exist three taps for draining off energy to be used for ATP formation. Since electrons removed from succinate enter the electron-transport chain at the level of cytochrome b or CoQ, two of these three must be located between CoQ and oxygen. One must be found between NAD and cytochrome b. From Table 1 it is possible to calculate the free energy release for each step of electron transfer. This is shown in Table 2.

At least 9 kcal energy must be available in order to form a molecule of ATP. Therefore we would expect to find the energy taps located (a) between NAD and its dehydrogenase, (b) between cytochrome b and cytochrome c and (c) between cytochrome a and oxygen. This conclusion has also been checked by an experimental approach of Chance and Williams called "a study of crossover points." The integration of the electron-transport chain with the chain leading to ATP formation is often so tight that if ADP is limiting, the slowdown in the phosphorylation chain leads to a slowdown in the electron transport or respiration chain. This phenomenon is called respiratory control. Chance and Williams found that as ADP became limiting, gaps in redox state became

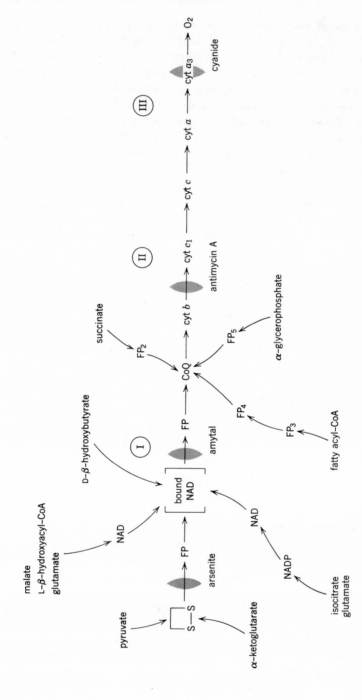

Figure 4 Schematic diagram of the respiratory chain. The scheme shows points of inhibition as well as sites of phosphorylation (Roman numerals). Redrawn from A. L. Lehninger, *The Mitochondrion*, Benjamin, New York, (1964).

Table 2   Thermodynamic relationships in respiratory chain

| Carrier | E_0' volts | $\Delta E_0'$ volts | ΔG, kcal | Phosphorylation sites | |
				From thermo-dynamic data	From cross-over points
NAD	−0.32				
Flavoprotein	−0.05	0.27	12.2	+	+
Cytochrome b	+0.04	0.09	4.05	0	0
Cytochrome c	+0.26	0.22	9.90	+	+
Cytochrome a	+0.29	0.03	1.35	0	0
Oxygen	+0.82	0.53	23.8	+	+

Reprinted with permission from A. L. Lehninger, *The Mitochondrion*, W. A. Benjamin, New York, (1964).

evident where the carrier on the substrate side of the chain became more reduced and the carrier on the side closer to oxygen more oxidized. These crossover points, which were influenced by the availability of ADP, corresponded to the three tap-off points indicated in Table 2 and Figure 4.

We have now reached the point where understanding drops precipitously off. Essentially, there are two questions and answers intertwined in the membranes of mitochondria and those forming the plasma boundaries of bacteria:

1. How can the energy released from the electron-transport chain be used to make ATP?

2. How are the enzymes for all of the substrate dehydrogenases, electron-transport components, and phosphorylating sequences integrated into the fabric of membranes to insure the smooth and efficient functioning of such a vital and complex system?

In regard to the first question, there are several observations which allow a general formulation that rests heavily on some unproven assumptions. The major assumption is that the high energy state is maintained in the form of chemical complexes of high reactivity. When the absence of ADP causes a backup in the phosphorylating sequence that in turn slows down the respiratory sequence, the agent 2,4 dinitrophenol (2,4-DNP) relieves the inhibition of respiration; but at the same time it blocks concomitant phosphorylation when ADP is added back. The ability of 2,4-DNP to stimulate respiration does not require inorganic phosphate. Furthermore, 2,4-DNP stimulates the breakdown of ATP. If a high-energy complex involving a member of the electron-transport chain reacted with a nonphosphorylated member of the phosphorylating chain and if this complex could be broken down by 2,4-DNP, the observations already stated could be explained.

For example:

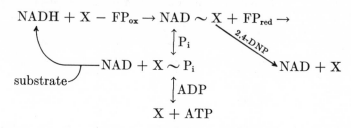

A further series of observations shows that phosphorylating tissue preparations catalyze an exchange of P_i and ATP (ATP-P_i exchange) only when ATP, ADP, and P_i are present but an ATP-ADP exchange occurs in the absence of P_i. Also, P_i is taken up first and then ADP is fixed. Other consistent observations involve the use of the agent oligomycin which (a) inhibits respiration linked to phosphorylation, (b) inhibits DNP-induced ATPase activity, (c) inhibits ATP-P_i exchange and the ATP-ADP exchange in phosphorylating preparations. When respiration is inhibited by oligomycin, 2,4-DNP releases the block and allows respiration unlinked to phosphorylation.

Wadkins and Lehninger linked these observations with some others concerning O^{18} exchanges between P_i and water and between ADP and ATP in the following type of formulation:

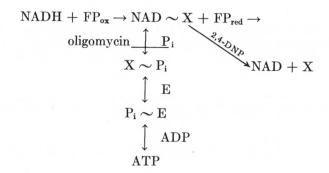

Many investigators studying these problems have formulated similar schemes but differ in certain details. All schemes of this kind assume the existence of chemical high energy derivatives involving members of the electron-transport chain and also other unidentified members of an as yet undiscovered phosphorylating chain.

From time to time evidence has been presented to support the idea that high energy compounds have been directly formed with various components of the electron-transport chain. For example, Boyer and his colleagues have described a labile phosphohistidine in a mitochondrial protein that is readily labeled from P^{32} labeled ATP. Griffiths and Chaplain described a phosphorylated derivative of NAD formed in mitochondria during oxidative phosphorylation. A similar compound has been described in bacteria by Pinchot.[1]

A most intriguing alternative possibility has been proposed by Mitchell (17). He suggests that the respiratory enzymes are oriented in the membrane surface such that during the oxidation of substrate they can direct the H^+ to one side of the membrane and electrons to the other. As the electrons are passed through the respiratory chain, they can cause the movement of $2H^+$ unidirectionally across the membrane at each phosphorylation site. The process is terminated when the electrons are finally passed to oxygen which is reduced to water in the process. Therefore, in the course of oxidation of NADH, six H^+ (corresponding to three phosphorylation sites) are moved from one side of the membrane to the other, causing an acidification on one side of the membrane and an alkalization on the other. The membrane is believed to be impermeable to H^+ and OH^-. The accumulation of such a double gradient rather than discrete chemical compounds is considered to represent the high-energy storage form of respiration. Since the enzyme catalyzing the reaction

$$ADP^{3-} + HPO_4^{2-} + H^+ \rightleftharpoons ATP^{4-} + H_2O$$

is also supposed to be asymmetrically situated in the membrane, H^+ and OH^- gradients can be used to force the equilibrium to ATP formation. That H^+ and OH^- can be extruded on opposite sides of the membrane has been observed as well as the ability of H^+ gradients to cause the formation of ATP in plant chloroplasts and rat liver mitochondria (18, 19).

That high-energy forms exist prior to ATP in this phosphorylating sequence is indicated by observations that in the presence of oligomycin

[1] Specific references and more detailed discussions of topics discussed in this section can be found in (15) and (16).

(an inhibitor of ATP formation and utilization by this enzymatic phosphorylation sequence) reactions that require energy can be powered by succinate oxidation. For example, electrons can be forced from succinate to NAD and then to acetoacetate, provided that succinate can produce high energy intermediates by being oxidized. Antimycin A which blocks the oxidation of succinate also blocks the reduction of NAD. ATP may also be used to force the reduction of NAD by succinate but in this case the reduction is blocked by oligomycin (20).

At the moment of writing, however, no one knows how energy from electron transport is really used to drive the formation of ATP.

The second problem stated on page 257 deals with the structure of the membrane and positioning of the numerous dehydrogenases and cofactors comprising the respiratory assembly and the phosphorylating or energy conservation enzyme sequence.

The individual carriers of the electron-transport chain have been found to occur in simple molar ratios to each other; fragmentation of mitochondria, even to very small dimensions, maintains these ratios among the members of the sequence. Therefore the concept of repeating respiratory assemblies seems valid. Estabrook and Holowinsky estimate that a single liver mitochondrion contains 17,000 molecules of cytochrome a and a heart mitochondrion may contain 50,000. If each assembly contains a single molecule of cytochrome a, then each mitochondrion must contain many thousands of respiratory assemblies.

With the known components of the respiratory chain and assuming (a) that the chain contains only one of each member and (b) that the *in vitro* measured molecular weights are valid measures for the *in vivo* state of the components, then as indicated in Table 3 a minimum molecular weight of 1,350,000 would be required. More likely, however, some structural nonfunctional components are present which would increase the minimal molecular weight figure. Another uncertainty is whether the respiratory assembly occurs alone or as part of a larger complex integrated with various specific dehydrogenases and coupling factors.

Many laboratories have subfractionated mitochondria to obtain fragments which possess electron-transport capabilities but which are inert in catalyzing oxidative phosphorylation. When another fraction is added, however, the two fractions unite to allow ATP formation coupled to oxidation. These factors, isolated from mitochondria which couple the electron-transport fragments to formation of ATP, are called coupling factors. Racker and Pullman have isolated and described several such factors. Sanadi and his colleagues have also described coupling factors which appear to be linked to site I phosphorylation associated with NAD oxidation.

Table 3   Components of the respiratory assembly*

Electron-transferring enzymes		Mol. wt.
NADH dehydrogenase		1,000,000
Succinate dehydrogenase		200,000
Cytochrome b		28,000
Cytochrome c_1		40,000
Cytochrome c		12,000
Cytochrome a monomer		70,000
	Subtotal	1,350,000
Coupling factors:		
Site I 2 × 30,000		60,000
Site II 2 × 30,000		60,000
Site III 2 × 30,000		60,000
	Subtotal	180,000
Auxiliary enzymes tightly bound to respiratory fragments:		
D-β-hydroxybutyrate dehydrogenase		60,000
α-glycerophosphate dehydrogenase		60,000
Fatty acyl-CoA dehydrogenases		60,000
Electron-transferring flavoprotein		60,000
Pyridine nucleotide transhydrogenase		60,000
	Subtotal	300,000
	Minimum assembly weight	1,350,000
	Maximum assembly weight	1,830,000

* The "minimum assembly" is one containing only the electron carriers; the "maxim mum assembly" also contains the coupling factors and other tightly bound dhydrogenases present in mitochondrial membrane. Reprinted with permission froe A. L.-Lehninger, *The Mitochondrion*, W. A. Benjamin, New York, (1964).

David E. Green is one biochemist who has appreciated the importance of the concept of membrane biochemistry as defined here. Metabolic sequences involving several enzymes, according to Green, imply the existence in cells of structures containing the enzymes in more or less fixed spatial frames. During the past twenty years, especially in regard to the enzymes catalyzing the citric acid oxidation cycle, electron transport, and oxidative phosphorylation, he has interpreted the results of his own laboratory and those of other groups in terms of integrated enzyme assemblies.

Green, in his enthusiastic appreciation of the potential biological significance of structural integration for complex enzymatic reaction sequences, has sometimes formulated his proposals in too much detail and based them on the rather limited available information at the time. Because of this, he has frequently found it necessary to modify extensively the specifics of the formulations. Nonetheless, Green has continually emphasized the importance of functional-structural relations, has stimulated other investigators to considerable efforts to prove or disprove various details of his proposals, and has contributed new findings and concepts in this area.

The history of these developments serves to illustrate one of the points made in the early chapters, that contemporary views of a field based on experimental data available at the time and interpreted with certain assumptions can produce plausible pictures which in later years are recognized to be quite inaccurate. Furthermore, the evolution of thinking in this area represents attempts to formulate biochemistry in terms of membrane structure and, although details have been often incorrect, it represents the early history of this new phase of cellular biochemistry. A knowledge of the background in this area should help informed students to evaluate the current status of the field with the orientation and perspective provided by the earlier developments. With these purposes in mind, I shall now trace these developments.

In 1947 Green and his collaborators were impressed with the apparent constancy of activity ratios for many constituent enzymes catalyzing the citric acid cycle in a particulate preparation obtained from rabbit kidney. They postulated that the enzymes were all bound to the same structural unit and named the hypothetical structure the cyclophorase system, emphasizing its functional nature (21, 22). The cyclophorase system was also capable of fatty acid oxidation, various synthetic reactions, electron transport, and oxidative phosphorylation.

At the same time, Claude and Hogeboom, Schneider, and others were perfecting techniques for isolating cell organelles, among which were included mitochondria. Schneider and Potter (23) and Lehninger and Kennedy (24, 25) independently showed that mitochondria contained all the enzymes which Green et al. attributed to the cyclophorase system. Therefore in 1949, the cyclophorase system as a functional unit became identified with the mitochondrion as a structural unit.

In 1951, Green felt that cyclophorase function required the intact structure of the entire mitochondrion (26). When mitochondria were fragmented by high speed, blendor-type homogenizers, it was believed that the different fragments possessed ability to carry out only isolated single-step parts of the over-all sequence. Continued blending broke the fragments into smaller units which at the time were incorrectly thought to be microsomes. Since microsomes are unable to carry out oxidative phosphorylation, this observation was interpreted to mean that the organized structure of the entire mitochondrion was required for oxidative phosphorylation and the loss of such organization accompanying the degradation of mitochondria to microsomes prevented the functioning of oxidative phosphorylation reactions. In fact, Green felt that the comminution of mitochondria into smaller fragments was incompatible with the interpretation of the structure being essentially like a cellule, which delineates its internal fluid contents from those of the cytoplasm by a membrane.

The concept of any free, soluble enzymes functioning in metabolism was questioned by Green many years before other biochemists were so concerned with enzyme-cell structure relations. As one argument, he pointed out that in a fixed structural arrangement, one coenzyme could serve the needs of an enzyme but in free solution hundreds of times as much coenzyme might be required to accommodate the demands of rapid kinetics.

In 1951 the concept of oxidative phosphorylation was somewhat different from our current ideas. The oxidation of citric acid cycle intermediates was believed to be coupled to the formation of some unknown pyrophosphate ester which could lead either to the accumulation of free inorganic pyrophosphate or, if AMP, glucose and hexokinase were provided, the AMP could be phosphorylated in two steps to form ATP which could then phosphorylate glucose. An alternative possibility was considered that the inorganic pyrophosphate formed during oxidation might result from the action of an ATPase which splits newly formed ATP into AMP and pyrophosphate.

Rather than think of mitochondria in terms of a membrane-bound organelle, Green referred to the system as cyclophorase-gel. Therefore when it was found that radioactive phosphate uptake accompanied oxidation, the interpretation was in terms of a specific incorporation into some substance of the gel involved in the phosphorylation sequence. Today it is known that during oxidation the mitochondrial membrane-transports many ions into the mitochondrion, one of which is phosphate.

In 1957 Green presented a lecture entitled "Studies in Organized Enzyme Systems" (27). During the interval of six years since his review on the cyclophorase system, the fine structure of mitochondria was more fully revealed by the studies of Sjöstrand and Palade (see pages 154–156). Once again Green emphasized the importance of structural integration for enzymes, even those classically thought of as soluble systems, such as those involved in glycolysis.

By this time Green's laboratory was preparing large quantities of mitochondria from beef heart. Operationally, a distinction was made between heavy and light mitochondria on the basis of their relative ease of sedimentation. The two fractions were also distinguished by their contents of bound pyridine nucleotides and abilities to function in oxidation and phosphorylation with different substrates. From the light mitochondrial fraction, by treatment with 15% alcohol and 0.15 M phosphate buffer at pH 7.4, two subfractions were isolated. The lighter fraction was called the electron-transport particles or simply ETP. A somewhat heavier fraction was called the phosphorylating electron-transport particles or PETP. ETP could oxidize succinate and NADH. PETP carried out Krebs cycle oxidations and oxidative phosphorylation. This work marks a significant departure from the concept expressed earlier that the entire mitochondrial structure was necessary for oxidative phosphorylation. Instead of the mitochondrion representing the smallest structural unit possessing all of the integrated reactions for oxidation and phosphorylation, the concept was offered that the mitochondrion contains a mosaic of integrated units which probably would be smaller in size than those particles isolated in the PETP fraction. Similar preparations could not be obtained from the heavy mitochondrial fraction.

During these isolation procedures, the electron microscope was used to trace the origin of the subfractions. The interpretation obtained in this fashion was that first the cristae of the mitochondria migrated to one side of the organelle, leaving a vast clear space bounded by the mitochondrial membrane(s?). During the subsequent treatment with alcohol and phosphate, the mitochondrial envelope was believed to undergo vesiculation to smaller spherical vesicles which were the units of the ETP fraction. The condensed cristae then formed separate vesicles which made up the PETP. Therefore the structural location within the intact mitochondria for oxidation and electron transport was the envelope and for integrated phosphorylation, the cristae.

A very important indicator of the importance of structural arrangement is the fact that NADH formed from oxidation of citric acid cycle intermediates is efficiently oxidized by mitochondria with concomitant phosphorylation; but when NADH is added, direct oxidative phosphorylation does not occur.

Since electrons can be fed into the electron-transport chain both by NADH and by succinate, anyone conscious of structural requirements would wonder whether two separate chains exist to carry out oxidation of NAD linked substrates and succinate or whether there exists a branched chain which meets at the level of cytochrome b.

By treating ETP with the surface active agent deoxycholate, a red particle and a green particle fraction could then be isolated. These subfractions differed markedly from one another in their ability to oxidize either NADH or succinate. All succinate activity was localized in the red particle, whereas NADH activity was carried out by both particles and the resulting soluble fraction. These observations were considered to be indicative of the existence of separate units in the ETP for NADH and succinate oxidation. It was felt, however, that the separate electron chains were in some way interconnected so that electrons inserted as NADH may ultimately be able to reduce any of the cytochrome a. This concession was dictated by the observation that both NADH and succinate could reduce more than their share of cytochrome a calculated on the basis of totally separate chains.

From the observation that various surface active agents could subdivide functional units from the mitochondria and its subfractions, the conclusion was reached that lipids or lipoproteins provide the structural links to hold the whole system together.

By May 1961 Green's formulation of the structure of the electron-transport particle was stated in much more detail (28). Existing biochemical and electron microscopic data were blended to support the same kind of structure. By two independent modes of calculation, one based on molecular weights of the electron-transport components and the other on the dimensions of electron micrographic images of preparations of ETP, a range of 3 to 4×10^6 was calculated as the molecular weight of the electron-transport unit. This figure was also supported by ultracentrifugal measurements. An estimate of the total number of repeating units of such particles in a single mitochondrion was also made by two different methods of calculation. For example, knowing the concentration of succinic dehydrogenase in moles/mg of mitochondrial protein and also the number of mitochondria per mg, he calculated that each mitochondrion would be able to house 17,000 ETP units. The second method was based on equating the volume of one ETP to the volume of the structured elements in a single mitochondrion (including cristae and outer limiting membrane). This calculation gave a figure of 18,000 units.

Green and Oda reviewed the work of Green et al. on a structural protein isolated from mitochondria and which is believed to account for 60% of its total protein (29). They proposed that the structural protein provided the framework on which the electron-transport chain is mounted. The cement for the structure was the lipid, which also provided an essential medium for the entire process.

Previous work had separated the electron-transport chain into four separate complexes as indicated in Table 4. The separate complexes could also be recombined to reconstruct the entire electron-transport chain. The formulation of these interactions is shown in Figure 5.

The electron micrographs showed a rectangular or cylindrical repeating unit in preparations of ETP. The dimensions were 450 A in length and 150 A in width. It was proposed that these units were the bricks from which all the cristae and limiting membranes of the mitochondria were constructed. It was calculated that 15,000 such bricks could build a whole mitochondrion. Electron micrographs of whole mitochondria showed a space between the outer and inner mitochondrial membranes

Table 4   The components of the four basic complexes of the electron transport system

I DPNH-Q reductase	flavoprotein (f_D)
	non-heme iron
	lipid
II Succinic-Q reductase	flavoprotein (fs)
	non-heme iron
	lipid
	cytochrome $b^{1)}$
III QH₂-cytochrome *c* reductase	cytochrome *b*
	cytochrome c_1
	non-heme iron
	lipid
IV Cytochrome oxidase	cytochrome *a*
	copper
	lipid

1) The association of cytochrome *b* with the succinic-Q reductase may be fortuitous. [Reprinted with permission from D. R. Green and T. Oda, *J. Biochem. (Tokyo)*, **49**, 742 (1961).]

and also within each crista. Since each crista membrane was given as 60 A thick and the space between them as 70 A, Green and Oda proposed that the part that looked clear actually contained lipid and was divided equally between ETP units back-to-back. Therefore, instead of being 60 A membranes separated by a 70 A space, there were two units each

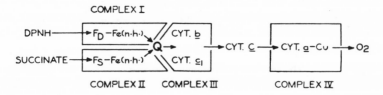

Figure 5   The arrangement of the four component complexes of the electron transport chain. Coenzyme Q is the link between succinic-Q reductase (II) and DPNH-Q reductase (I) on the one side and QH₂-c reductase (III) on the other. Cytochrome c is the link between III and reduced cytochrome c oxidase (IV). From D. E. Green and T. Oda, *J. Biochem. (Tokyo)*, **49**, 743 (1961).

with a width of 95 A. One final important point was that it was essential for phosphorylation that the blocks remain back-to-back with lipid in between. Coenzyme Q and cytochrome c, both lipid soluble, would circulate in this space between units. ETP particles not capable of phosphorylation were visualized as separated from the apposing ETP configuration and consequently lacking the lipid channel. On the other hand, a particle called ETPH and capable of oxidative phosphorylation retained the paired unit relationship.

At a symposium on the plasma membrane held in New York City in December, 1961, Fernández-Morán described a newly discovered feature of the fine structure of mitochondria isolated from the inner segment of frog retinal rods and beef heart muscle (30). A repeating unit seen as a round or polyhedral particle 80 to 100 A in diameter covered the surface of the inner mitochondrial membranes (see page 181). The name elementary particle (EP) was used to describe these bodies which Fernández-Morán believed represented the culmination of the long search for the cytological unit of electron-transport. He estimated a molecular weight of 500,000 to 750,000 for these particles. Since Green and Oda estimated 3 to 4 $\times$ 10^6 as the minimum molecular weight for the required enzyme sequence, somewhat of a problem was presented. Fernández-Morán referred to a paper in press from Green's laboratory where particles resembling EP were isolated and found to have high concentrations of the electron-transport carriers. The older ETP was believed to be a composite of EP integrated with structural protein (SP).

In two extensive papers by Fernández-Morán, Oda, Blair, and Green, the proponents of the EP concept came to grips with the apparent discrepancy based on molecular weight of the particle and the assumed molecular weight of the electron-transport sequence (31, 32). They were able to cut the original estimate more than half by considering that a major portion of the originally estimated weight included structural protein. A newer look, involving reconstruction of the electron transfer chain from the four purified complexes (see page 266) yielded an estimated weight of 1.4 $\times$ 10^6. This figure was supported by calculations based on the content of cytochrome a in EP preparations. Furthermore, the EP was enlarged to include a stalk and basepiece in addition to the 80 to 100 A head. The total volume of such a unit, assuming a density of 1.25, yielded a calculated molecular weight of 1.3 $\times$ 10^6. Another calculation of the minimal molecular weight of an electron-transport chain was made from the molecular weights of purified preparations of the components. Assuming no structural protein or lipid, this calculated value was 0.9 $\times$ 10^6. Allowing for 30% lipid, a value of 1.3 $\times$ 10^6 was obtained.

A tentative assignment of positions in the EP was offered on the basis of some plausible correlative assumptions. Thus, Complexes I and II (DPNH-coenzyme Q-reductase and succinic-coenzyme Q-reductase) were placed in the basepiece. Complex III (QH$_2$-cytochrome c-reductase) was visualized in the stalk and Complex IV (cytochrome oxidase) was given the headpiece or sphere.

By 1963 Green had sufficiently resolved the size dilemma to announce that the molecular weight of each elementary particle was 2 to 3 × 10^6 and that the isolated unit of electron transfer had a size of under 1 × 10^6 on a protein basis. Therefore he proposed that the elementary particle probably contained, in addition to a complete electron transfer chain, the enzymes and factors for coupling oxidation to phosphorylation, the translocation of ions, and a contractile protein (33).

Green's detailed structural and functional speculations did not go unchallenged. Chance and Parsons (34) considered two alternatives of Green's findings relating to elementary particles. They introduced the term "oxysome" for a hypothetical unit that carries out all the functions of oxidative phosphorylation. If all of the components are in the one particle, it would be a lumped oxysome, which was one of Green's views of the elementary particle. If the various components are distributed in different units, then the term "distributed oxysomes" would apply. The total molecular weight and sizes for cytochrome c and cytochrome oxidase $(a + a_3)$ are so large that if they are contained in the respiratory unit termed elementary particle, then they must account for a major fraction of its size. If mitochondria deficient in cytochromes are examined then, according to the lumped oxysome or elementary particle concept, the intramitochondrial subunits (IMS) should be smaller or absent altogether. According to the distributed oxysome concept, some of the IMS should disappear or shrink while others remained unchanged. Chance and Parsons compared the size and frequency of distribution for visible repeating units (IMS) to natural differences in content of cytochromes for mitochondria from the thoracic muscle of young and adult honey bees *Apir melliferce* and from the longitudinal muscle of the worm *Ascaris lumbricoides.* Although cytochromes a, a_3 and c_1 were virtually missing from *Ascaris,* the size of its IMS was unchanged. Differences in content of cytochromes between young and adult bees likewise did not result in IMS units of smaller dimensions. A decreased frequency of IMS, however, could be correlated with the diminished cytochrome contents. Therefore the concept of distributed oxysomes might be valid but the results seemed incompatible with the stated picture for elementary particles.

An even more damaging blow was dealt to Green's view of the

EP in a later paper by Chance, Parsons, and Williams (35). Mitochondria from rat liver were stripped of IMS particles by ultrasonic irradiation. The stripped membranes were found to have an enhanced concentration of cytochrome c and cytochrome a on a protein basis. Some cytochrome c_1 and b were lost, however, and may be contained in the subunits. The authors concluded that cytochromes a_3, a, and c were absent from the IMS and that this was also true for DPNH and succinate dehydrogenases.

In another investigation (36) Racker, Chance, and Parsons presented a correlated biochemical and electron microscopic study which identified the headpieces of the elementary particle with a water soluble ATPase and coupling factor F_1 previously described by Racker and his coworkers (37).

By 1965 Green's formulation of the nature of the elementary structural unit for electron transport had gone through a drastic revision. On the other hand, the general concept of repeating units of structure and function comprising the basic architecture of membranes was extended from mitochondria to all other kinds of membranes that seem to be the sites for metabolic sequences (see page 233).

In regard to the mitochondrion, Green and Perdue admitted that repeating units were chemically not identical, although their physical dimensions seem to be the same (38). They further stated that in their own laboratory, membranes free of the repeating knobs (headpieces) were fully active for electron transport. At this point there is no disagreement on one principle. The components of the electron-transport chain are contained in the membranes of the cristae and inner mitochondrial membranes and not in the projecting elementary particles (EP). Green and Perdue suggested that the active components are contained in the basepieces of the elementary particles. They calculated that the basepiece could accommodate a mass of 400,000 molecular weight. They previously found that the whole electron-transport unit free of structural protein had a mass of 1.5×10^6. Since this could be separated into four complexes (see page 266), then each basepiece could conceivably accommodate one of the four individual complexes of the whole chain. They claimed that the basepieces could be made to aggregate into membranes (a process they call vesicularization) and again be deaggregated by use of bile salts and the removal and addition of phospholipids. Vesicularization could be accomplished with each of the four complexes. Structural protein would not vesicularize and was not involved in either the vesicularization or function of any of the complexes of the electron-transport chain. All of the enzymes concerned in fatty acid oxidation, fatty acid elongation, citric acid cycle oxidations, and oxidation of

β-hydroxybutyrate and amino acids were assigned to the outer membrane with the implication that some or all would be localized in a repeating unit assumed to be characteristic of this membrane (see page 183). The prediction was made that other cellular membranes would eventually be found to fit the same general pattern involving functional and structural subunits.

This altered view of the nature of the elementary repeating units of electron transfer function and membrane structure was again stated and further defined in 1966 by Green and Tzagaloff (39). Each of the four complexes, which together formed the complete electron-transport chain, was fixed *in theory* at about 300,000 to 400,000 in molecular weight and said to contain about 8 to 10 molecules of protein, about 14 molecules of phospholipid, and no structural protein.[1] The basepiece of the tripartite elementary particle (basepiece-stem-headpiece) was calculated to comprise a molecular weight of 300,000 to 400,000, and so only one part of a complete electron-transport chain was accommodated in each basepiece. Since individual complexes could vesicularize to form membranes, there was no automatic formation of complete membranes with unique arrangements involving the four complexes. Green and Tzagaloff resolved the difficulty of visualizing an integration among such a randomly reconstructed membrane by postulating that lipid-soluble coenzyme Q and cytochrome c were mobile in the membrane and could connect the electron transfer sequence among the different complexes (see Figure 5).

This quite obviously leaves the story very much in the middle and in the air. Future histories will tell how these studies have developed.

FUNCTION OF LYSOSOMES

The finding of collections of highly destructive lytic enzymes in membrane-bound vesicles (lysosomes, see pages 166–168) in virtually all cell types has presented a mystery to biochemists. De Duve and Wattiaux present an exhaustive survey of the literature up to September, 1965 (40). Many speculations concerning possible functions of lysosomes are reviewed but it must be admitted that a satisfying general explanation for the role of these organelles is not available at this time. An excellent summation of some of the main observations concerning lysosomes is presented by the authors and is reproduced in an abridged form here.

---

[1] Complex I as isolated actually had a molecular weight of 1.17 x 10⁶ and complex IV, 260,000.

1. Lysosomes are involved in a wide variety of animal cells in the storage, processing and digestion of extracellular materials taken in by the cells; these materials range all the way from whole cells and particles of cellular size to small molecules and ultramicrodroplets of fluid.

2. When newly taken in, the exogenous material is frequently bound within a membrane-lined vacuole devoid of acid hydrolase activity. Such a structure is termed phagosome, or, if the revised classification suggested in this review is adopted, heterophagosomes.

3. Exposure of the material to the lysosomal hydrolases occurs through fusion of the phagosome with a lysosome, resulting in the formation of a secondary lysosome of the heterolysosome type; the lysosome in this process may be either a primary or a secondary lysosome.

4. Heterophagosomes and heterolysosomes are involved in numerous fusion and fission processes. As a result of these dynamic exchanges, the contents of the individual members of this system are continuously being redistributed among the whole population of vacuoles, with phagosomes bringing new exogenous material, and primary lysosomes bringing new enzymes to the digestive pool. These exchanges are probably not very rapid and may be restricted to a greater or lesser extent, depending on the polarity and main traffic flow characteristics of each cell type; they are quite intensive in the liver.

As a further aid to understanding some of the dynamic exchanges involving phagosomes and lysosomes, the authors have designed a very useful illustration which is reproduced in Figure 6.

The lysosomes, it is believed, may be formed by the rough-surfaced endoplasmic reticulum and Golgi apparatus in a manner similar to zymogen granule formation in guinea pig pancreas (see pages 272–273). Material destined for digestion may enter the cell from outside (heterophagy) or be normal cell constituents that become encapsulated with membrane (autophagy). The latter process may be involved in the normal turnover of cell constituents or during differentiation. These phagosomes, before receiving hydrolytic enzymes by fusing with lysosomes, are called prelysosomes. Lysosomes contain hydrolases and before fusion with a phagosome they are primary lysosomes. The particles containing substrates undergoing digestion are called secondary lysosomes. After digestion has progressed as far as it can, the remaining material is held in a particle called a post-lysosome. The elimination of these waste products by a process involving the fusion of the membrane of the postlysosome with that of the cell membrane is called defecation. Some cells lack this faculty and are said to be constipated. Severe constipation may result

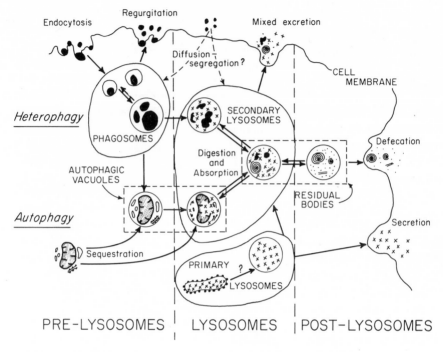

Figure 6 Synthetic diagram illustrating the various forms of lysosomes and related particles and the different types of interaction they may exhibit with each other and with the cell membrane. Each cell type is believed to be the site of one or more of the circuits shown but not necessarily of all. Crosses symbolize acid hydrolases. From C. de Duve and R. Wattiaux, *Ann. Revs. Physiol.*, **28**, 435 (1966).

in the aging and death of the cell. A future Nobel prize may be awarded to the discoverer of cell laxatives.

FUNCTIONS OF THE GOLGI APPARATUS

The Golgi apparatus has been suspected for some time of playing an important role in cellular function. Although current knowledge implicates the structure in the concentration, packaging and secretion of enzymes, there still remains an aura of mystery and uncertainty about other key functions that may be discharged via this organelle. Our appreciation of its function in enzyme secretion is due largely to the efforts of Palade and his coworkers (41, 42).

This information has been amassed mainly in a study of the exocrine cells of various animal pancreases. Originally, most of the conclusions were based on the interpretation of electron micrographs taken at various time intervals after feeding starved guinea pigs. These studies were correlated with a biochemical investigation of the labeling of the pancreatic zymogen α-chymotrypsinogen. Radioactive amino acid was administered to intact animals which were then sacrificed in a timed sequence and the pancreas was removed and subjected to tissue fractionation techniques. The accumulated observations supported the concept of a five-stage process for the secretion of pancreatic zymogens. Stage 1 is the release of the completed enzyme from a membrane-associated ribosome. Stage 2 is represented by the accumulation of newly synthesized enzymes (zymogens) in the intracisternal space of the rough-surfaced endoplasmic reticulum (er). Originally Palade maintained that the zymogens were contained in intracisternal zymogen granules which lacked a limiting membrane. The occurrence of these granules, however, seemed to be limited mainly to guinea pig pancreas. Later studies by Palade and his coworkers led them to modify the original view on the essentiality of intracisternal granules in the secretion process and they concluded that more likely the zymogens at this stage were in a nonparticulate form and generally dispersed in the intracisternal space. In stage 3 the zymogens are found in smooth-surfaced vacuoles of the Golgi region. The membrane from these vacuoles is believed to be supplied by the Golgi body. The vacuoles gradually fill until all of their space is occupied with zymogen, at which time the membrane-bounded organelle becomes a zymogen granule and thus stage 4 of the process is reached. The final stage is the emptying of the granule contents into the lumen or extracellular space by a process of reverse pinocytosis where the membrane of the granule fuses with the plasma membrane. From the first to the fourth stage approximately 45 min are required. The fourth stage, which represents a storage condition until the enzymes are needed for the digestive process, may be maintained for hours.

The scheme outlined above was further refined and substantiated by elegant radiographic procedures carried out by Caro, Jamieson, and Palade (43, 44). By using tritiated amino acids and the electron microscope, the radioautographic procedure allowed resolution to 0.1 μ. This work completely corroborated the over-all scheme. For the transition from stage 2 to stage 3, it was found that transitional elements which were partly rough and partly smooth-surfaced er seemed to act as a shuttle, transporting packets of zymogen from the cavities of the rough-surfaced er to the condensing vacuoles of the Golgi complex.

The Golgi apparatus was shown to function in a different but related fashion in root-cap cells of wheat by Northcoate and Pickett-Heaps (45). These authors also used tritiated precursors, electron microscopy, radioautography, tissue fractionation, and chemical analysis to trace the sequence of events following the path through the cell of administered glucose. For the external root-cap cells they found that glucose was quickly concentrated in the Golgi region where it was incorporated into high molecular weight polysaccharides, both as glucose and its metabolically converted product, galactose. Vesicles from the Golgi area then moved to the cell membrane, where the two membrane systems fused and the newly-synthesized polysaccharide was transferred to the cell wall, which seemed to be renewed at these points. An external slime layer also received radioactive polysaccharide by this process. Therefore the Golgi apparatus functioned in the synthesis of polysaccharide and its transfer in membrane-packaged units to the outside surface of the cell. Northcoate and Pickett-Heaps reviewed some other related observations involving Golgi function. Thus the Golgi has been experimentally implicated in the packaging and transfer to the extracellular space of proteins, mucopolysaccharides and even spicules and scales produced in certain flagellates. Undoubtedly future work will allow us a better appreciation of the full extent of participation of the Golgi body in cellular affairs.

Thus, in this chapter I have tried to show how several of the less well understood important metabolic activities of cells have led us to a growing appreciation of the importance of cell membranes in cellular activities. I would like to emphasize that this chapter barely introduces this subject, which may grow to include countless numbers of individual oriented enzymes and integrated enzyme sequences. The list of membrane associated enzyme activities is already too large to attempt an enumeration. The study of biochemical processes occurring in or on membranes comprises an important area of biochemistry that should be known as membrane biochemistry.

REFERENCES

1. T. Rosenberg and W. J. Wilbrandt, *Gen. Physiol.,* **41,** 289 (1959).
2. E. Heinz and P. M. Walsh, *J. Biol. Chem.,* **233,** 1488 (1958).
3. G. N. Cohen and J. Monod, *Bact. Revs.,* **21,** 169 (1957).
4. R. J. Britten and F. T. McClure, in *Amino Acid Pools,* Ed., J. Holden, Elsevier, Amsterdam, 1962, p. 595.
5. E. Lundsgaard, *Biochem. Z.,* **264,** 221 (1933).
6. R. K. Crane, *Physiol. Rev.,* **40,** 789 (1960).
7. A. S. Keston, *Science,* **120,** 355 (1954).
8. R. K. Crane, *Fed. Proc.,* **24,** 1000 (1965).
9. C. F. Fox and E. P. Kennedy, *Proc. Nat. Acad. Sci.,* **54,** 891 (1965).
10. A. B. Pardee, L. S. Prestidge, M. B. Whipple, and J. Dreyfuss, *J. Biol. Chem.,* **241,** 3962 (1966).

11. H. C. Neu and L. A. Heppel, *J. Biol. Chem.*, **240**, 3685 (1965).
12. W. Kundig, F. D. Kundig, B. Anderson, and S. Roseman, *J. Biol. Chem.*, **241**, 3243 (1966).
13. J. R.. Piperno and D. L. Oxender, *J. Biol. Chem.*, **241**, 5732 (1966).
14. H. C. Neu, D. F. Ashman, and T. D. Price, *Biochem. Biophys. Res. Comm.*, **25**, 615 (1966).
15. A. L. Lehninger, *The Mitochondrion*, W. A. Benjamin, New York, 1964.
16. E. Racker, *Mechanisms in Bioenergetics*, Academic Press, New York, 1965.
17. P. Mitchell and J. Moyle, in *Regulation of Metabolic Processes in Mitochondria*, Eds., E. Quagliarello, E. C. Slater, S. Papa, and J. M. Tager, Vol. 7, Biochim. Biophys. Acta Library, Amer. Elsevier, New York, 1966, p. 65.
18. A. T. Jagendorf and E. Uribe, *Proc. Nat. Acad. Sci.*, **55**, 170 (1966).
19. R. A. Reid, J. Moyle, and P. Mitchell, *Nature*, **212**, 257 (1966).
20. L. Ernster, *Symp. on Intracellular Respiration*, Vol. V, (5th International Cong. Biochem. Moscow 1961) Pergamon, Oxford, 1963, p. 115.
21. D. E. Green, W. F. Loomis, S. R. Dickman, V. H. Auerbach, and B. Noyce, *Abs. of Papers 11th Meeting Amer. Chem. Soc.*, (1947), p. 26B.
22. D. E. Green, W. F. Loomis, and V. H. Auerbach, *J. Biol. Chem.*, **172**, 389 (1948).
23. W. C. Schneider and V. R. Potter, *J. Biol. Chem.*, **177**, 893 (1949).
24. A. L. Lehninger and E. P. Kennedy, *J. Biol. Chem.*, **172**, 847 (1948).
25. E. P. Kennedy and A. L. Lehninger, *J. Biol. Chem.*, **179**, 957 (1949).
26. D. E. Green, *Biol. Revs.* (Cambridge Phil. Soc.), **26**, 410 (1951).
27. D. E. Green, *The Harvey Lectures*, **52**, 177, Acad. Press, N.Y., 1958.
28. D. E. Green and T. Oda, *J. Biochem.* (Japan), **49**, 742 (1961).
29. D. E. Green, H. D. Tisdale, R. S. Criddle, P. Y. Chen, and R. M. Bock, *Biochem. Biophys. Res. Comm.*, **5**, 109 (1961).
30. H. Fernández-Morán, *Circulation*, **26**, 1039 (1963).
31. P. V. Blair, T. Oda, D. E. Green, and H. Fernández-Morán, *Biochem.*, **2**, 756 (1963).
32. H. Fernández-Morán, T. Oda, P. V. Blair, and D. E. Green, *J. Cell Biol.*, **22**, 63 (1964).
33. D. E. Green, in *Intracellular Membranous Structure*, Eds., S. Seno and E. V. Cowdry, Japan Soc. Cell. Biol., Chugoku Press, Okayama, 1965, p. 89.
34. B. Chance and D. F. Parsons, *Science*, **142**, 1176 (1963).
35. B. Chance, D. F. Parsons, and G. R. Williams, *Science*, **143**, 136 (1964).
36. E. Racker, B. Chance, and D. F. Parsons, *Fed. Proc.*, **23**, 431 (1964).
37. M. E. Pullman, H. S. Penefsky, A. Dotta, and E. Racker, *J. Biol. Chem.*, **235**, 3322 (1960).
38. D. E. Green and J. F. Perdue, *Ann. N.Y. Acad. Sci.*, **137**, 667 (1966).
39. D. E. Green and A. Tzagaloff, *Arch. Biochem. and Biophys.*, **116**, 293 (1966).
40. C. de Duve and R. Wattiaux, *Ann. Rev. of Physiol.*, **28**, 435 (1966).
41. G. E. Palade, P. Siekevitz, and L. G. Caro, in *Ciba Foundation Symp. on the Exocrine Pancreas*, Eds., A. V. S. Reuck and M. B. Cameron, J. London, and A. Churchill, 1962, p. 23.
42. G. E. Palade, *J. Amer. Med. Soc.*, **198**, 143 (1966).
43. L. G. Caro and G. E. Palade, *J. Cell Biol.*, **20**, 473 (1964).
44. J. D. Jamieson and G. E. Palade, *Proc. Nat. Acad. Sci.*, **55**, 424 (1966).
45. D. H. Northcoate and J. D. Pickett-Heaps, *Biochem. J.*, **98**, 159 (1966).

Chapter 7

HORMONES AND MEMBRANES

Study of the mechanism of action of hormones poses some particularly difficult problems. In the living animal a given hormone may be secreted at a level such that slow progressive changes occur over a period of time often measured in months or years. Moreover, the ultimate effects of a given hormone are probably the result of complex and subtle interactions with equally small doses of other hormones that can aid or modify the response of individual tissues or enzyme interactions. Frequently, in order to amplify the response of an animal to a particular hormone, the gland which normally secretes the hormone is surgically removed. The animal then enters an abnormal state in which the tissues are forced to adapt themselves to the severe hormone imbalance. The response elicited by a dose of the hormone under these conditions may not reproduce the normal physiological situation.

Although the natural effects that are the result of the action of a particular hormone may be accomplished in a period of months, the investigator attempts to study the mechanisms in experiments of short duration by administering a rather large dose of the hormone. Thus it is reasonable to consider that such an excess of hormone may affect targets in cells which may not be influenced by the hormone under normal conditions. There is the further uncertainty as to precisely what dose administered under *in vitro* conditions would approximate the tissue's normal exposure to hormone in the intact animal.

Another problem of major proportions is whether a particular phenomenon under study is directly influenced by a given hormone or whether some unrecognized primary event has influenced the observed reaction, as well as other interactions in the cell.

Attempts to explain the mode of action of hormones have always reflected the prevailing climate of research. This view was ably expressed by Hechter and Halkerston (1a) as follows:

"The prevailing conceptual fashions in biochemistry and biophysics have dominated the experimental approach to elucidation of hormone action; as the former have changed, corresponding changes have occurred in the latter. An atomistic biochemistry—concerned almost exclusively with energetics and powered by the then new enzymology, which systematically unraveled the classical problems of intermediary metabolism—gave rise to the idea that the physiological effects produced by steroids (hormones) are ultimately the expression of metabolic changes in responsive cell types in characteristic "target" tissues and organs. In recognition of the knowledge that most of the chemical reactions in the cell are catalyzed by specific enzymes there was almost universal acceptance of the view that hormones produce their characteristic effects by acting (directly or indirectly) to regulate the activity of enzymes, with widespread secondary effects upon cell function."

Today's ideas for all kinds of regulatory phenomena are dominated by thoughts of nucleic acids and their role in the regulation of protein synthesis in microorganisms, particularly $E.$ $coli.$ Work in the field of hormone research has been directed largely to seeking and finding explanations of hormone action in terms of RNA, particularly, mRNA. The field in the last few years has grown much too large to be reviewed in detail. In this chapter, three hormones—growth hormone, insulin, and thyroid hormone—are reviewed in some detail and other hormones are briefly discussed.

It is my feeling that we are about to enter a most fruitful period in biochemistry where it will be recognized that the structure of cells and their membranes exert a vital influence and control on most of the integrated reaction sequences of the cell. A greater understanding of hormone action should result from a consideration of the cell's enzymatic capabilities in relation to its structural organization. Similar ideas have been mentioned in the past (Hechter and Peters) but the time was not ripe for them to be heard (1b, 2).

Tata has recently reviewed the effects of hormones on the synthesis and utilization of ribonucleic acids (3). After considering the various lines of experimental evidence pointing to effects of hormones on RNA metabolism and the possibility of explaining hormone action in terms of the processes of transcription (copying a nucleotide sequence in DNA by an RNA polymerase which produces mRNA) and translation (producing a protein to correspond to the sequence of nucleotides in mRNA), Tata stated, "However, it is well known that almost every hormone that has a marked effect on RNA metabolism also has a potent effect on intracellular lipid and phospholipid synthesis and turnover." In his concluding remarks Tata observed, "Perhaps we have not yet fully realized the full implications of cellular architecture in the control of biosynthetic activity, and the observation that intracellular membranes are metabolically and functionally not static structures is of great importance."

The coverage given to considerations of hormone action in this chapter is admittedly incomplete. I have selected only some of the observations which lend themselves to interpretations involving cellular structure. There are many other excellent studies that bear on this point that I have not been able to include as well as important studies that are not obviously integrated into the concept of hormonal-induced structural modifications of the cell. I hope that I make my point clear, however, with the examples chosen for this chapter.

Growth Hormone

In an extensive review of his and other studies with growth hormone, Korner proposed that the hormone exerts its anabolic influence through a regulation of RNA biosynthesis (4). He pointed out that removal of the anterior lobe of the pituitary of rats resulted in loss of weight and protein content of the animal. Acellular liver systems prepared from such animals exhibited diminished amino acid-incorporation ability. Treatment of the animal with growth hormone repaired the *in vivo* and *in vitro* biosynthetic deficiencies. Examination of the components of the acellular amino acid-incorporating system indicated that the ribosomal components of the microsomes were the units responsible for the diminished incorporating ability in the deficient animals. Growth hormone added *in vitro* was not effective in this system.

After reviewing the concepts of mRNA regulation of protein synthesis in *E. coli*, Korner acknowledged that the evidence of a comparable control system in mammalian tissues was scanty. His own experiments with hemoglobin formation from liver ribosomes mixed with reticulocyte supernatant fluids (possibly containing hemoglobin mRNA) gave a *false positive* result which was apparent only after several rigorous control experiments. He cited the existence of active polysomes in mammalian cells as an argument for the existence of functional mRNA. This view was strengthened by a review of reported cases where actinomycin D reduced amino acid-incorporation in such systems (see Chapter 3). The disagreement between his interpretations and those drawn by Revel and Hiatt regarding the significance of actinomycin D effects in mammalian cells (page 117) was noted. It may be recalled that these authors questioned that actinomycin D was acting principally through an effect on mRNA.

In Korner's experiments 100 μg of growth hormone was administered 12 hr before sacrificing the animal. An important limitation of interpretation in these studies is the uncertainty of what other initial action the hormone may take before producing the effects leading to a stimulated acellular liver amino acid-incorporating system. When an acellular system was isolated from hypophysectomized animals it was found that free ribosomes were more affected than polysomes in terms of a diminished incorporating ability. He concluded, however, that polysomes were still the important element of the incorporating system but that hypophysectomy reduced their number and not the incorporating ability of the remaining ones. This possibility was borne out by a reduced recovery of polysomes from the operated animals. This leaves unanswered the question of why the free ribosomes were less active on a per-unit basis in the system obtained from the operated animals.

Some clarification of this problem can be found in the work of Garren, Richardson, Jr., and Crocco (5). These authors confirmed that less ribosomal material was present in the livers of hypophysectomized animals. However, the distribution of ribosomes between polysomes and monomers was not altered and the uptake of labeled orotic acid into the polysomes was not impaired by hypophysectomy. The amino acid-incorporating ability of polysomes from hypophysectomized animals was less than that of polysomes from normal animals. Ribosomes from hypophysectomized animals were less efficient in using polyuridylic acid to stimulate phenylalanine incorporation than were normal ribosomes. The authors therefore concluded that the ribosomes themselves are able to influence the rate of protein synthesis independent of mRNA and that one effect of pituitary hormone is to influence the state of the ribosome.

Another crucial experiment tested whether the injection of actinomycin D would abolish the stimulatory effect of growth hormone. If growth hormone works by stimulating the formation of mRNA, then actinomycin D might abolish its effect. It was found that actinomycin D led to diminished rates of amino acid-incorporation in the acellular system but the stimulatory effect of growth hormone was still present even in the heaviest actinomycin treatment. The same picture was obtained for the labeling of both nuclear and cytoplasmic RNA with orotic acid as stimulated by growth hormone in the presence and absence of actinomycin D (i.e., growth hormone was still able to stimulate RNA synthesis in the presence of actinomycin D) (4).

Analysis by sucrose density gradient centrifugation of the RNA labeled normally, after hypophysectomy, and after treatment with growth hormone or actinomycin D showed that all classes of RNA—ribosomal, messenger, and transfer RNA—were affected, not just mRNA. Korner concluded that growth hormone stimulated the synthesis of all forms of RNA which then caused stimulation of protein synthesis. Since actinomycin D is supposed to prevent mRNA formation, his findings of stimulation of synthesis of protein in the presence of actinomycin D requires additional assumptions for any theory placing a key role on mRNA in the regulatory process. Furthermore, as Korner acknowledges, the enhanced rate of RNA synthesis following growth hormone administration could itself be a result of some other general stimulatory effect rather than being a primary response.

Although Korner presents a possible mode of operation of growth hormone via the RNA's of the cell, not much evidence could be marshalled in direct support of this idea. Furthermore, in the discussion following Korner's presentation, Knobil reported that the addition of growth hormone stimulated amino acid-incorporation in isolated rat diaphragm but there was no concomitant stimulation of adenylate incorporation into RNA (6). The same kind of negative finding for the stimulation of RNA labeling by growth hormone in diaphragm preparations was reported by Dawson (7).

Kostyo studied the separation of the effects produced by growth hormone on amino acid transport and protein synthesis in muscle (8). He reviewed a number of reports in the literature which indicate that growth hormone stimulates the transport of amino acids in muscle cells as well as the incorporation of amino acids into muscle protein. It is possible that a primary stimulation of amino acid transport could result in the stimulation of protein synthesis. The idea that the primary stimulation for protein synthesis is exerted through RNA seems to be refuted by the fact that the effects on amino acid transport and protein synthesis are rapid, whereas the effects on RNA metabolism are frequently evident only after hours or days following hormone administration. Kostyo took advantage of recent findings linking amino acid transport to the presence of sodium in the medium and found that by replacing sodium with choline, growth hormone no longer stimulated glycine accumulation in the rat diaphragm although the hormone did stimulate glycine incorporation into the protein. These experiments show that the stimulation of protein synthesis does not result from an increased free amino acid pool.

Nonetheless, the studies demonstrate that one hormone can simultaneously affect a membrane process involving amino acid transport and also the incorporation of the amino acid into protein. Kostyo reviewed some of the literature concerning diaphragm tissue, which strongly indicates that external amino acid has a direct route for incorporation into protein, not involving its passage through the intracellular pool (see Chapter 8).

These studies of Kostyo and others could be explained by the view that growth hormone interacts with a single receptor site at the cell membrane where it affects an early stage common to protein synthesis and amino acid accumulation. The absence of sodium would then affect a subsequent stage concerned with pool formation.

Martin and Young noted that the *in vitro* addition of growth hormone to isolated diaphragm from hypophysectomized rats stimulates the uptake of glucose, the accumulation of amino acids, and the incorporation of amino acids into protein (9). They found that in the presence of 10 μg/ml of actinomycin D, growth hormone still stimulated glycine incorporation into protein even though orotic acid incorporation into RNA was reduced by 90 to 95%. Growth hormone administered in the absence of actinomycin D did *not* stimulate RNA synthesis as measured by orotic acid incorporation.

Insulin

In 1949, Levine, Goldstein, Klein, and Huddlestun proposed that the action of insulin might be centered on the transmembrane transport of glucose, rather than on any of the known steps involved in its subsequent metabolism (10). They demonstrated an effect of insulin on the uptake of galactose from the blood in nephrectomized dogs. For several years afterwards, this idea became a part of the general scientific thinking on the problem of the mechanism of insulin action. Various reports demonstrating effects of insulin on the uptake of sugars added weight to the theory. Other known manifestations of insulin's effects could be fitted into this explanation if glucose entry was a rate-limiting factor for various of its further metabolic conversions. Even as late as 1955 and 1957, metabolic effects of insulin, such as increases of glucose oxidation, glycogen synthesis, formation of fat, esterification of fatty acids and protein-sparing action, were explained in terms of an effect on glucose entry (11, 12).

After 1955 some difficulties arose for the idea of such a single primary site of action (e.g., there was the effect of insulin on liver metabolism, where glucose transport is not a rate-limiting process). Reviewing the status of the problem in 1965, Levine admitted that subsequently studied effects of the hormone in a variety of tissues made it difficult or well nigh impossible to deduce them from the effect on glucose transport (13). For example, the influx of substances unrelated chemically to sugars is stimulated by insulin in the *absence* of added external glucose. Such substances include nonutilizable and natural L-amino acids, long-chain fatty acids, and potassium.

After considering some of these problems at length, Krahl, in 1957, proposed that a single site of action of insulin at an "extracellular-intra-cellular" barrier might initiate a series of unspecified intermolecular rearrangements which are propagated along the boundary and into the cell interior to influence many different paths of cellular metabolism (14). His discussion and proposed solution in the research climate of 1957 was admittedly speculative and without much experimental support. However, in the research climate of a decade later the suggestion seems much more timely.

Levine reemphasized the possibility that a primary site for insulin action may exist in the structure of the cell membrane and that all other actions result from the involvement of the cell membrane in various processes (13).

Insulin has been shown to increase the incorporation of labeled amino acids into tissue proteins in the absence of medium glucose (15). According to prevalent ideas on processes of protein synthesis in 1965, in order to preserve the concept of a single primary membrane site of action for insulin, it would be useful to show that the effect on protein synthesis could result from the increased amino acid pool brought about by the insulin effect on the membrane. Wool has devoted consider-able effort to this problem and these studies are discussed on page 283.

Current thinking on the regulation of protein synthesis involves the sequence DNA → mRNA → ribosome complex. It was therefore most natural that Wool explored the possibility of an effect of insulin on nucleic acid synthesis. He found that insulin enhanced the incorpora-tion of labeled adenine and glucose into RNA of isolated rat diaphragm (16). No similar effect was found for the labeling of DNA. Insulin also brought about a net synthesis of RNA. The general significance of these results in terms of the hormone's action on mRNA synthesis was explored by Wool.

The explanation of insulin action on protein synthesis through gene transcription (mRNA) was short-lived. Eboué-Bonis, Chambaut, Volfin, and Clauser found that when actinomycin D completely inhibited the labeling of RNA by C^{14}-adenine in rat diaphragm, the incorporation of labeled amino acids into protein was unchanged and the 100% stimu-lation of amino acid-incorporation evoked by insulin was likewise insen-sitive to the nearly total inhibition of RNA labeling (17). They also showed that the uptake of glucose and the labeling of ATP were not dependent on protein or RNA synthesis, which processes could be manip-ulated by use of puromycin or actinomycin D, respectively. These obser-vations were confirmed by Wool and Moyer (18). Although these authors acknowledged this finding as a blow to the explanation of insulin action on protein synthesis through mRNA transcription, they discussed other possibilities such as the small percentage of actinomycin D-resistant RNA synthesis representing a small number of crucial messengers or insulin exerting its effect on the utilization of available mRNA.

Although Levine recognized the difficulty posed by the effect of insulin on protein synthesis (which, according to contemporary ideas, is an activity carried on primarily in the cell sap) to his idea of a primary locus of action at the membrane surface (13), he added weight to the idea by citing the studies of Rieser and Rieser (19) and Rieser and Roberts (20), who used proteolytic enzymes to alter membrane structure and obtain insulin-like effects. He also cited the studies of Rodbell, who with the same purpose, used phospholipase C in the isolated fat cell in order to modify the membrane and elicit insulin-like activity on glucose transport, fatty acid synthesis, and protein synthesis (21).

The possibility of reconciling the effects of insulin on both the transport of L-amino acids and protein synthesis in terms of a single site of action was vigorously explored by I. G. Wool. The point of reference for all of Wool's extensive studies and discussions was that the incorporation of radioactive amino acid into protein must proceed from an internal amino acid pool and must be a reaction distal or subsequent to the reactions which place the amino acid into this internal pool.

The separation of effects of insulin on glucose entry and protein synthesis was evident from studies showing that protein synthesis in muscle could be increased by insulin in the absence of extracellular glucose (22). To separate the processes of transport of external amino acid into the cell and its incorporation into protein, several different approaches were used. Thus C^{14}-amino acid was injected into an animal prior to isolation of its diaphragm. One hemidiaphragm served as a control and the other was incubated with or without insulin. Under these conditions it was shown that insulin stimulated the labeling of protein from amino acids already in the cells. The same conclusion was reached by another elegant approach: amino acids were synthesized intracellularly from labeled precursors such as pyruvate, α-ketoglutarate and bicarbonate. In every case insulin still exerted its effect and so the penetration of external amino acid was not the site of insulin's action leading to a stimulation of protein synthesis (23).

Rampersad and Wool went on to prepare a conventional amino acid-incorporating system from a deoxycholate treated rat heart muscle homogenate (24). They found that insulin added in vitro to this system did not effect the incorporation of radioactivity from phenylalanyl-sRNA into protein in the presence or absence of polyuridylic acid (poly-U) (22, 25). Ribosomes isolated from a diabetic animal, however, were decreased in activity even in the presence of added poly-U and so a defect in the ribosomes was indicated. Insulin added *in vitro* was without effect but the administration of insulin 1 hr before sacrificing the animals exerted a profound effect on the activity of the subsequently isolated ribosomal system. From these and related studies, Wool concluded that the locus of the action of insulin on protein synthesis is on the ribosomes themselves. However, a vitally important qualification was not emphasized; that the action of insulin on the ribosomes must be secondary to some other cellular effect since insulin had no effect on the isolated ribosomes.

That insulin does influence the transport of natural amino acids across the cell membrane was amply demonstrated by studies of Scharff and Wool (26). Wool discussed the interesting finding of Scharff and himself that the increased amount of amino acid transported under the influence of insulin was frequently located in newly synthesized protein and could be found in the free amino acid pool only in the presence of puromycin, which inhibited protein synthesis. Wool's interpretation of all of his studies with his collaborators was that insulin can enhance protein synthesis in muscle by an action at a site distal to amino acid transport, presumably by an effect on some intracytoplasmic process. This conclusion is not the only or even the best that can be drawn from the mass of important observations obtained by Wool and his collaborators. Before discussing an alternative interpretation I would like to cite another effect of insulin on a process known to be membrane-bound: the formation of high energy compounds.

Clauser, Volfin, and Eboué-Bonis studied the effect of insulin on P^{32}-phosphate uptake and labeling of high energy compounds in the isolated diaphragm of normal and hypophysectomized rats (27). Insulin had no effect on the rate of penetration of inorganic phosphorous into the intracellular space but promoted in every case a striking enhancement of the labeling of organic phosphates, particularly ATP, in the presence or absence of added glucose. In the studies of Wool already cited, the ATP supply was not limiting and therefore the insulin effect on protein synthesis was not mediated through ATP forming reactions. The effect of insulin on ATP formation, however, represents another action of the hormone which must be considered as either a separate effect or as one that could result from a primary site of action. Since oxidative phosphorylation is a membrane-located metabolic sequence, a primary site of action of insulin at a membrane surface would be consistent with these observations.

Returning now to the dilemma pondered by Wool on the necessity of considering two different sites of action for insulin to explain the effects on amino acid transport and protein synthesis, it is important to take note of the many published studies which indicate that for various types of tissue, including muscle, extracellular amino acids destined for incorporation into protein do not necessarily pass through the total or expandable free amino acid pool but instead can be incorporated directly into protein (see pages 298–300). In other words, the incorporation of external amino acids into cellular protein need not be considered as a reaction distal to the formation of the intracellular pool. On the basis of such studies as these and many others, I suggested a model for protein synthesis which proposed that the process of protein synthesis is membrane-oriented (28). An external amino acid, on entering the cell membrane, could either be directly incorporated into protein by membrane-bound ribosomes (at the cell membrane or by a continuous membrane channel to the endoplasmic reticulum) or, if the rate of entry exceeded the requirements or capacity of the protein synthetic machinery, it could be deposited in an intracellular pool. Therefore the entry of an amino acid into the cell membrane may represent a reaction common to the processes of protein synthesis and amino acid transport. If the process of protein synthesis is blocked (e.g., by the addition of puromycin) then the amino acid would be directly deposited in the pool. Insulin, by reacting at a site on the cell membrane, could enhance the penetration of the membrane by amino acid, glucose, potassium, or other substances. It could possibly also stimulate ATP formation, the synthesis of fats, esterification of fatty acids, glucose oxidation, and glycogen synthesis. The finding of many apparently unrelated effects of insulin action on cellular metabolism could indicate many different sites of action or, in the face of independent evidence linking these many activities to the integrity of the membrane, these various manifestations of insulin action could point to the important control that cell membranes exert over many metabolic pathways of the cell.

Thyroid Hormones

The biological activities of thyroid hormones present another example of a multitude of apparently unrelated biochemical events under the influence of a single active hormone substance. The fundamental question confronting endocrinologists and biochemists is how to trace several different and important metabolic consequences to the action of a single administered hormone.

This dilemma was originally viewed at the whole animal level where basal metabolic rate, growth, development, and physiological function of organs are dramatically influenced by the thyroid state. More recent concern has been devoted to an apparent multiplicity of action at the molecular level.

In 1951 it was reported by several groups that thyroxin added to isolated liver mitochondria will uncouple oxidative phosphorylation (29–31). Although similar observations have been reported many times, the influence of thyroid hormones on respiration and phosphorylation has been the subject of much confusion and disagreement. In some instances oxygen uptake has been increased with no change in P:O ratio, the P:O ratio has been lowered with no change in oxygen uptake, and in "digitonin" preparations of mitochondria, thyroxine, and tri-iodothyronine have not uncoupled oxidative phosphorylation whereas in sonicated preparations they have. (For references to these observations see Tata [32].)

In 1954 Hoch and Lipmann demonstrated that liver mitochondria from thyrotoxic hamsters were in a state of loose coupling (33) (i.e., in the absence of ADP, respiration continued at a high rate and the addition of ADP did little to alter this rate). Thyroxine and related compounds added *in vitro* inhibited both the phosphate-ATP and ADP-ATP exchange reactions in rat liver mitochondrial preparations (34, 35). These exchange reactions are believed to represent the terminal steps of the oxidative phosphorylation process. Tata feels that the uncoupling of oxidative phosphorylation by thyroid hormones occurs only at excessively high concentrations *in vitro* and questions how much of the direct effects of the hormones *in vivo* at physiologic concentrations may be traced to such an action.

Thyroxine is able to influence the structure and stability of mitochondria and under various conditions can lead to either a swelling or contraction of these organelles (32). Activities of mitochondrial enzyme systems such as succinoxidase and cytochrome oxidase can be elevated both by *in vivo* and *in vitro* administration of the hormone. The activity of microsomal TPNH-cytochrome *c*-reductase of thyrotoxic rats was found to be greatly increased by Phillips and Langdon (36) and this effect has been confirmed by others. Bronk has emphasized the direct stimulatory effect of thyroid hormones on the functional capacity of the electron transport and oxidative phosphorylation systems. Stimulation of protein synthesis could then represent a result of the greater availability of energy (37, 38). This interpretation is opposite to that of Tata *et al.* (39) and of Sokoloff and Kaufman (40), who have suggested that the increased demand for energy by thyroid-stimulated protein synthesis causes an increase in oxidative function. Microsomal glucose-6-phosphatase is likewise increased after administration of thyroid hormones. Various other effects of the hormones on isolated enzyme activities have been reported from many laboratories and some of these actions are reviewed by Tata (32). The effects of thyroid preparations on blood cholesterol and lipid metabolism are comprehensive subjects which by themselves could fill entire chapters. Furthermore, profound effects on cellular levels and turnover of carbohydrates, water, electrolytes, vitamins, and coenzymes have also been reported. One pronounced and well recognized effect is the depletion of tissue glycogen levels. From the point of view of this monograph, however, one of the most interesting facets of thyroid action is its pronounced effect on protein synthesis.

This story begins in 1952 when Dutoit reported that the incorporation of radioactive alanine into protein by liver slices was depressed after thyroidectomy of the animal and stimulated after *in vivo* pretreatment of the rat with large doses of thyroxine (41). Further evidence of a reduced rate of protein biosynthesis in hypothyroidism and its return to normal after effective thyroid therapy has been obtained for intact man by Crispell et al. (42). The ability of thyroid hormone to stimulate the synthesis of serum proteins in intact man is demonstrated by the work of Schwartz (43), Rothschild, Bauman, Yalow, and Berson (44) and Lewallen, Rall, and Berman (45). In 1959, Sokoloff and his collaborators instituted a series of studies on the effects of thyroxine and related compounds on amino acid-incorporation in acellular systems.

Sokoloff and Kaufman found that thyroxine at a concentration of 1.3×10^{-5} M increased the rate of incorporation of DL-C^{14}-leucine into the protein of homogenates of rat liver by 19% (40). Pretreatment of rats with thyroxine allowed the production of liver homogenates with a 46% higher rate for leucine incorporation. In order to localize the source of enhanced activity, mitochondria, microsomes, and supernatant fractions were prepared from normal and thyroxine treated rats, and homogenates were reconstituted in all possible combinations. It was found that the increased activity was encountered only in the combinations which included the mitochondria from the hyperthyroid animals. At low concentrations of the hormone effects on protein synthesis preceded those on oxidative phosphorylation but this situation was reversed at extremely high levels of the hormone (above 10^{-3} M).

In further studies of the phenomenon, Sokoloff and Kaufman reported that the stimulation of *in vitro* incorporation of amino acid into protein was dependent on the nature of the oxidizable substrate present, being best with α-ketoglutarate (67%) and quite erratic with succinate (46). Although only the natural isomer, L-thyroxine, was effective *in vivo*, D-thyroxine stimulated the *in vitro* system. This seemingly paradoxical situation also applied to the uncoupling of oxidative phosphorylation. The dependence of the stimulating activity of thyroxine on the presence of mitochondria was once again confirmed and it was further shown that an oxidizable substrate was also required. Data were presented which indicated that all of the increased amino acid-incorporation was into the microsomal fraction, but later work showed the mitochondrial fraction itself to account for an appreciable fraction of this enhanced level of incorporation. A lag phase of about 25 min duration prior to the appearance of the stimulated level of incorporation seemed to exist but this lag could be eliminated (or diminished) by preincubation of the system with oxidizable substrate and thyroxine.

Sokoloff, Kaufman, and Gelboin next reported that thyroxine stimulated the incorporation of C^{14}-valyl-sRNA as effectively as that of free C^{14}-valine. A large quantity of C^{12}-valine added to the flask did not alter the results with labeled valyl-sRNA. Under the same conditions added GTP did not enhance incorporation and so the possibility that thyroxine was acting to increase a limiting level of GTP seems to be eliminated. The following study by Sokoloff, Kaufman, Campbell, Francis, and Gelboin further established that the essential step that was stimulated by thyroxine was the transfer of the labeled amino acid from sRNA to protein (48). This stimulation seemed to require a lag phase of about 5 min. It was again shown that the thyroxine effect seemed to be a stimulation of amino acid-incorporation rather than an increase in a limiting quantity of either GTP or ATP. The composite picture that emerged was that the incubation of thyroxine with mitochondria and an oxidizable substrate caused the formation and release of a substance which stimulated the transfer of amino acid from sRNA to protein. The exciting information was given that unpublished experiments of Sokoloff, Campbell, and Francis demonstrated that the soluble supernatant fraction from a preincubated mitochondrial system could replace the mitochondrial requirement for a thyroxine stimulation of amino acid-incorporation into microsomal protein.

Following the original suggestion of Sokoloff and Kaufman (40) that the acceleration of metabolic rate characteristic of thyroxine action may be secondary to a stimulation of energy requiring processes such as protein synthesis, Weiss and Sokoloff studied the effects of the drug puromycin which inhibits protein synthesis (49). They found that when protein synthesis was inhibited, the hypermetabolic state of rats induced by prior treatment with thyroxine was completely reversed and the consumption of oxygen was restored to normal levels. They concluded that the calorogenic effect of thyroxine is secondary to its effect on protein synthesis.

Michels, Cason, and Sokoloff demonstrated that L-thyroxine increased the incorporation of radioactive amino acids *in vivo* in liver, kidney, and heart, but not in spleen, testis, or brain (50). This is similar to its effect on oxidative metabolism in the respective organs. Thus they suggested the stimulation of protein synthesis is a physiological action of thyroid hormone.

Sokoloff, Campbell, Francis, and Klee demonstrated that thyroxine stimulation of amino acid-incorporation *in vitro* could be obtained with ribosomes as well as with microsomes; this seems to exclude the possibility that the effect is secondary to some alteration in the structure or permeability of the microsomal membrane (51). These results were at variance with those of Roche, Michels, and Kamei who reported that a thyroxine stimulation *in vitro* of amino acid-incorporation into microsomes was eliminated when the microsomes were replaced with ribosomes (52). Furthermore, the actively incorporating units of Sokoloff et al. could be ribosomes retaining portions of membrane.

Further studies of Sokoloff and Klee (53, 54) explored the correlation of effects of thyroxine on metabolic rate and protein synthesis in brain tissue of immature rats and the absence of such effects in brain tissue from older animals. All of the observations reported above for rat liver were found also for the case involving immature brain tissue, although the actual percentage of stimulation was frequently much smaller. In mixed reconstituted homogenates a stimulation of amino acid-incorporation by thyroxine could be obtained with mitochondria from immature brain or liver, but not from adult brain. In these studies incorporation of amino acid into the mitochondria from immature brain was about 50% higher than into the microsomal and cell sap fractions.

In the contemporary research climate all regulatory effects on protein synthesis are attributed sooner or later to some interaction with mRNA or its synthesis. Thyroid hormones are no exception to this rule. Widnell and Tata reported that RNA-polymerase activity in nuclei prepared from livers of thyroidectomized rats is low, and that thyroid hormone administration to such animals raises the activity toward normal levels before the effect on amino acid-incorporation into protein is observed (55). They therefore proposed that thyroxine must exert its influence at the gene level in the process of transcription. Sokoloff, Francis, and Campbell refuted this suggestion by four different types of experiments (56). First, they found that the addition of thyroxine *in vitro* to cell free, normal rat liver homogenates resulted in a stimulation of amino acid-incorporation into microsomal protein in the absence of any detectable effect on the incorporation of RNA precursors into RNA. Second, the thyroxine stimulation of amino acid-incorporation into protein was present even when nuclei containing RNA-polymerase activity and mRNA were added in optimal amounts. Third, inhibition of DNA-dependent RNA-polymerase activity by actinomycin D or DNAase had no effect on amino acid-incorporation into protein or the thyroxine stimulation of this incorporation. Finally, thyroxine stimulated the poly-U directed incorporation of phenlyalanine into polyphenylalanine. Sokoloff et al. tried to resolve their differences with Widnell and Tata by suggesting that the hypothyroid state of the animals in the latter's experiments led to degenerative changes in the livers, the repair of which may be facilitated by thyroxine by a mechanism different from the usual type of direct hormone stimulation of normal processes.

The stimulatory effect of thyroxine on C^{14}-leucine and C^{14}-valine incorporation in a rabbit reticulocyte lysate was shown by Krause and Sokoloff to lead to an increased labeling of the α and β chains of isolated homoglobin (57). Thus, the hormone has been shown to be capable of influencing the formation of a normal cellular protein in the *in vitro* system.

Tata, Ernster, Lindberg, Arrhenius, Pedersen, and Hedman, in evaluating the multifaceted response of tissues to thyroid hormones at the cellular level, stated that their results indicated that the enzymes or functions linked firmly to membranous subcellular structure are more markedly affected during the early phase of thyroid hormone action than activities not dependent on structural integrity (39). Similarly, Lee, Takemori, and Lardy found a thyroid stimulated activity of mitochondrial bound glycerol-1-phosphate dehydrogenase but no such stimulation for the soluble form of the enzyme (58). The microsomal membrane-bound reactions which incorporate acetate into cholesterol were similarly susceptible to stimulation (59). These observations indicate the regulatory importance of membranes and are similar to ideas of Hechter and Peters linking hormone action to effects on the structural organization or cytoskeleton of the cell which, according to Peters, is the master coordinator for the cell's basic functions (1b, 2).

Tata and Widnell studied in some detail consecutive stimulatory effects in liver tissue of a single injection of tri-iodothyronine to thyroidectomized rats (60). DNA-dependent RNA polymerase activities of two types were followed. Although a stimulation of the rapid labeling of nuclear RNA by C^{14}-orotic acid was evident in 3 to 4 hr, the Mg^{2+}-activated RNA-polymerase (for which the product is mainly ribosomal type of RNA) began to be stimulated at 10 to 12 hr after treatment and reached its peak of activity at 24 to 48 hr. The Mn^{2+} ammonium sulfate-activated RNA-polymerase (for which the product is more DNA-like) was unaffected for 24 hr after hormone administration and stimulated by 30 to 40% at about 50 hr. These enzymes were not stimulated by *in vitro* addition of the hormone to isolated nuclei. Stimulation of RNA synthesis was evident for mitochondrial, microsomal, and soluble RNA. The actual amount of ribosomal RNA per gram of liver was increased by about 50%, 35 to 45 hr after injection. A very significant observation was that *the newly formed ribosomes (and possibly attached mRNA) were more firmly bound to microsomal membranes after hormone treatment.* Although data were not presented in this work, previous studies would indicate that the enhanced level of C^{14}-amino acid-incorporation into microsomes would become evident about 17 hr after hormone treatment and reach a peak at about 45 hr.

Steroid Hormones and Other Considerations

Hechter and Halkerston critically reviewed the accumulating evidence which seeks to explain steroid hormone action in terms of a hormone-gene interaction (1a). Their discussion concerned the hormones ecdysone, aldosterone, estradiol, and cortisol. For the many details of their arguments the reader should consult the original review.

They concluded, that in only one case—the effect of aldosterone on sodium transport in toad bladder—is there evidence for a direct action of steroid at the gene locus of a responsive cell type; the evidence here is suggestive, not conclusive. "In summary," they stated, "the attractive hormone-gene thesis appears to be wholly inadequate to provide a unitary basis for steroid hormone action at the molecular level."

Contemporary thinking in terms of regulatory processes has also been influenced by other ideas of Monod and his colleagues. Monod, Changeux, and Jacob introduced the concept of "allosteric transition" (61). This hypothesis assumes that proteins may possess two distinct types of receptor sites: (a) the active enzyme site which binds the substrate, and (b) the allosteric site which has a different affinity and which occurs at another area of the molecule. The manner of folding of the enzyme or protein and its spatial configuration are influenced by the binding of some small molecule to the allosteric site. In such fashion the biological activity of the macromolecule is under the regulatory control of the small molecule. Although no clear case of hormone action in terms of an allosteric transition has been presented, the idea is an extremely valuable addition to our concept of possible cellular regulatory mechanisms.

Another proposal for a unifying concept of hormone activity is that of Sutherland and Rall (62). Sutherland and Robison have recently reviewed the model (63). Briefly stated, it is proposed that the catechol amine hormones (epinephrine and norepinephrine) in particular owe their metabolic ability to their interaction with the membrane-bound enzyme adenyl cyclase. This enzyme is present in all animal cells examined to date with the exception of nonnucleated erythrocytes. ATP is the substrate for the reaction, and in the presence of Mg^{2+} the enzyme catalyzes the formation of cyclic 3′,5′-AMP and inorganic pyrophosphate. The biological effects of the hormones are said to be caused by the direct action of cyclic AMP on some active protein or enzyme system.

There is an accumulation of observations consistent with such a sequence of events in various tissues. In liver, epinephrine reacts with the membrane adenyl cyclase system and leads to an elevation of the intracellular concentration of cyclic 3′,5′-AMP. This in turn results in an increase in the amount of the phosphorylated, or active, form of phosphorylase, which in the liver is the rate-limiting enzyme in the conversion of glycogen to glucose. An impressive list of enzymes and metabolic processes in different tissues known to be influenced by cyclic 3′,5′-AMP was compiled by Sutherland and Robison. The theory was extended to include other hormones and active substances such as glucagon, ACTH, vasopressin, luteinizing hormone, thyroid-stimulating hormone, serotonin, acetylcholine, and histamine. Details of this broad concept are discussed by Sutherland, Øye and Butcher (64). Hormones which have not been shown to influence the level of cyclic 3′,5′-AMP, such as insulin, growth hormone, and oxytocin, could fit into the same general scheme by assuming another such critical enzyme in place of adenyl cyclase and another regulatory intermediary in place of cyclic 3′,5′-AMP. But with this vagueness we are back to other less clearly stated ideas of indirect hormone action.

One less attractive feature of the generalization of the concept of Sutherland and colleagues is that this simplification in terms of mechanism of action of various hormones introduces complexities of another kind. The impressive specificity of action of various hormones in terms of different tissues, different cells, and different metabolic events within a given cell is lost by the necessity of exerting their influence through a single intermediary substance. Therefore we must consider different forms for the same enzyme or different enzymes for producing the same substance so that a given target tissue will respond to a given hormone in a unique manner. When more than one hormone can effect a particular cell in different ways or in an additive fashion at saturation levels for a similar response, we must consider compartments within the cell where the cyclic AMP produced by one hormone cannot reach sites that normally would be responsive to cyclic AMP produced by another hormone.

Although only three hormones are considered in detail in this chapter, many other hormones present the same picture of multiple cellular responses; for example, ACTH stimulates protein synthesis *in vivo* (65) and *in vitro* (66), as well as enhancing the rate of synthesis of rapidly labeled RNA (67) and also of stable RNA (68). It also influences the permeability of adrenal cortical slices to ascorbic acid (69), sugars, and amino acids (70). Similar multiplicities of response affecting protein synthesis, RNA synthesis, and permeability are produced by thyroid stimulating hormone, testosterone, estrogens, cortisone, and aldosterone (3, 70, 71). Although most students of hormone action have felt compelled to consider that the various manifestations of hormone regulation should be traceable to some primary event, most have thrown their hands up in despair when forced to consider the effects on some obvious membrane-bound system such as permeability with those on protein synthesis. Examples of this can be seen in discussions by Hechter and Lester (70), Levine (13), and Wool (22), to mention only a few. This dilemma in the year 1967 emphasized the prevailing satisfaction with a view of protein synthesis involving only messengers, ribosomes, and soluble factors. At the 152nd Annual Meeting of the American Chemical Society held in New York City in September, 1966 a symposium was held on the current state of knowledge of the synthesis, structure, and function of ribosomes. The leading experts in these fields were assembled in panels for a review of the current state of knowledge and to indicate the direction for future developments. The word membrane was not mentioned once in the whole symposium. Free ribosomes were discussed in terms of the binding of messengers and sRNA, but not membranes. Protein synthesis as carried out by the ribosomes was discussed only in terms of the soluble factors necessary to catalyze peptide bond formation. It is not surprising, then, that the finding of a hormone influence on protein synthesis would seem to dictate the necessity of a second target to be found somewhere in the intracytoplasmic space. This may actually turn out to be the case, but it is not the only or necessarily the best explanation.

Hormones are seen to affect such processes as permeability, electron transport, oxidative phosphorylation, and lipid biosynthesis—processes which are thought of as membrane-associated. They also affect RNA and protein synthesis. As has been the emphasis of this monograph, there is considerable reason for considering macromolecular biosynthesis in terms of the involvement of cellular structure in general and membranes in particular. Although future work may establish primary non-membrane-linked targets for some or many hormones, the simple concept of a primary site of action—*on membrane integrity*—is of fundamental importance. Rather than interpreting the effects on protein synthesis against such a possibility (of a primary membrane target), it might be more profitable to explore the kind of regulation that membrane components may exert on protein and RNA synthesis. Regulation at the cellular level could be achieved via cell membranes as well as through the various important forms of RNA that have recently been so thoroughly considered.

The cytoskeleton, as briefly indicated by Peters, and the structural integrity of its parts may provide the area for our most fruitful phase in biochemical investigation of regulatory mechanisms.

REFERENCES

1a. O. Hechter and I. D. K. Halkerston, *Ann. Rev. Physiol.*, **27**, 133 (1965).

1b. O. Hechter, *Vitamins and Horm.*, **13**, 293 (1955).

2. R. A. Peters, *Nature*, **177**, 426 (1956).

3. J. R. Tata, *Progress in Nucleic Acid Res. and Mol. Biol.*, **5**, 191 (1966).

4. A. Korner, *Rec. Prog. Horm. Res.*, **21**, 205 (1965).

5. L. D. Garren, A. P. Richardson, Jr., and R. M. Crocco, *J. Biol. Chem.*, **242**, 650 (1967).

6. E. Knobil, *Rec. Prog. Horm. Res.*, **21**, 236 (1965).

7. K. G. Dawson, *Rec. Prog. Horm. Res.*, **21**, 238 (1965).

8. J. L. Kostyo, *Endocrinology*, **75**, 113 (1964).

9. T. E. Martin and F. G. Young, *Nature*, **208**, 684 (1965).

10. R. Levine, M. Goldstein, S. Klein, and B. Huddlestun, *J. Biol. Chem.*, **179**, 985 (1949).

11. R. Levine and M. S. Goldstein, *Rec. Prog. Horm. Res.*, **11**, 343 (1955).

12. R. Levine, *Surv. Biol. Progr.*, **3**, 185 (1957).

13. R. Levine, *Fed. Proc.*, **24**, 1071 (1965).

14. M. E. Krahl, *Perspectives in Biol. and Med.*, **1**, 69 (1957).

15. I. G. Wool and M. E. Krahl, *Nature*, **183**, 1399 (1959).

16. I. G. Wool, *Biochim. Biophys. Acta*, **68**, 28 (1963).

17. D. Eboué-Bonis, A. M. Chambaut, P. Volfin, and H. Clauser, *Nature*, **199**, 1183 (1963).

18. I. G. Wool and A. N. Moyer, *Biochim. Biophys. Acta*, **91**, 248 (1964).

19. P. Rieser and C. Rieser, *Fed. Proc.*, **23**, 410 (1964).

20. P. Rieser and M. D. Roberts, *Fed. Proc.*, **24**, 577 (1965).

21. M. Rodbell, *J. Biol. Chem.*, **239**, 375 (1964).

22. I. G. Wool, *Fed. Proc.*, **24**, 1060 (1965).

23. K. L. Manchester and M. E. Krahl, *J. Biol. Chem.*, **234**, 2938 (1959).

24. O. Rampersad and I. G. Wool, *Fed. Proc.*, **23**, 316 (1964).

25. O. Rampersad and I. G. Wool, *Fed. Proc.*, **24**, 511 (1965).

26. R. Scharff and I. G. Wool, *Biochem. J.*, **97**, 272 (1965).

27. H. Clauser, P. Volfin, and D. Eboué-Bonis, *Gen. Comp. Endocrinol.*, **2**, 369 (1962).

28. R. W. Hendler, *Nature*, **193**, 821 (1962).

29. M. Niemeyer, R. K. Crane, E. P. Kennedy, and F. Lipmann, *Fed. Proc.*, **10**, 229 (1951).

30. H. A. Lardy and G. Feldott, *Ann. N.Y. Acad. Sci.*, **54**, 636 (1951).

31. C. Martius and B. Hess, *Arch. Biochem. Biophys.*, **33**, 486, (1951).

32. J. R. Tata, in *Actions of Hormones on Molecular Processes*, Eds., G. Litwack and D. Kritchevsky, John Wiley and Sons, 1964, p. 58.

33. F. Hoch and F. Lipmann, *Proc. Nat. Acad. Sci.*, **40**, 909 (1954).

34. O. Lindberg, H. Low, T. E. Conover, and L. Ernster, in *Biological Structure*

and Function, Vol. 2, Eds., T. W. Goodwin and O. Lindberg, Academic Press, New York, 1961, p. 3.

35. J. R. Bronk, *Biochim. Biophys. Acta.*, **69**, 375 (1963).
36. A. H. Phillips and R. G. Langdon, *Biochim. Biophys. Acta*, **19**, 380 (1956).
37. J. R. Bronk, *Science*, **141**, 816 (1963).
38. J. R. Bronk and D. S. Parsons, *J. Physiol.* (London), **184**, 942 (1966).
39. J. R. Tata, L. Ernster, O. Lindberg, E. Arrhenius, S. Pedersen, and R. Hedman, *Biochem. J.*, **86**, 408 (1963).
40. L. Sokoloff and S. Kaufman, *Science*, **129**, 569 (1959).
41. C. H. Dutoit, in *Phosphorus Metabolism*, Vol. II, Eds., W. D. McElroy and B. Glass, Johns Hopkins Press, Baltimore, 1952, p. 597.
42. K. R. Crispell, W. Parson, G. Hollifield, and S. Brent, *J. Clin. Invest.*, **35**, 164 (1956).
43. E. Schwartz, *J. Lab. Clin. Med.*, **45**, 340 (1955).
44. M. A. Rothchild, A. Bauman, R. S. Yalow, and S. A. Berson, *J. Clin. Invest.* **36**, 422 (1957).
45. C. G. Lewallen, J. E. Rall, and M. Berman, *J. Clin. Invest.*, **38**, 88 (1959).
46. L. Sokoloff and S. Kaufman, *J. Biol. Chem.*, **236**, 795 (1960).
47. L. Sokoloff, S. Kaufman, and H. V. Gelboin, *Biochim. Biophys. Acta*, **52**, 410 (1961).
48. L. Sokoloff, S. Kaufman, P. L. Campbell, C. M. Francis, and H. V. Gelboin, *J. Biol. Chem.*, **238**, 1432 (1963).
49. W. P. Weiss and L. Sokoloff, *Science*, **140**, 1324 (1963).
50. R. Michels, J. Cason, and L. Sokoloff, *Science*, **140**, 1417 (1963).
51. L. Sokoloff, P. L. Campbell, C. M. Francis, and C. B. Klee, *Biochim. Biophys. Acta*, **76**, 329 (1963).
52. J. Roche, R. Michels, and T. Kamei, *Biochim. Biophys. Acta*, **61**, 647 (1962).
53. L. Sokoloff and C. B. Klee, in *Proc. Assn. for Res. in Nerv. and Ment. Dis.*, Vol. 43, Williams and Wilkins, Baltimore, 1963, p. 371.
54. C. B. Klee and L. Sokoloff, *J. Neurochem.*, **11**, 709 (1964).
55. C. C. Widnell and J. R. Tata, *Biochim. Biophys. Acta*, **72**, 506 (1963).
56. L. Sokoloff, C. M. Francis, and P. L. Campbell, *Proc. Nat. Acad. Sci.*, **52**, 728 (1964).
57. R. L. Krause and L. Sokoloff, *Biochim. Biophys. Acta*, **108**, 165 (1965).
58. Y. P. Lee, A. E. Takemori, and H. Lardy, *J. Biol. Chem.*, **234**, 3051 (1959).
59. K. Fletcher and N. B. Myant, *J. Physiol.*, **154**, 145 (1960).
60. J. R. Tata and C. C. Widnell, *Biochem. J.*, **98**, 604 (1966).
61. J. Monod, J. P. Changeux, and F. Jacob, *J. Mol. Biol.*, **6**, 306 (1963).
62. E. W. Sutherland and T. W. Rall, *Pharmacol. Rev.*, **12**, 265 (1960).
63. E. W. Sutherland and G. A. Robison, *Pharmacol. Rev.*, **18**, 145 (1966).
64. E. W. Sutherland, I. Øye, and R. W. Butcher, *Rec. Prog. Horm. Res.*, **21**, 623 (1965).
65. R. V. Farese and W. Reddy, *Biochim. Biophys. Acta*, **76**, 145 (1963).
66. R. V. Farese, *Endocrinol.*, **76**, 795 (1965).
67. E. D. Bransome and E. Chargaff, *Biochim. Biophys. Acta*, **91**, 180 (1964).
68. R. C. Imrie, T. R. Ramaiah, F. Antoni, and W. C. Hutchison, *J. Endocrinol.*, **32**, 303 (1965).
69. S. K. Sharma, R. M. Johnstone, and J. H. Quastel, *Can. J. Biochem. Physiol.*, **41**, 597 (1963).
70. O. Hechter and G. Lester, *Rec. Prog. Horm. Res.*, **16**, 139 (1960).
71. T. Z. Csáky, *Ann. Rev. Physiol.*, **27**, 415 (1965).

Chapter 8

PROTEIN SYNTHESIS IN TERMS
OF MEMBRANE BIOCHEMISTRY

If someone wanted to emphasize the spectacular advances and achievements of biochemistry during the past ten years he would most likely turn to the study of protein synthesis. This phase of biochemical investigation has grown to the point where it and the study of related macromolecules have become recognized under a new and impressive name—molecular biology. Young graduate students and recent Ph.D.'s, when confronted with the contemporary view of protein synthesis, could easily gain the impression that this phase of biochemical development is almost complete.

Since these achievements have been accomplished without much regard to a consideration of membranes and since it is not necessary to invoke membranes in presenting this contemporary view, one may properly wonder why it is worth thinking of protein synthesis in terms of cell membranes at all. The answer to this question can be presented in two ways. First I would like to point out the parts of the contemporary picture which are missing. Secondly, I would like to show where existing data in the literature point to the membranes of cells for the answers to the problems that still exist.

The general problem of protein synthesis is discussed at length in Chapter 1. It is important at this point to reemphasize that what we seek is an understanding of the manner by which cells turn out a spectrum of unique proteins with great efficiency in terms of the quantities of protein produced in a unit time and particularly in terms of the fidelity of the process. As we examine the present nature of our information about protein synthesis several aspects become evident. We know a great deal about how existing machinery in a cell can be used to make peptide bonds. We know also how to artificially influence these processes to cause certain desired amino acids to participate in peptide bond formation.

It has been known for a long time that when a cell is destroyed in order to obtain an acellular amino acid-incorporating system, about 99% of the quantitative ability to incorporate amino acids is lost (1), (i.e., the number of micromoles of amino acid incorporated per milligram of tissue is decreased by this amount). It is difficult to be certain of the exact value for this loss since in the organized tissue an isolated amino acid precursor pool may exist that has an indeterminate specific activity. Nonetheless, where sufficient data has been obtained to allow a calculation, the loss is seen to be very high (about 99%). It is possible that the loss may simply be due to the dilution of some crucial cofactor and that the residual process that is measured is truly representative of the 99% that is lost. On the other hand, the residual process may represent a minor cellular pathway which can withstand the severity of the isolation procedure and the acellular environment. There is a third possibility—that under the conditions of the acellular experiment new and abnormal processes may arise. There are amply documented cases of this kind in the literature (see pages 23–25).

For the most part, the surviving activity in acellular systems is not characteristic of the intact cell's ability to make its normal proteins (i.e., either all or most of the radioactive protein formed is different from the proteins which are characteristic of the tissue). There are some notable exceptions to this generalization. The best known acellular system capable of forming an authentic product is derived from reticulocytes and produces hemoglobin. Using sensitive immunological procedures it has also been shown that acellular liver systems can produce small amounts of normal proteins (2, 3). It is entirely possible that the authentic synthesis may represent a statistically small fraction of the preparation that has retained a sufficient degree of structural integrity to be representative of the original cellular condition. With bacterial ribosomes from either *E. coli* or *Euglena gracilis* it has been shown that when RNA from f-2 phage is supplied the protein product produced does appear to be authentic coat protein of f-2 phage (4, 5).

In Chapter 3 (pages 60–74) the rate of passage of amino acid through the cellular soluble RNA fraction was discussed at length. The conclusion reached was that in all of the reported cases it appeared that these rates were too low to handle the amount of amino acid which was becoming incorporated into protein. On the other hand, the evidence implicating the participation of amino acyl-tRNA in the cellular process of protein synthesis by this time seems incontestable. Perhaps the kinetics of passage of amino acid through the cell's soluble RNA pool does not measure the tRNA pool actually used by the cell. If protein synthesis occurs at a cell membrane surface there may exist a pool of tRNA at the membrane which possesses the proper kinetic ability to handle the required flux of amino acid entering protein.

The missing parts of the complete elucidation of the mechanism of protein synthesis may be found in the structural debris of tissue that is usually discarded in the course of preparation of acellular systems. At least this is a possibility worth exploring. There are many published papers which point to an important role for cell membranes in the process of protein synthesis. The further study of the intimate details involved in these processes may involve the considerations used to define a new aspect of biochemistry: membrane biochemistry (Chapter 6).

In the pages that follow I have tried to organize some of the pertinent reported experimental work into a pattern that illustrates the suspected vital role of membranes in the cellular process of protein synthesis.

MEMBRANES HAVE ACCESS TO ADEQUATE SUPPLIES OF AMINO ACIDS FOR PROTEIN SYNTHESIS

A rather curious observation has been made many times under different conditions for different tissue systems. External amino acids seem to have direct access to the sites of protein synthesis and they do not have to pass through the internal storage pools of amino acids. Halvorson and Cohen studied the accumulation of amino acids in yeast and the use of internal and external amino acids for protein synthesis (6). Two identical cultures of yeast were allowed to accumulate C^{14}-phenylalanine and C^{12}-phenylalanine, respectively, for 10 min. Then after centrifugation and washing, the suspensions were transferred quantitatively to media containing, respectively, the same concentrations of C^{12}-phenylalanine and C^{14}-phenylalanine. They followed the changes in the incorporation rate of radioactive phenylalanine and compared the experimental curves with the theoretically predictable curves, assuming that the pool is a necessary intermediary. In both cases it was found that exogenous amino acid was used for protein synthesis in preference to phenylalanine accumulated in the internal pool. It was also found that exogenous valine reduced exogenous phenylalanine uptake into the internal pool by more than 90% but the incorporation rate into protein was cut by only 50%.

In two brief communications Kipnis and Reiss described their studies on the effects of prior hypophysectomy and growth hormone administration on amino acid uptake into cellular pools and incorporation into protein of isolated rat diaphragms (7, 8). Hypophysectomy reduced incorporation into protein by 50%. Growth hormone restored normal levels of incorporation. Under both conditions, however, the pool sizes and specific activities of individual amino acids were unchanged. Furthermore, the radioactivity appeared in the protein as a linear function of time, whereas specific activity of the amino acid pool approached equilibrium exponentially (this point is further discussed in the following). The hormone, it seems, may have stimulated the direct incorporation of external amino acid into protein.

In a detailed kinetic study of the uptake of labeled amino acids into internal pools and proteins of intact rat diaphragm and isolated guinea pig and rabbit lymph node cells, Kipnis, Reiss, and Helmreich presented strong evidence in support of the idea that a direct route exists from external amino acid to the protein synthesizing sites of these tissues (9). If the external amino acid had to pass through the internal pool *en route* to incorporation into protein, then the curve showing the rate of incorporation of radioactivity into protein should show a continuous increase in slope reflecting the continuous rise in specific activity of the internal pool. After equilibration of the internal pool is achieved, the incorporation rate should become constant. If the external amino acid could be directly incorporated into protein, then a straight line curve of constant slope (extrapolating back through the origin) should represent the rate of uptake into protein. In all cases the data ruled out the necessity for external amino acid to equilibrate with the internal pool prior to incorporation into protein.

In a later study, Rosenberg, Berman, and Segal added the refinements of studying the rate of oxidation in addition to the rates of uptake into the pools and proteins, and the use of a computer to evaluate the data in terms of various possible models (10). They studied the uptake of glycine and the oxidation of lysine, and its uptake into kidney-cortex slices of rats. In both cases the incorporation into protein was linear and not consistent with the necessity for prior equilibration of external amino acid with the internal pool. The appearance of radioactive $C^{14}O_2$ from lysine, however, showed the increasing rate demanded for a process requiring the external amino acid to enter the internal free amino acid pool. The authors emphasized the compatibility of these data with a model for protein synthesis which placed the synthesizing site in the vicinity of the cellular membranes.

Pichler came to similar conclusions as a result of his studies with mouse Ehrlich ascites tumor cells (11). He found that oxidation of glutamic acid required the amino acid to first pass through the internal pool, whereas incorporation into protein was by a direct route. He considered these findings in terms of the involvement of the cell membranes in protein synthesis.

The earlier findings of Halvorson and Cohen in yeast and similar studies by Kempner and Cowie in *E. coli* (12) were originally discussed in terms of functionally distinct pools of amino acid existing inside the cell. These and many other published works, however, in addition to my own studies with lipoamino acids, led me to propose a model for protein synthesis which placed the protein synthesizing ribosomes at membrane sites (1).

In my early studies with the albumin-secreting cells of the hen oviduct, I was very surprised to find that radioactive amino acids acquired a lipid solubility at a much greater rate than their uptake into nucleic acids. Furthermore, when the specific activity of the medium was lowered, the amino acids lost their lipid solubility faster than they were able to break their association with nucleic acids. The specific lipid hydrolyzing enzyme α-lecithinase was a potent inhibitor of protein synthesis at extremely low concentrations (13). Surface active agents were also potent inhibitors of amino acid-incorporation, and lipotropic agents such as CoA and CTP were stimulatory. The lipid-soluble forms of the amino acid could be fractionated in organic solvent systems to produce chromatographic peaks characteristic of true lipid substances. Both functional groups of the amino acid seemed to be involved in linkages with nonpolar substances and strong hydrolysis could liberate free amino acids. Every amino acid tested behaved in a similar fashion. Similar observations for amino acids in a wide variety of tissues were reported from other laboratories. Whatever the true function of these substances proves to be, it is at least obvious that amino acids are readily available to the lipoidal environment provided by cell membranes (97).

CELL MEMBRANES PROVIDE ACTIVE SITES FOR PROTEIN SYNTHESIS

Gale and Folkes in 1955 reported on the characteristics of an active amino acid-incorporating system isolated from *Staphylococcus aureus.* Electron microscopic analysis of the active fraction revealed it to be composed mainly of cell membranes (14).

Hunter, Crathorn, and Butler in 1957 showed that when protoplasts of *B. megaterium* incorporated amino acids into protein and were subsequently lysed and fractionated, the membrane fraction appeared to con-

tain the primary sites of protein synthesis (15). This fraction reached
a plateau level of labeling after the first few minutes of incubation.
The cytoplasmic proteins continued to show an increased level of radioactivity and after about 5 min this level surpassed that of the membrane
fraction. A more thorough study by Butler et al. was reported the following year (16). They showed by labeling the protoplasts first and fractionating afterward or by carrying out the incorporation studies on the
isolated fractions that the membrane fraction was the most active both
in vivo and *in vitro*. Proteins labeled in the membrane fraction were
subsequently released as soluble proteins during continued incubation.
In 1961 Godson, Hunter, and Butler reported that 50% of the cell's
RNA could be isolated with the membrane fraction if lysis was accomplished in a concentrated buffer (17). Washing with a buffer of low
ionic strength removed ribosomal subunits from the membrane.

Wachsmann, Fukuhara, and Nisman reported on studies with *B.
megaterium* which once again established the isolated membrane fraction
as the most active (compared to free ribosomes and the soluble fraction)
in terms of amino acid-incorporation ability (18).

Beljanski and Ochoa described experiments with an acellular system
derived from *Alcaligenes faecalis* (19). The active fraction consisted
mostly of cell membrane fragments. The authors claimed that the amino
acids were being activated for incorporation into protein by a means
other than the conventional amino acid activating enzymes since there
was virtually no amino acid-dependent PP^{32} exchange[1] observable with
ATP.

Spiegelman described the results of detailed studies on protein and
nucleic acid synthesizing abilities of acellular fractions isolated from
penicillin produced spheroplasts of *E. coli* (20). When the spheroplasts
were pulse-labeled and quickly fractionated, the membrane fraction was
the most active both for amino acid and uridine incorporation ability.
Experiments were also carried out on the isolated fractions. Although
a $100,000 \times g$ supernatant fraction showed some incorporation ability,
the addition of a mixture of other amino acids did not augment this
incorporation. This description also applies to the ribosome fraction.
When the ribosomes were studied together with the high speed supernatant fraction, the resultant behavior was just slightly better than
would be expected from the sum of their individual activities. The membrane fragments were the only fraction found to possess significant
capacity to synthesize protein as measured by either amino acid-incorporation or induced enzyme synthesis. The amino acid-incorporating

[1] P^{32}-labeled pyrophosphate

ability of the membrane fraction exceeded by a factor of 100 that observed with the other fractions when examined under comparable conditions. The presence of other amino acids significantly augmented the incorporation of leucine which was almost inert in their absence. This fraction was also very active in the synthesis of RNA.

Schachtschabel and Zillig fractionated intact *E. coli* type B after cell destruction accomplished by shaking the cells with fine glass beads in a vibrating homogenizer. They found that both the membrane-containing fraction (cell debris) and the free ribosome fraction were capable of amino acid-incorporation when properly supplemented (21). The cell-debris fraction contained about 5% of the cellular RNA and about 10% of the total ribosomes. Based on a per-ribosome amino acid-incorporation ability, the debris system exceeded that of the ribosome system by several powers of ten. Rigorous precautions were taken to decrease the possibility that the observed activity was due to contaminating whole cells in the debris fraction.

That ribosomes occur intimately associated with cellular membranes has been amply demonstrated for a variety of cell types and seems to be a general cytological fact. Siekevitz and Palade showed that the microsomes of biochemical investigations were derived from the ribonucleoprotein (ribosome)-studded endoplasmic reticulum of liver and pancreas (22, 23). Sjöstrand and Elfvin have gone farther than most other cytologists in suggesting that the association of ribosomes with membranes is so intimate that the former forms a smooth carpet that extensively covers the latter (24). The appearance of attached ribosomes as discrete balls which spot the membrane surface, according to predominant contemporary view, is attributed to an artifact of fixation, staining, and dehydration. Sjöstrand and Elfvin used a technique of rapid freezing to prepare specimens for electron microscopy. They believe that their interpretation is more accurate for the living cell. This is an argument that will have to be evaluated by further work but it does tend to emphasize the strong association of ribosomes and membranes.

The association of ribosomes and ribosomal clusters with bacterial membranes has been reported several times. Thus, Fitz-James showed ribosomal clusters (polysomes ?) associated with the membranes of sucrose-lysozyme-produced protoplasts from *B. megaterium* (25). Pfister and Lundgren subjected cells of *B. cereus* to freezing and thawing to rupture cell walls and release most of the cytoplasm. The partially emptied cells revealed clusters of 10 to 55 ribosomes associated with cell membranes (26). Van Iterson proposed that ribosomes in bacteria are all connected to, or are a part of, a reticular fibrillar network that extends throughout the cytoplasm and is continuous from the nuclear region to the plasma membrane (27). Electron micrographs supporting this thesis were shown for preparations from *E. coli, B. subtilis* and *Proteus L* form cells. The fibrils were of the order of 20 to 35 A in diameter and were not digested by either DNAase of RNAase. Similar conclusions were independently reached by Schlessinger, Marchesi, and Kwan, who presented an electron microscopic study of *B. megaterium* (28). They found that the binding of ribosomes to membranes and/or reticulum was strongly influenced by Mg^{2+} concentration and at 0.02 M Mg^{2+}, 60% of the ribosomes were bound. Treatment with ribonuclease or deoxyribonuclease did not digest the reticulum connecting and holding the ribosomes. Treatment with 0.2% deoxycholate for 10 min did release about 60% of the bound RNA. The authors suggested that although the bacterial reticulum does not morphologically resemble the endoplasmic reticulum of mammalian cells, it might serve a similar functional purpose.

In view of the accumulated information in 1963 linking membrane sites to protein synthesis, I wondered whether two distinctly different types of protein synthesis were handled by free ribosomes and membrane-bound ribosomes, or whether active ribosomes were mainly located within cells at membrane sites—and techniques of cell disruption artificially translocated active ribosomes from a bound to a free form. Furthermore, if ribosomes function more efficiently in a bound form, why did actively protein synthesizing cells like *E. coli* seem to have only free ribosomes? The idea to be tested was whether gentle techniques of cell disruption left active ribosomes associated with membranes, whereas the more usual, more violent techniques of cell disruption would cause bound ribosomes to become free. In collaboration, J. Tani and I studied the pulse-labeling with radioactive amino acids of spheroplasts of *E. coli* type K-12 followed by cell disruption by mild and severe methods (29). Membrane and free ribosome fractions were examined for the formation of nascent (radioactive) protein. When the cells were disrupted by osmotic shock, the membrane fraction was consistently more radioactive than the free ribosome fraction and on a per-unit RNA basis bound ribosomes were about ten times more active than free ribosomes. Breaking the same cells in a tissue press yielded free ribosome fractions that were more radioactive than the membrane fractions. Grinding the cells with alumina also reduced the radioactivity isolated with the membrane fraction by as much as fivefold but the released activity was mostly hydrolyzed and, instead of appearing with the ribosomes, was found mainly in the supernatant fraction. When precautions were taken to limit hydrolytic digestion, the ribosomes obtained after alumina grinding were more radioactive than the membrane fraction.

Another series of studies was carried out with intact cells of *E. coli* type B (30). The cells were broken by violent shaking with glass beads. The isolated ribosome fraction was consistently more radioactive than the membrane fraction. Treatment of the membranes with deoxycholate, however, released ribosomes that were of manyfold higher specific activity than those of the free ribosome fraction; although both ribosome populations showed the same kinetics for uptake of radioactivity and loss of radioactivity following a chase with C^{12}-amino acids.

The biochemical studies with *E. coli* already outlined were correlated with electron microscopic and ultracentrifugal examinations of the intact cells, and derived membrane and ribosome fractions (31). It was shown that ribosomes appear to be attached to the internal surface of the cell membrane in K-12 spheroplasts and the isolated membranes from these cells. The ribosomes obtained from the membranes of *E. coli* type B by treatment with deoxycholate were shown to be morphologically similar to free ribosomes treated with the same concentration of deoxycholate. The sedimentation behavior of the two ribosome populations was similar although the ribosomes liberated from the membranes appeared to be more dissociated into subunits than the DOC-treated free ribosomes.

Reports describing the greater protein synthetic ability of membrane-bound ribosomes as compared to free ribosomes isolated from the same tissue have been appearing in the literature with increasing frequency during the past six years. This has been true for both microbial systems and those derived from higher organisms.

Hauge and Halvorson pulse-labeled intact yeast suspensions with $S^{35}O_4$ and then broke the cells with a French tissue press (32). The most active fraction was found to sediment between $7000 \times g$ and $60,000 \times g$. The lower speed centrifugation insured the absence of whole cells in the active fraction. The isolated preparation contained 52% protein, 27% lipid, and 21% RNA. An electron microscopic examination revealed mitochondria, mitochondrial fragments, and membrane vesicles with attached ribosomes. Expressed on a per-unit RNA basis, the lipid-rich fraction incorporated radioactivity 5 to 7.5 times faster than the free ribosomal particles.

Schlessinger stated that at least 25% of the ribosomes of *B. megaterium* occur in clusters of four or more, bound to membranes in crude cell lysates made by alumina grinding or lysozyme treatment of cells in 10^{-2} M Mg^{2+} (33). These bound ribosomes are usually discarded for studies of protein synthesis in acellular systems. The membrane-bound ribosomes or polysomes incorporated radioactive amino acids into protein about three times more efficiently than did unattached ribosomes and/or polysomes in acellular incubations. Part of this enhanced level of incorporation was due to the greater duration of active incorporation that was maintained by the bound particles. When polysomes and ribosomes were released from the membranes by lowering the Mg^{2+} concentration they lost their advantage in protein synthesis and functioned with the efficiency of the unattached particles. Schlessinger speculated that polyribosomes were protected by membranes from nuclease attack and that in the organized bacterial cell the bulk of protein synthesis occurs on membrane-bound polysomes.

Hallberg and Hauge studied the incorporation of $S^{35}O_4$ into the protein of cells of B. anitratum (34). Cells were broken in a French pressure cell, and preliminary centrifugation at 20,000 $\times g$ for two 10 min periods removed whole cells and large structural fragments. Free ribosomes were separated from membranes in the supernatant fraction by application of sucrose gradient centrifugation techniques. Per-unit of 260 mμ-absorbing material (nucleic acid) the membrane fraction accumulated 13 times as much radioactivity as the free ribosome fraction in a 3 min labeling period.

Moore and Umbreit prepared membrane and free ribosome fractions from S. faecalis (35). On the basis of their respective RNA contents, the membrane fraction was more active than the ribosomal fraction by a factor of 3 to 5. Both systems were stimulated by providing a source of energy and were markedly inhibited by ribonuclease. The labeled products produced in each system were similar in their electrophoretic behavior. Both the membrane and free ribosome fractions contained phospholipid. Intermediary fractions obtained by successive washing of the membrane fraction also contained phospholipids. All phospholipid extracts behaved qualitatively similarly when examined by thin-layer chromatography. The incorporation ability of each fraction could be linearly correlated with its phospholipid content. Phospholipid is present only in membranes of S. faecalis and is absent from cell walls and cytoplasm. Therefore, the data indicate that a membrane constituent is the limiting factor in protein synthesis.

It was observed that during isolation of the membranes, 30s ribosomal subunits were lost from the membrane fraction and became included in the free ribosomal fraction. The authors suggest that this may account for the apparent lesser extent of amino acid incorporation per unit of RNA in the free ribosome fraction since the 30s subunits by themselves would contribute to RNA content but not to amino acid-incorporating ability. By the same token, however, the residual excess of 50s subunits in the membrane would contribute to its RNA content without functioning in amino acid incorporation. The net effect of these two factors is not easy to evaluate. The sum total of the observations are consistent with the emerging pattern of functionally active membrane-ribosome structural units.

Haywood and Sinsheimer used a novel approach to ascertain the site of synthesis of viral protein in phage-infected *E. coli* speroplasts (36). The spheroplasts were treated with actinomycin D to suppress DNA-dependent host specific protein synthesis. An RNA-containing coliphage, MS_2, was introduced and the infected culture was labeled with an H^3-amino acid. An uninfected actinomycin-treated spheroplast preparation was then labeled with the corresponding C^{14}-amino acid. The two cultures were lysed, combined, and fractionated. The greatest excess of H^3-radioactivity was found in a 13,000 $\times$ g sediment. Deoxycholate and EDTA treatment released from this sediment a radioactive particle which sedimented with a velocity constant of 41s. This component possessed solubility properties which were described by Weissman, Borst, Burdon, Billeter, and Ochoa (37) to be characteristic for the viral RNA synthetase. If this component is actually the RNA synthetase, Haywood and Sinsheimer concluded that the cell membrane might represent the site of synthesis of the virus. Other possible interpretations, however, are not yet excluded. This situation is somewhat similar to one described by Penman, Becker, and Darnell, who studied the formation of poliovirus in infected HeLa cells treated with actinomycin (38). It was found that most of the protein and RNA synthesis in these cells occurred in a large, easily sedimentable cell fragment which could be disrupted by treatment with deoxycholate. This structural unit also contained the major amount of viral RNA-polymerase and seems to be the site of formation and assembly for the entire virus.

Protein synthesis in the organism *B. stearothermophilus* was studied by Bubela and Holdsworth (39). After short-term labeling periods with amino acids, protoplasts of the cells were lysed and fractionated. The 5000 $\times$ g pellet was the most radioactive fraction and its protein was of higher specific activity than that of the intact cells. A membrane preparation was isolated from the 5000 $\times$ g pellet and this fraction was of higher specific activity than the pellet itself. An isolated membrane fraction actively incorporated amino acids into protein and also into RNA and lipid-containing fractions. The order of labeling was first into lipid then into RNA and then into protein. The membranes contained amino acid activating enzymes and these were more stable to the high temperatures (60–65°) which were optimal for the operation of protein synthetic reactions in the system than were the activating enzymes of the soluble extract.

Nisman and his coworkers have studied an easily sedimentable fraction obtained from digitonin-induced lysates of *E. coli* and believe their results implicate an active membrane involvement in synthetic reactions (40, 41). This preparation, which Nisman refers to as a membrane preparation, is capable of synthetic operations which are unique among subcellular fractions. It can be induced to form specific enzymes such as β-galactosidase and alkaline phosphatase. It is specifically dependent on DNA from cells that can be induced to form these enzymes, it is stimulated by the enzyme polynucleotide phosphorylase, and it requires nucleoside triphosphates and an energy source. Since the system is so complex in its requirements and capabilities, it seems that the responsible structures might be intact or nearly intact whole spheroplasts that could be present. The fact that spheroplasts would acquire some characteristics that might distinguish them from untreated spheroplasts during the preparation of the lysate is in itself understandable and such observed differences as stimulation of the treated preparation by nucleic acids and a differential response to particular inducers cannot be used as arguments in favor of the interpretation that the treated preparation is much less complex than whole spheroplasts. Therefore the importance of these studies in terms of delineating the synthetic capabilities of isolated membrane preparations must await a better characterization of the nature of the active cellular constituents that are present.

In 1956 Palade and Siekevitz showed the existence and preponderance of membrane-bound ribosomes in rat liver by a correlated electron microscopic and biochemical study (42). Siekevitz and Palade separated membrane-bound from free ribosomes of guinea pig pancreas by differential centrifugation and presented evidence that *in vivo* the membrane-bound particles are more important in protein synthesis (43). Henshaw, Bojarski, and Hiatt studied the relative participation of free and bound ribosomes in protein synthesis in rat liver both *in vivo* and *in vitro* (44). By use of sucrose gradient ultracentrifugation techniques, the crude microsome fraction was resolved into a membrane-containing subfraction and a free ribosomal subfraction. Twenty minutes after an intraperitoneal injection of C^{14}-arginine into a rat, the liver was isolated and fractionated. Nearly all of the incorporated radioactivity was found in the membrane-bound ribosomal fraction. If a shorter time interval were used, this experiment would be more significant since during 20 min *in vivo* an appreciable quantity of fully synthesized protein could accumulate at sites other than where it was synthesized. For *in vitro* experiments, the subfractions were obtained from rats that had not been given radioactive amino acid. The membrane-bound ribosomes were very active for the incorporation of C^{14}-phenylalanine, whereas the free ribosome fraction was virtually inert. When poly-U was added, however, both fractions were raised to the same high level of incorporating ability. This stimulation of total incorporation represented 320% for the membrane-bound ribosomes and 2860% for the free ribosomes. For the unstimulated ribosomes the specific activity of the bound variety was five times that of the free ribosomes.

Howell, Loeb, and Tomkins also studied the function of free and membrane-bound ribosomes in rat liver (45). They confirmed the finding of Palade and Siekevitz that over two thirds of the ribosomes in this tissue were membrane-bound (42). Treatment of these membrane-rich fractions with deoxycholate released the bound ribosomes predominantly as dimers. After a 10-sec in vivo perfusion of C^{14}-valine into the portal vein of an anaesthetized rat, the liver was removed and fractionated. Free ribosomes and dimers had the lowest specific activities. Larger aggregates and membrane-bound dimers had the highest specific activities and the bulk of the incorporated activity was contained in the membrane-bound form. When ribosomes were released from the larger membrane fragments by DOC treatment, the ribosomes were relatively inert for amino acid-incorporation. Ribosomes which were isolated from the post mitochondrial supernatant fraction were not so sensitive to DOC treatment. When the membrane-bound ribosomes were tested without prior DOC treatment, the ability to incorporate amino acids (per unit of RNA) was the same for free and bound ribosomes. Since most of the ribosomes in liver are membrane-bound, this variety accounts for most of the in vivo protein synthesis.

Campbell, Cooper, and Hicks compared free and membrane-bound ribosomes from normal and regenerating rat liver (46). There was an enhanced level of incorporation for microsomes from regenerating liver but not for the free ribosomes. The stimulatory effects of poly-U for phenylalanine incorporation by rough-surfaced and smooth-surfaced endoplasmic reticulum and free ribonucleoprotein particles was studied as a function of Mg^{2+} concentration. The conclusion of these authors was that the major protein synthesizing activity of the liver cell was associated with the ribosomes attached to the endoplasmic reticulum. The free ribosomes seemed to be comparatively inactive but could be stimulated to synthesize protein by combination with an RNA fraction possessing messenger-like characteristics (i.e., polyuridylic acid).

The following year, Campbell, Serck-Hanssen, and Lowe were disturbed by a situation which to them posed a dilemma (47). Previous studies from their own laboratory and others showed that membrane-bound ribosomes of liver accounted for most of the protein synthesis of that organ, and free ribosomes were comparatively inactive. Many electron microscopists, however, suggested that membrane-bound ribosomes synthesize protein for export from the tissue, whereas free ribosomes make the proteins for intracellular purposes. Campbell *et al.* felt, therefore, that evidence should be obtained in support of the protein synthetic capabilities of free ribosomes. Since the evidence used by the microscopists was electron micrographs of fixed tissue, I see no dilemma. It is not possible to tell what a ribosome is doing from its picture in a micrograph. The electron microscopists simply proposed a possible interpretation of the cytological distribution of free and bound ribosomes. There was no conflict of any experimental data.

Very young rats have relatively more free ribosomes than adult rats but the total microsome fraction (free plus bound ribosomes) from each, incorporated similar amounts of radioactive phenylalanine without poly-U and also were similarly stimulated by the presence of poly-U. Since the young rats have many free ribosomes, Campbell et al. concluded that the free ribosomes must be active. This, of course, is only one of several possibilities. It could be that the fewer membrane-bound ribosomes of the young animal were more active than their counterparts in the adult animal. When light and heavy microsome fractions were incubated with radioactive leucine or phenylalanine and then fractionated by sucrose gradient centrifugation techniques, it was found that most of the radioactivity was located in a region intermediate between free ribosomes and the heavy-bound fraction. Although the authors felt that this fraction probably contained some associated endoplasmic reticulum, they believe that it was essentially a free polysome fraction and that the finding of amino acid-incorporating ability by such free polysomes can resolve the paradox between earlier biochemical studies and the feelings of the electron microscopists about the segregated synthetic functions of free and bound polysomes or ribosomes.

The protein synthetic capabilities of free and membrane-bound ribosomes of rat liver were also studied by Hallinan and Munro (48). Time course studies on the distribution of radioactive amino acid in rough-surfaced vesicles, smooth-surfaced vesicles, and free ribosomes isolated from the liver after intraperitoneal injection of the amino acid showed that the first fraction to be labeled was the rough-surfaced fraction. This fraction also attained the highest levels of incorporated radioactivity. *In vitro* experiments were also carried out on the isolated fractions. It was found that the energy-dependent uptake of leucine by the rough-surfaced vesicles was considerably greater than by the free-ribosomes or smooth-surfaced vesicles. On the other hand, when radioactive RNA precursors were injected into the rat it was found that free ribosomes became labeled prior to the membrane-attached variety. This situation could mean that newly formed free ribosomes become attached to the membrane in order to begin their protein synthetic function.

Tobacco cells grown in a suspension culture were the source of tissue used by O'Neal Nicolson and Flamm to study protein synthesis by free and membrane-bound ribosomes (49). After incubations of the intact cells with C^{14}-labeled algal protein hydrolysate, it was found that the bound ribosomes incorporated three times as much radioactivity as the free ribosomes in the first 20 sec of incubation. At 15 min, the ratio of activities was down to 2:1. The bound ribosomes also incorporated three times as much uridine into their 25s RNA as did the free ribosomes. During aging, as the protein synthetic ability decreased, there was a loss of total ribosomes accompanied by an increase in the percentage and actual number of free ribosomes.

THE PRESENCE OF RAPIDLY LABELED RNA, ACTIVATING ENZYMES AND tRNA (OR sRNA) IN MEMBRANES

In addition to an adequate supply of amino acids for protein synthesis and active ribosomes located at cell membranes, it is important to know if other parts, assumed to be vital to the protein synthetic apparatus, are also present.

Rapidly labeled RNA is believed to function partly as a precursor to ribosomes and partly as the carrier of genetic information from DNA to the cytoplasmic ribosomes. Most interpretations of the latter type can trace their origin back to experiments of Volkin and Astrachan who used P^{32} to label the rapidly formed RNA in T_{2r}-phage-infected *E. coli* (50). They found that the base ratio of this RNA seemed to reflect the base ratio for DNA. Although the bulk of this RNA was not membrane-bound, the fraction with the highest specific activity (less than 5% of the total) was found in the membrane-containing fraction. Very similar observations were reported by Nomura, Hall, and Spiegelman in a more complete study of the same system (51). The presence of small amounts of rapidly labeled RNA in bacterial membranes from uninfected cultures has also been reported by Spiegelman (20). Suit found that after a 5 min pulse of P^{32}, membranes isolated from spheroplasts of *E. coli* possessed RNA more than three times as radioactive as the supernatant RNA (52). The supernatant RNA in these experiments included the ribosomes. Diaminopimelic acid concentration was used to control cell wall synthesis in a mutant. Since the membrane-RNA was still rapidly formed when cell wall synthesis was inhibited and its rate of turnover did not increase when cell wall synthesis was resumed, it was concluded that the membrane-RNA was not particularly involved with cell wall synthesis. *S. faecalis* membranes were isolated from cells that had been briefly labeled with P^{32} by Abrams, Nielsen, and Thaemert. Nucleotides obtained after hydrolysis of the RNA isolated from these thoroughly washed membranes were four to six times more radioactive than corresponding nucleotides isolated after hydrolysis from the ribosomal RNA (53). Since the base composition of this newly synthesized and membrane-bound RNA resembled that of ribosomal RNA and since accompanying electron micrographs showed ribosomal material in association with cell membranes, Abrams *et al.* believed that these observations tended to relate to newly synthesized ribosomes at the membrane surface. A similar conclusion for mammalian tissues was reached by Tata on the basis of his studies with thyroxine (see page 291). Suit, in a later communication, studied the rapidly labeled membrane-RNA in normal *E. coli* that had been "stepped-down" and also in T_2-infected *E. coli* (54). When rapidly growing *E. coli* is transferred to a minimal medium from a rich medium, ribosome synthesis is decreased and the newly formed RNA is DNA-like (step-down conditions). In both cases the P^{32}-pulse-labeled membrane-RNA was five to fifteen times higher in specific activity than was the ribosomal RNA. Furthermore, this newly synthesized RNA of the membrane fraction was DNA-like in its composition.

Yudkin and Davis found that membrane-associated RNA of *B. megaterium* was ribosome-like in its base composition but the rapidly labeled RNA of the membranes was DNA-like in base composition and was more active in uptake of P^{32} than was the ribosomal RNA (55). The interesting observation was made that while actinomycin D treatment (10 μg/ml) inhibited the labeling of cytoplasmic RNA by 70%, the labeling of membrane RNA was actually increased by 30%. The authors interpreted these findings as indicating that ribosomes physiologically attached to the protoplast membrane have a high affinity for messenger RNA, which suggests that such ribosomes are of unusual importance in the synthesis of protein. Furthermore, the association of mRNA with membrane-bound ribosomes seems to protect it from hydrolysis.

Evidence of the association of rapidly labeled RNA with biological membranes in intact cells has also been obtained by Brown and Littna in developing embryos of *Xenopus laevis,* the South African Clawed Toad (56). Although electron micrographs of embryos demonstrate the ribosomes to be free in the cytoplasm, the addition of deoxycholate (DOC) not only releases ribosome particles from that portion of the embryo homogenates that forms pellets in low centrifugal fields, but these particles released by DOC contain heterogeneous rapidly labeled RNA that is absent from ribosomes which are isolated in the absence of DOC. Rapidly labeled RNA was specifically associated with that fraction of ribosomes which was released by DOC.

Petrović, Bećarević, and Petrović claim that the usual extraction procedures for RNA do not remove all of the RNA from the microsomes of rat liver and that, in fact, the species that is incompletely removed is characterized by a rapid turnover (57). The technique to insure a more complete extraction involved the use of a high concentration of sodium dodecylsulphate and low ionic strength on membranes whose structure was destroyed by treatment with phenol. These observations are almost identical to those of Kuff and Hymer who reported that conventional extraction procedures applied to microsomes of rat liver did not extract a small percentage of RNA that was quite rapidly labeled (58). By a two step procedure using prior treatment with lauryl trimethyl ammonium chloride and $MgSO_4$ followed by extraction with sodium dodecylsulfate and EDTA, a more complete recovery was accomplished.

The suspicion of the existence of newly formed mRNA in the membranes of cells suggests the possibility that genetic information in the form of DNA must also be present. In 1963, Jacob, Brenner, and Cuzin proposed a model for chromosomal duplication in bacteria which implicated a membrane DNA point of replication (59). There have been various isolated reports in the literature pointing to the existence of DNA in association with easily sedimentable particulate fractions from cell homogenates as well as electron micrographs of DNA associated with the cell membrane. Ganesan and Lederbug studied the formation of nascent DNA in growing *B. subtilis* (60). They found that the nascent DNA was predominantly associated with the membrane-rich fraction. The membrane-rich fraction contained DNA-polymerase activity which accounted for approximately 25% of the cell content.

It is essential for protein synthesizing sites to have a continuous and adequate supply of energy. Current views on this problem have been formulated from studies with isolated ribosomal systems. Thus ATP can activate amino acids under catalysis of activating enzymes to form enzyme-bound amino acyl-AMP and free pyrophosphate. The active amino acid can next be transferred to tRNA or sRNA to form an active amino acyl-RNA which is the energized form of amino acid that can be complexed with the mRNA-ribosomal unit in order to undergo peptide bond formation (see page 85). It is important to know if such components of the protein synthetic apparatus are contained in membranes. Other possibilities for funneling energy into protein synthetic reactions will also be considered later in this chapter (see pages 319–323).

McCorquodale and Zillig prepared from *E. coli* type B a membrane-rich fraction, a ribosomal fraction, a ribosome-free and DNA-free supernatant fraction and nucleotide-free protein fraction produced by electrophoresis of the supernatant fraction (61). The ability of added amino acids to catalyze P^{32}-pyrophosphate-ATP exchange was studied and taken as a measure of amino acid activating enzymes. Virtually no such ability was found in the ribosomal fraction. The original supernatant fraction possessed activity for only cysteine, serine, histidine, tryptophan, tyrosine, leucine, isoleucine, threonine, valine, and methionine. The electrophoresed supernatant fraction lost its ability to respond with serine and methionine. The membrane-rich fraction, however, gave positive responses for all 21 amino acids tested.

Hunter, Brookes, Crathorn, and Butler reported that the membranes isolated from *B. megaterium* possessed enzymes that catalyzed a P^{32}-pyrophosphate-ATP exchange when amino acids were added and, furthermore, that radioactive amino acid-containing RNA could be extracted from membranes that were incubated with amino acids (62). Although the amino acid could be removed by mild alkali and RNAase treatment and seemed to be transferable to protein in a subsequent incubation, its specific radioactivity was lower than that of the protein. Therefore the role of such an intermediate in the membrane preparation was not established.

Wachsmann, Fukuhara, and Nisman were not able to find evidence for amino acid catalyzed PP-ATP exchange in a membrane-containing fraction that was very active for amino acid-incorporation (63). They did find, however, that an RNAase-stable bound form of amino acid was released by treatment with 1 M NH_2OH at pH 7.0. These conditions have been frequently used to split high energy metabolic intermediates. Although no further clarification of this finding has appeared, the authors speculated on the possible occurrence of activated amino acids other than amino acyl-sRNA in the membrane.

A different approach to the problem of finding the localization of amino acid activating enzymes in *E. coli* was taken by Roberts, Bensch, and Carter, who concluded that these enzymes may be situated in the living bacteria at their cell membranes (64). These workers found that intact cells of *E. coli*, after freezing or treatment with EDTA or polymyxin B, could catalyze an extracellular amino acid-dependent P^{32}-pyrophosphate-ATP exchange. The exchange was catalyzed by the cells since the supernatant fluid separated from the cells was inactive and the amount of exchange was proportional to the amount of cells. When either C^{14}-ATP or P^{32}-(PP) was used it was found that essentially all of the radioactivity remained outside of the cells. Electron microscopic examination of EDTA-treated active cells showed no evidence of lysis or protoplast formation and, in fact, showed no difference from untreated control cells. The authors believed that these observations were consistent with the localization of amino acid activating enzymes at the cell membrane.

In the studies of Bubela and Holdsworth cited earlier, amino acid activating enzymes were determined by measuring amino acid hydroxamate-forming ability (39). Activating enzymes were found both in the soluble and membrane fractions. The membrane-associated enzymes were much more stable to heat denaturation than the free enzymes. The membrane preparation, which as already discussed was the most active cellular fraction for protein synthesis, also catalyzed the formation of lipoamino acids and membrane-bound amino acyl-tRNA. Formation of lipoamino acid preceded that of amino acyl-tRNA which in turn preceded that of incorporation into the protein.

Whole chick embryos were the subjects of experimentation of Norton, Key, and Scholes (65). These investigators examined isolated microsomes and particle-free supernatant fractions for their relative abilities to catalyze amino acid-dependent pyrophosphate-ATP exchange. Their microsome fraction also contained free ribosomes. It was found that for 12 of the 13 amino acids tested, a substantial portion of the enzymatic activity was associated with the microsomal fraction; tryptophan was the exception. The microsome-associated activities could be removed or inactivated by extensive washing or treatment with deoxycholate. The extent of association or absence of such enzymes in the membrane-containing fraction could well reflect the destructive nature of the procedures used to disrupt the cells. In this study a 1-min period of sonication at 0° increased the proportion of enzyme activity in the soluble fraction. The authors concluded with this statement:

"It is reasonable to believe that in the intact cell of the chick embryo, the amino acid activating enzymes having a role in protein synthesis (amino acyl-RNA synthetases) are bound to the microsomes, possibly to the membranes of the endoplasmic reticulum."

THE PRESENCE OF NASCENT PROTEIN IN MEMBRANES OF CELLS

For many years, Peters at the Mary Imogene Bassett Hospital in Cooperstown, New York, has been studying the formation and secretion of serum albumin in the liver. By the combined use of labeled amino acids with biochemical and immunological procedures, he has traced the passage of amino acid from the blood through various fractions of the liver and finally to circulating serum albumin (66–68). The appearance of radioactive albumin in the soluble fraction of the liver cell after incubation with labeled amino acid shows a lag of about 15 min. No such lag is seen in the labeling of albumin held in association with the microsomal fraction. The associated early labeled albumin can be released from the microsomes by treatment with deoxycholate.

In *in vivo* experiments with rats, when time points were taken only a few minutes after injection of radioactive amino acid, and the isolated liver microsome fraction was separated into rough-surfaced and smooth-surfaced elements, it was found that the first appearance of newly made albumin was in the rough-surfaced elements of the endoplasmic reticulum (*er*). In the next few minutes after injection, however, the level of radioactive albumin in the rough *er* decreased while the level in the smooth *er* rose to a peak.

A similar situation has been described by Ogata, Omori, and Hirokawa, who studied the synthesis of antibodies by spleen preparations *in vitro* (69). The spleen microsomes contained newly made antibody protein which could be released by deoxycholate and which seems to represent the precursor to extracellular antibody.

Another early indication of the association of newly made protein with membranous organelles of the cell was reported by Douglas and Munro, who studied amylase formation in the pancreas of live pigeons (70). At all times after injection of C^{14}-glycine the microsomes represented the most radioactive fraction. The major fraction of this radioactive protein was released by treatment with deoxycholate. The microsome fraction normally contains a bound and inactive form of amylase which can be activated by mechanically shaking the structures with small glass beads (ballotini). Although this treatment destroys the structural integrity of the closed vesicle system, it does not release the bound amylase. Therefore an association of the enzyme with the membranous structure is indicated. Since amylase is a secretory product, the implication of these studies is that the enzyme is synthesized at a microsomal (membrane) location, although more definitive experiments to firmly prove this point were not performed in this study.

In 1960 Siekevitz and Palade clearly showed that the synthesis of pancreatic α-chymotrypsinogen in the living guinea pig occurred at the rough-surfaced endoplasmic reticulum (43). An extension of this investigation involving the apparent synthesis of amylase by pigeon pancreatic microsomes in an *in vitro* system was reported by Redman, Siekevitz, and Palade (71). By a procedure which partially subfractionated the microsomes into ribosomes, solubilized membranes, and vesicle contents (72), the ribosomes were shown to be the first component to possess radioactive amylase. The procedure which differentiates membrane components from cisternal contents did not yield clear-cut results in this system but the indication was that newly synthesized amylase leaves the attached ribosome and is immediately transferred into the cisternal space.

The concept of synthesis of secretory protein on membrane-attached ribosomes, with the growing peptide chain oriented toward the membrane surface, and the subsequent unidirectional discharge of finished protein across the endoplasmic reticulum into the cisternal space receives strong support from experiments of Redman and Sabatini (73). During amino acid-incorporation by a liver microsomal system, most of the incorporated activity (about 80%) remains associated with the ribosomal particles rather than the DOC solubilized components. It is well known that puromycin causes premature release of nascent peptides from their attachment to ribosomes (see pages 92–96). In free ribosomal systems the abortively released chains can be recovered from the aqueous medium in the form of peptidyl-puromycin molecules. In the microsomal system it was found that puromycin treatment also caused the membrane-attached ribosomes to release their growing peptide chains; but instead of releasing the chains into the aqueous medium they were released into the membrane and cisternal content portions of the microsomes. By use of the Ernster, Siekevitz, and Palade procedure (72) which yields a fraction soluble in 0.25% DOC believed to represent the cisternal contents and an insoluble membrane-enriched fraction, the authors concluded that the abortively released peptide chains were contained within the microsomal vesicles rather than in association with the membranes. Not only do these studies establish the formation of nascent protein by membrane-attached ribosomes and the direct secretion of protein into the intracisternal space, but they show that at all times the growing peptide chain is oriented on the side of the ribosome apposing the membrane or at a ribosomal membrane interface. This situation was posulated in the model I proposed for protein synthesis, which was based on other considerations (1).

The distribution of newly synthesized amylase in microsomal subfractions isolated from guinea pig pancreas after *in vivo* injections of C^{14}-leucine was studied by Siekevitz and Palade (74). By use of a very low concentration of DOC (0.04%), the microsomes could be separated by centrifugation in a sucrose gradient into a pellet consisting of membrane vesicles with attached ribosomal granules (about 50% of the total), a heavy fraction consisting of closed vesicles with few attached ribosomes (less than 1% of the total) and a free ribosome peak containing about 50% of the total ribosomes. The distribution of amylase activity was 40%, 44%, and 16%, respectively, in these three fractions. After a 5 to 15 min labeling period, however, the total amounts of radioactive amylase in these three fractions were 76%, 20%, and 4%. The relative specific activities, therefore, were 7.6, 1.8, and 1.0. A higher DOC concentration (0.2%) produced a pellet which contained only one tenth of the total RNA and one seventh of the total amylase activity, but which still retained most of its radioactive amylase. It thus appears that the ribosomes which are most difficult to detach from the membranes are the ones that have associated nascent enzyme molecules. The results also indicate that the newly synthesized enzyme molecule may be held tightly by the membrane itself.

A similar situation in rat liver was described by Sabatini, Tashiro, and Palade (75). After *in vivo* labeling of ribosomal protein, they solubilized by EDTA treatment about 70% of the ribosomal material of a microsomal preparation, but only 20 to 30% of the radioactive ribosomal protein, and concluded that the small proportion of ribosomes which remained attached to microsomal membranes contained the greatest proportion of newly synthesized protein.

Since the presence of an extensive intracellular membrane system with attached ribosomes (rough-surfaced endoplasmic reticulum) is characteristic of cells active in the secretion of proteins, a prevailing view among electron microscopists is that the importance of the attachment of ribosomes to a membrane lies in the secretory rather than synthetic process. The studies described in this section which deal with the location of newly synthesized proteins on membrane-attached particles have been carried out with such kinds of secretory cells. It should not be overlooked, however, that another common feature of all of these cells is the capacity for extensive protein synthesis. The fact that membrane-attached ribosomes have been found to be active in synthesizing secretory proteins is compatible with different situations. It could mean that attached ribosomes make exportable proteins and free ribosomes make endogenous proteins. It could also mean that membrane-attached ribosomes are more active for making all proteins of the cell. In the case of microbial systems where secretory proteins are not the principal product of protein synthesis it has been shown in many instances that membrane-bound ribosomes are the most active for general protein synthesis. The dichotomy of function for the two ribosome populations suggested by the electron microscopists may indeed turn out to be true, but at the present time there is scant biochemical evidence to support it. The fact that many rapidly growing tissues (such as from young animals, tumors, and bacteria) have high proportions of free ribosomes and small amounts of membrane-bound ribosomes, does not necessarily mean that free ribosomes function alone since a small proportion of bound ribosomes in *in vivo* may be extremely active and since small fragments of attached membranous material may function adequately for ribosomes not actually attached to continuous membranes in these systems. The question of the relative importance of the two ribosomal populations is certainly unresolved at this point. The ability of free polysomes to incorporate radioactive amino acids in the intact cell is indicated, at least for the plasma-cell tumor of mice, by studies of Kuff, Hymer, Shelton, and Roberts (76).

An additional word of caution should be sounded in these considerations. In nearly all cell fractionations there is a blending of different cellular components for any given fraction. Thus very large polysomes may be found in fractions containing either smooth or rough-surfaced membranous elements. Therefore in the absence of other criteria such as enzymic assays, chemical characterization, or electron microscopic examinations, a given centrifugal fraction should not be accepted as representing either polysomes or membrane-bound ribosomes alone.

OTHER RELEVANT CONSIDERATIONS BEARING ON THE INVOLVEMENT OF MEMBRANES IN PROTEIN SYNTHESIS

There are various other isolated studies which serve to indicate a relationship between cell membranes and protein synthesis and even this chapter which is designed to emphasize this relation, does not present a full catalog of such relevant observations.

An example of the kind of information available is taken from the observation of Dubin, Hancock, and Davis that an early consequence of streptomycin treatment with growing *E. coli* involves an impairment of both protein synthetic ability and membrane integrity (77). A site of inhibition of protein synthesis has been shown to be the 30s subribosomal component (78, 79). Either two separate targets are involved or one sensitive site can affect both ribosomal and membrane function.

Nomura, in a recent investigation of the mechanism of killing action of antibiotic colicines, has also presented evidence in favor of a functional coordination of macromolecular synthesis and cell membranes (80). Thus several of the colicines inhibit, either together or separately, the processes of RNA and protein synthesis, oxidative phosphorylation, and maintenance of DNA. The addition of trypsin reverses the action of the colicines. Since trypsin does not penetrate the bacteria, the colicines presumably act at the cell surface. Nomura discussed the possibility that reactions involved in macromolecular synthesis are integrated in the cell membrane.

In chapter 7 the mechanism of action of several hormones was considered. Although early investigations with hormones were prefaced with the idea that a single target of action in the cell for a given hormone would present the simplest biological explanation for a mechanism of action, some apparent difficulties with this concept seem to have arisen. Thus early consequences of several hormones seemed to simultaneously affect the permeation of various metabolites, oxidation of substrates, formation of ATP, and metabolism of fatty substances. Present biochemical concepts would accommodate all of these effects as consequences of an alteration in membrane structure, since these reactions are known to be membrane-associated. On finding, however, that protein synthesis is also affected shortly after hormone administration, many investigators reluctantly abandoned the simpler theory in favor of a more complicated explanation involving both membrane and intracytoplasmic targets. The alternative possibility that effects on protein synthesis follow from initial modifications of membrane configuration have, unfortunately, not been seriously enough considered.

The concept of protein synthesis in terms of membrane biochemistry offers other possibilities. Classical biochemistry solves the problems of energy supply by the intermediacy of ATP, which is produced as a result of a membrane-bound and oriented enzyme system. If protein synthesis itself is a membrane-oriented system, could there not be a more direct relationship between membrane-bound energy producing systems and those of protein synthesis?

The concept that high energy intermediates which are normally generated in the cell by respiration may be directly used for work instead of for ATP formation was proposed by Slater (81). The first experimental demonstration of this possibility was reported by Chance and Hollunger approximately seven years later (82, 83). They found that succinate added to mitochondria in the absence of ADP could cause the rapid and extensive reduction of mitochondrial pyridine nucleotide (an energy requiring reaction). Many other groups have since confirmed and extended these findings. Succinate can similarly cause the reduction of acetoacetate (presumably via NAD) in the absence of phosphate when a portion of the succinate could be oxidized (84, 85). The energy for reversing electron flow, which normally would be from NADH to cytochrome b could be supplied by the oxidation of succinate or by adding ATP. Oligomycin, an agent which blocks energy conversion between the oxidative phosphorylation chain and ATP, will not inhibit the reversal of electron flow when energy is obtained by oxidation of succinate, but will inhibit when energy is supplied by ATP. This means that high energy intermediates normally used to form ATP and which can be regenerated from ATP, may be directly used for work. In the case discussed, the work is reversal of electron transport. In a similar manner it has been shown that high energy intermediates can be directly used for ion accumulation by mitochondria (86, 87). In 1963, in the preface to the recorded proceedings for the first colloquium of the Johnson Research Foundation on the subject of energy-linked functions of mitochondria (88), Chance wrote:

"The key point that has emerged with greater force and clarity from these discussions than was possible in the independent . . . and excellent . . . contributions of the participants prior to this meeting is a change in the concept of the role of ATP as the sole energetic reaction product of the mitochondrial reactions. While a good deal of evidence for high energy precursors of mitochondria had been accumulated, much of it by the participants of this colloquium, the closing ring of evidence for the function of these high energy intermediates not only in the energy-linked reduction of DPN and TPN, but also in the transport of ions across membranes, now leads to a new and important concept in mitochondrial energy relations, and focuses our attention on a new class of nonphosphorylated high energy compounds."

Protein synthesis requires enormous amounts of energy and as early as 1961 there were strong indications that some protein synthesis may occur at membrane surfaces. At that time I suggested the possibility that the cell may have a way of directly linking the membrane-bound energy producing reactions to some of the membrane-bound protein synthetic reactions, without using the ATP pool dissolved in the cytoplasm (89). This possibility has also been seriously considered by several other workers.

Bronk, using rat liver mitochondria with succinate as energy source, showed that leucine incorporation into protein was virtually eliminated by antimycin A (blocks reduction of cytochrome c by cytochrome b) and 2,4-dinitrophenol (causes abortive hydrolysis of a nonphosphorylated high energy intermediate) but it was not inhibited by oligomycin at concentrations which block ATP formation (90). The addition of ADP or AMP also inhibited incorporation, and it was felt that they may be competing with protein synthesis for the free energy produced by succinate oxidation. Furthermore, tetraiodothyroacetic acid, a potent inhibitor of phosphorylation, virtually doubled the rate of leucine incorporation. Although Kroon made similar observations with beef heart and rat liver mitochondria, regarding the insensitivity of amino acid-incorporation to oligomycin, he was more cautious in his interpretation (91). Kroon pointed out that the magnitude of incorporation of amino acids by mitochondrial protein in the systems studied was quite low.

If ATP levels were not limiting, the small level of oligomycin-insensitive ATP production might be sufficient to drive the amino acid-incorporation; but even if this were true, we would expect some effect on amino acid-incorporation after cutting off most of ATP production. He also cautioned that secondary effects such as high energy-induced uptake of certain ions may be the cause of enhanced protein synthesis. Campbell, Mahler, Moore, and Tewari reexamined the question in a system with a much higher level of protein synthesis, namely, rat brain mitochondria, and found the same indication of a possible coupling for the systems of electron transport and protein synthesis (92); that is, the addition of ADP inhibited protein synthesis, oligomycin either stimulated or had no effect, and 2,4-dinitrophenol, azide, cyanide, and antimycin A inhibited protein synthesis.

For protein synthesis in mitochondria, it is conceptually easy to visualize a direct linkage between the systems of electron-transport and those of protein synthesis. But if energy is made available in the mitochondria, how can it be directly shuttled to the protein synthetic sites of the endoplasmic reticulum? There is no evidence that such a shift is accomplished by any means other than ATP. There are, of course, several other possibilities. In the membranes of the endoplasmic reticulum are contained fragments of an electron-transport system. There is both a TPNH-cytochrome c-reductase and a DPNH-cytochrome c-reductase. Cytochrome b_5 and coenzyme Q are also present in microsomes. The TPNH-cytochrome-reductase seems to be involved in drug metabolism. There is no convincing explanation of the role of the DPNH-cytochrome-reductase. If it is functional, what happens to the energy produced in the oxidation? What functions as the final electron sink? There is no evidence of cytochrome oxidase or cytochrome c being present. If there is a direct, albeit transient, association between mitochondria and endoplasmic reticulum, can the mitochondria complete the electron-transport sequence by oxidizing microsomal cytochrome, or in some other way communicate with the endoplasmic reticulum? There is absolutely no solid evidence to support any of these possibilities but they are intriguing.

In microbial systems membrane-bound reactions of oxidative phosphorylation are integrated in the cell membrane and invaginations of these membranes that frequently occur in gram-positive organisms (mesosomes). There is an impressive accumulation of evidence that protein synthesis also occurs with great efficiency at bacterial membrane locations (pages 300–306). Therefore microbial systems would present the logical place in which to seek a structural and functional relationship between these two reaction sequences.

In 1947 Spiegelman (93), and Reiner, and Spiegelman (94) reported on a curious phenomenon for which they could offer no explanation. The formation of adaptive enzymes in anaerobic yeast was inhibited by the two classic uncouplers of oxidative phosphorylation—2,4-dinitrophenol (DNP) and sodium azide. Energy for protein synthesis was provided by glucose fermentation and it was shown that the poisons did not inhibit these fermentative reactions. These same findings were reported by Kovac and Istenesova 17 years later, but this time the authors suggested that the results indicated that the same high energy intermediates that are involved in ATP formation might be involved in protein synthetic reactions (95). Although I was fully prepared to accept such a possibility, these experiments could easily have been explained by a number of other possibilities. DNP is a notorious stimulator of ATPase activity and, although azide is principally an ATPase inhibitor, it can, under certain conditions, induce a transient ATPase activity. A reduction of ATP levels by these poisons could equally well account for the results. Also, if the poisons interfered with the formation of any amino acid from glucose or any nucleoside triphosphate, or the formation of "messengers" or sRNA, or the reactions involving these or any of the other factors in the conventional scheme of protein synthesis, the inhibition of induced-enzyme formation would follow.

Dr. Leonard Jarett, a postdoctorate research associate in my laboratory from 1964 to 1966, and I set out to re-examine this system and to run all of the pertinent controls. We found that dinitrophenol and azide, at concentrations where they normally uncouple oxidative phosphorylation in yeast, would inhibit anaerobic growth, enzyme induction, and the incorporation of radioactive amino acids under growing and nongrowing conditions. During the period of inhibition, ATP pools were maintained or elevated. The poisons did not reduce cellular levels of any amino acid or the nucleoside triphosphates. All of the reactions of *in vitro* protein synthesis as well as the stimulation of protein synthesis by exogenous polynucleotide messages were insensitive to the poisons in concentrations much higher than were effective in inhibiting cellular protein synthesis. Polysomes did not appear to be any more sensitive to these poisons than were free ribosomes and so a targeted sensitivity of mRNA seems to be eliminated. It was similarly found that although DNA-dependent RNA-polymerase was insensitive to the poisons, RNA labeling in the intact cell was inhibited. This inhibition did not precede that of protein synthesis, both effects being simultaneous. Therefore the possibility that some RNA synthesis may also draw directly on a pool of high energy intermediates must also be considered (96).

The aforementioned results do not prove the suspected direct relationship between reactions of energy metabolism and protein synthesis. In spite of the lack of inhibitory effects of 2,4-dinitrophenol and azide on all known reactions of protein synthesis, there is always the possibility of a key unknown reaction or set of conditions that was not tested and which could explain the observations by more conventional considerations. The above studies merely serve to indicate that it is still worth pursuing the possibility of a direct biochemical and structural linkage for these complex membrane-associated and interdependent metabolic networks.

A more convincing proof would be the isolation of a structural fragment which (a) possessed the necessary parts of the oxidative phosphorylation sequence and (b) which could catalyze the ATP-independent incorporation of amino acid in the presence of oxygen and an oxidizable substrate and (c) which under anaerobic conditions depended on ATP and was sensitive to oligomycin. In our laboratory, we are endeavoring to isolate such a fragment from *E. coli* spheroplasts.

In Chapter 3 it was emphasized that although many facts point to the ability of sRNA to participate in the synthesis of protein, when this role of sRNA in intact cells was investigated it invariably proved kinetically inadequate to its task. One possible resolution to this quandary was suggested in a model (Figure 1) I proposed several years ago (1). This model also explained the loss of quantitative and qualitative ability of a cell to synthesize protein after homogenization, the apparent accessibility of sites of protein synthesis in the intact cell to external amino acids, the observations of acellular incorporating systems that were not dependent on free sRNA, and the observed high incorporating ability of membrane attached ribosomes. The model proposed that in the living cell active ribosomes occur in an extended form in intimate association with a cell membrane. Nascent protein is formed at an interface between the ribosomal and membrane surfaces. External amino acid at the first stage of metabolic contact with a cell enters its external membrane. From this location it can either be directly incorporated into protein at an attached ribosomal site or it can be stored in the internal pool. Internal pool amino acids can also become incorporated into protein by approaching the ribosome-membrane interface from the aqueous side using reactions involving free sRNA or by re-entering the membrane directly to join the path used by external amino acids. When cells are broken by homogenizing techniques, the continuity of the membrane route may be compromised and the relative contribution of the aqueous route involving free sRNA unduly emphasized.

Since the time that this model was proposed, many new observations have accumulated that could be explained by it. I have discussed these,

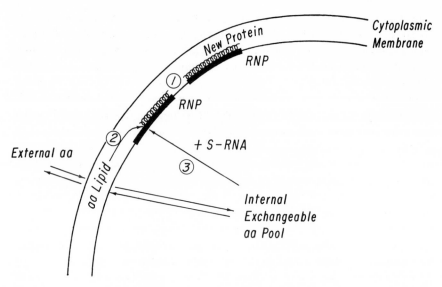

Figure 1 A model for protein synthesis. 1. Protein is synthesized at an interface between a lipoprotein membrane and a ribonucleoprotein film. 2. Amino acid preferentially approaches the interface from the lipoprotein membrane side, either from outside or inside the cell. 3. Amino acid from the interior pool could approach from the RNA side utilizing sRNA reactions. RNP = ribonucleoprotein or ribosome; sRNA = soluble ribonucleic acid. Cytoplasmic membrane is representative of any cell membrane such as endoplasmic reticulum or nuclear membrane. From R. W. Hendler, *Nature,* **193,** 821 (1962).

particularly in the last two chapters. The newer work of Palade, Siekevitz, Redman, Sabatini, and Tashiro add further support (see pages 315–317) to the model. In this chapter, I have emphasized that evidence exists for amino acid activation and the occurrence of sRNA in membranes. The possibility of a direct use of membrane-bound high energy intermediates has been considered.

There is one further addition that can be made to the original model that may resolve the dilemma about the slowness of the kinetics of passage of amino acid through free sRNA in the intact cell. If a species of sRNA (or tRNA) exists associated with the membrane at the ribosomal area of attachment, then it might have the anticipated kinetic behavior in terms of binding amino acid and passing it to the ribosomal site of synthesis. In fact, if sRNA can be charged with amino acid while it is already fixed to the template, delays resulting from competition between charged and uncharged varieties for the template site would be eliminated. It has been found that uncharged sRNA is efficiently bound by ribosomes (see page 94). If the uncharged pool of active sRNA's is already in the vicinity of the template and synthetic machinery, then the difficulty of diffusion and selection times for the comparatively bulky free sRNA species in finding their proper location on the ribosomal template can be overcome.

In short, a membrane-bound pool of tRNA could offer the means of resolving many of the remaining problems that have arisen in the attempt to blend the acellular and cellular observations.

Whichever way future experiments may lead, the most important message of this book has been that concepts of the cell are in a state of continual evolution and, in spite of the fact that strong evidence seems to support our contemporary view of protein synthesis, we must not close our eyes to possibilities that do not at this moment fit into the general scheme. I will even go further by predicting that our future understanding of the cells' processes for protein synthesis will be pursued and found in studies involving cellular membranes—their structure and function. This same route will hopefully lead to a clarification of other of the less well understood cellular processes such as active transport, oxidative phosphorylation, and regulation of metabolic activities, to mention just a few examples.

REFERENCES

1. R. W. Hendler, *Nature,* **193,** 821 (1962).
2. M. C. Ganoza, C. A. Williams, and F. Lipmann, *Proc. Nat. Acad. Sci.,* **53,** 619 (1965).
3. P. N. Campbell and B. A. Kernot, *Biochem. J.,* **82,** 262 (1962).
4. D. Nathans, G. Notani, J. H. Schwartz, and N. D. Zinder, *Proc. Nat. Acad. Sci.,* **48,** 1424 (1962).
5. J. H. Schwartz, J. M. Eisenstadt, G. Brawerman, and N. D. Zinder, *Proc. Nat. Acad. Sci.,* **53,** 195 (1965).
6. H. O. Halvorson and G. N. Cohen, *Ann. Inst. Pasteur,* **95,** 73 (1958).
7. D. M. Kipnis and E. Reiss, *J. Clin. Invest.,* **39,** 1002 (1960).
8. E. Reiss and D. M. Kipnis, *J. Lab. Clin. Med.,* **54,** 937 (1959).
9. D. M. Kipnis, E. Reiss, and E. Helmreich, *Biochim. Biophys. Acta,* **51,** 519 (1961).
10. L. E. Rosenberg, M. Berman, and S. Segal, *Biochim. Biophys. Acta,* **71,** 664 (1963).
11. A. G. Pichler, *Biochim. Biophys. Acta,* **104,** 104 (1965).
12. E. Kempner and D. B. Cowie, *Biochim. Biophys. Acta.,* **42,** 401 (1960).
13. R. W. Hendler, *Science,* **128,** 143 (1958).
14. E. F. Gale and J. P. Folkes, *Biochem. J.,* **59,** 661 (1955).
15. G. D. Hunter, A. R. Crathorn, and J. A. V. Butler, *Nature,* **180,** 383 (1957).
16. J. A. V. Butler, A. R. Crathorn, and G. D. Hunter, *Biochem. J.,* **69,** 544 (1958).
17. G. N. Godson, G. D. Hunter, and J. A. V. Butler, *Biochem. J.,* **81,** 59 (1961).

18. J. T. Wachsmann, H. Fukuhara, and B. Nisman, *Biochim. Biophys. Acta*, **42**, 388 (1960).
19. M. Beljanski, and S. Ochoa, *Proc. Nat. Acad. Sci.*, **44**, 494 (1958).
20. S. Spiegelman in *Recent Progress in Microbiology* (7th Int. Cong. Microbiol. Symp.), Almquist and Wiksells, Uppsala, 1958, p. 81.
21. D. Schachtschabel and W. Zillig, *Hoppe-Seyler Z. Physiol. Chem.*, **314**, 262 (1959).
22. G. E. Palade and P. Siekevitz, *J. Biophys. Biochem. Cyt.*, **2**, 171 (1956).
23. G. E. Palade and P. Siekevitz, *J. Biophys. Biochem. Cyt.*, **2**, 671 (1956).
24. F. S. Sjöstrand and L. G. Elfvin, *J. Ultrastructure Res.*, **10**, 263 (1964).
25. P. C. Fitz-James, *Canad. J. Microbiol.*, **10**, 92 (1964).
26. R. M. Pfister and D. G. Lundgren, *J. Bact.*, **88**, 1119 (1964).
27. W. van Iterson, *J. Cell. Biol.*, **28**, 563 (1966).
28. D. Schlessinger, V. T. Marchesi, and B. C. K. Kwan, *J. Bact.*, **90**, 456 (1965).
29. J. Tani and R. W. Hendler, *Biochim. Biophys. Acta*, **80**, 279 (1964).
30. R. W. Hendler, and J. Tani, *Biochim. Biophys. Acta*, **80**, 294 (1964).
31. R. W. Hendler, W. G. Banfield, J. Tani, and E. L. Kuff, *Biochim. Biophys. Acta*, **80**, 307 (1964).
32. J. G. Hauge and H. O. Halvorson, *Biochim. Biophys. Acta*, **61**, 101 (1962).
33. D. Schlessinger, *J. Mol. Biol.*, **7**, 569 (1963).
34. P. A. Hallberg and J. G. Hauge, *Biochim. Biophys. Acta*, **95**, 80 (1965).
35. L. D. Moore and W. W. Umbreit, *Biochim. Biophys. Acta*, **103**, 466 (1965).
36. A. M. Haywood and R. L. Sinsheimer, *J. Mol. Biol.*, **14**, 305 (1965).
37. C. Weissman, P. Borst, R. H. Burdon, M. A. Billeter, and S. Ochoa, *Proc. Nat. Acad. Sci.*, **51**, 890 (1964).
38. S. Penman, Y. Becker, and J. E. Darnell, *J. Mol. Biol.*, **8**, 541 (1964).
39. B. Bubela and E. S. Holdsworth, *Biochim. Biophys. Acta.*, 376 (1966).
40. B. Nisman, H. Fukuhara, J. Demailly, and C. Genin, *Biochim. Biophys. Acta*, **55**, 704 (1962).
41. D. H. L. Bishop, C. Roche, and B. Nisman, *Biochem. J.*, **90**, 378 (1964).
42. G. E. Palade and P. Siekevitz, *J. Biophys. Biochem, Cytol.*, **2**, 171 (1956).
43. P. Siekevitz and G. E. Palade, *J. Biophys. Biochem. Cytol.*, **7**, 619 (1960).
44. E. C. Henshaw, T. B. Bojarski, and H. H. Hiatt, *J. Mol. Biol.*, 7, 122 (1963).
45. R. R. Howell, J. N. Loeb, and G. M. Tomkins, *Proc Nat. Acad. Sci.*, **52**, 1241 (1964).
46. P. N. Campbell, C. Cooper, and M. Hicks, *Biochem. J.*, **92**, 225 (1964).
47. P. N. Campbell, G. Serck-Hanssen, and E. Lowe, *Biochem. J.*, **7**, 422 (1965).
48. T. Hallinan and H. N. Munro, *Biochim. Biophys. Acta*, **8**, 285 (1965).
49. M. O'Neal Nicolson and W. G. Flamm, *Biochim. Biophys. Acta*, **108**, 266 (1966).
50. E. Volkin and L. Astrachan, *Virology*, **2**, 433 (1956).
51. M. Nomura, B. D. Hall, and S. Spiegelman, *J. Mol. Biol.*, **2**, 306 (1960).
52. J. C. Suit, *J. Bact.*, **84**, 1061 (1962).
53. A. Abrams, L. Nielsen, and J. Thaemert, *Biochim. Biophys. Acta*, **80**, 325 (1964).
54. J. C. Suit, *Biochim. Biophys. Acta*, **72**, 488 (1963).
55. M. D. Yudkin and B. Davis, *J. Mol. Biol.*, **12**, 193 (1965).
56. D. D. Brown and E. Littna, *J. Mol. Biol.*, **8**, 669 (1964).
57. E. Petrović, A. Bećarević, and J. Petrović, *Biochim. Biophys. Acta*, **95**, 518 (1965).
58. E. L. Kuff and W. C. Hymer, *Biochemistry*, **5**, 959 (1966).
59. F. Jacob, S. Brenner, and F. Cuzin, *Cold Spring Harbor Sym. Quant. Biol.*, **28**, 329 (1963).

60. A. T. Ganesan and J. Lederburg, *Biochem. Biophys. Res. Com.*, **18**, 824 (1965).
61. D. J. McCorquodale and W. Zillig, *Hoppe-Seyler Z. Physiol. Chem.*, **315**, 86. (1959).
62. G. D. Hunter, P. Brookes, A. R. Crathorn, and J. A. V. Butler, *Biochem. J.*, **73**, 369 (1959).
63. J. T. Wachsmann, H. Fukuhara, and B. Nisman, *Biochim. Biophys. Acta*, **42**, 388 (1960).
64. L. E. Roberts, K. Bensch, and C. E. Carter, *Biochim. Biophys. Acta*, **90**, 291 (1964).
65. S. J. Norton, M. D. Key, and S. W. Scholes, *Arch. Biochem. Biophys.*, **109**, 7 (1965).
66. T. Peters, *J. Biol. Chem.*, **229**, 659 (1957).
67. T. Peters, *J. Biol. Chem.*, **237**, 1181 (1962).
68. T. Peters, *J. Biol. Chem.*, **237**, 1186 (1962).
69. K. Ogata, S. Omori, and R. Hirokawa, *J. Biochem.* (Tokyo), **49**, 660 (1961).
70. T. A. Douglas and H. N. Munro, *Exptl. Cell. Res.*, **16**, 148 (1959).
71. C. M. Redman, P. Siekevitz, and G. E. Palade, *J. Biol. Chem.*, **241**, 1150 (1966).
72. L. Ernster, P. Siekevitz, G. E. Palade, *J. Cell. Biol.*, **15**, 541 (1962).
73. C. M. Redman and D. D. Sabatini, *Proc. Nat. Acad. Sci.*, **56**, 608 (1966).
74. P. Siekevitz and G. E. Palade, *J. Cell. Biol.*, **30**, 519 (1966).
75. D. Sabatini, Y. Tashiro, and G. E. Palade, *J. Mol. Biol.*, **19**, 503 (1966).
76. E. L. Kuff, W. C. Hymer, E. Shelton, and N. E. Roberts, *J. Cell. Biol.*, **29**, 63 (1966).
77. D. T. Dubin, R. Hancock, and B. D. Davis, *Biochim. Biophys. Acta*, **74**, 476 (1963).
78. J. E. Davies, *Proc. Nat. Acad. Sci.*, **51**, 659 (1964).
79. E. C. Cox, J. R. White, and J. G. Flaks, *Proc. Nat. Acad. Sci.*, **51**, 703 (1964).
80. M. Nomura, *Proc. Nat. Acad. Sci.*, **52**, 1514 (1964).
81. E. C. Slater, *Nature*, **172**, 975 (1953).
82. B. Chance and G. Hollunger, *Fed. Proc.*, **16**, 163 (1957).
83. B. Chance and G. Hollunger, *Nature*, **185**, 666 (1960).
84. L. Ernster, *Proc. Symp. on Intracellular Respiration*, Vol. 5, Ed., E. C. Slater, (5th Int. Congr. Bioch., Moscow, 1961) Pergamon Press, London, 1963, p. 115.
85. A. M. Snoswell, *Biochim. Biophys. Acta*, **60**, 143 (1962).
86. J. B. Chappell, M. Cohn, and G. D. Greville, in *Energy-Linked Functions of Mitochondria*, Ed., B. Chance (1st Colloq. of the Johnson Res. Found.) Academic Press, New York, 1963, p. 219.
87. G. P. Brierly, *ibid.*, p. 237.
88. B. Chance, *loc. cit.*
89. R. W. Hendler, *Biochim. Biophys. Acta*, **49**, 297 (1961).
90. J. R. Bronk, *Proc. Nat. Acad. Sci.*, **50**, 524 (1963).
91. A. M. Kroon, *Biochim. Biophys. Acta*, **91**, 145 (1964).
92. M. K. Campbell, H. R. Mahler, W. J. Moore, and S. Tewari, *Biochem.*, **5**, 1174 (1966).
93. S. Spiegelman, *J. Cell. and Comp. Physiol.*, **30**, 315 (1947).
94. J. M. Reiner and S. Spiegelman, *J. Cell. and Comp. Physiol.*, **30**, 347 (1947).
95. L. Kovac and A. Istenesova, *Biochim. Biophys. Acta*, **82**, 162 (1964).
96. L. Jarett and R. W. Hendler, *Biochem.*, **6**, 1693 (1967).
97. R. W. Hendler, *The Encyclopedia of Biochemistry*, Eds., R. J. Williams and E. M. Lansford, Jr., Reinhold Pub. Corp., N. Y., 1967, p. 491.

AUTHOR INDEX

329

SUBJECT INDEX